T0324742

Global Culture and Sport Series

Series Editors: Stephen Wagg, Leeds Beckett University, UK, and David Andrews, University of Maryland, USA The Global Culture and Sport series aims to contribute to and advance the debate about sport and globalization through engaging with various aspects of sport culture as a vehicle for critically excavating the tensions between the global and the local, transformation and tradition and sameness and difference. With studies ranging from snowboarding bodies, the globalization of rugby and the Olympics, to sport and migration, issues of racism and gender, and sport in the Arab world, this series showcases the range of exciting, pioneering research being developed in the field of sport sociology.

More information about this series at
http://www.springer.com/series/15008

Jon Dart • Stephen Wagg
Editors

Sport, Protest and Globalisation

Stopping Play

Editors
Jon Dart
School of Sport
Leeds Beckett University
Leeds, United Kingdom

Stephen Wagg
Carnegie Faculty
Leeds Beckett University
Leeds, United Kingdom

Global Culture and Sport Series
ISBN 978-1-137-46491-0   ISBN 978-1-137-46492-7   (eBook)
DOI 10.1057/978-1-137-46492-7

Library of Congress Control Number: 2016956277

Cover illustration: Trinity Mirror / Mirrorpix / Alamy Stock Photo

Printed on acid-free paper

This Palgrave Macmillan imprint is published by Springer Nature
The registered company is Macmillan Publishers Ltd. London

For everyone seeking to challenge injustice, in or via sport

Acknowledgements

Thanks to all the contributors to this book for their enthusiasm and support.

Contents

Contributors

Celeste González de Bustamante is an Associate Professor in the School of Journalism at the University of Arizona, USA.

Jon Dart teaches sports policy and sociology in the Carnegie Faculty of Leeds Beckett University.

Russell Field is Assistant Professor in the Faculty of Kinesiology and Recreation Management at the University of Manitoba in Winnipeg, Canada.

Christopher Gaffney teaches geography at the University of Zurich, Switzerland.

David Clay Large retired as Professor of History at Montana State University, USA, and now teaches at the University of San Francisco. He is the author of *Nazi Games: The Olympics of 1936* (New York, 2007).

Joshua Large teaches history at Universidad EAFIT in Medellin, Colombia.

Helen Jefferson Lenskyj is Professor Emerita at the University of Toronto and a leading critic of the Olympic industry. Her books include *Inside the Olympic Industry* (2000), *The Best Olympics Ever? Social Impacts of Sydney 2000* (2002), *Olympic Industry Resistance* (2008), *Gender Politics and the Olympic Industry* (2012) and *Sexual Diversity and the Sochi 2014 Olympics* (2014) (the last two books were published with Palgrave Macmillan).

John Minto is a teacher and political activist based in Christchurch, New Zealand. In a television documentary called *New Zealand's Top 100 History Makers* shown in the country in October 2005, John was placed 89th.

Christine O'Bonsawin is an associate professor of History and Indigenous Studies at the University of Victoria, Canada. She is a specialist in Canadian sport history, Olympic history and the history of indigenous peoples.

Carol A. Osborne teaches the history, sociology and philosophy of sport at Leeds Beckett University, UK. She is a former chair of the British Society of Sport History.

Anita Pleumarom is a geographer and member of the Tourism Investigation & Monitoring Team based in Bangkok, Thailand. She has written extensively on tourism and the environment.

Stephen Wagg is a professor in the Carnegie Faculty in Leeds Beckett University. He is co-editor of *The Palgrave Handbook of Olympic Studies* (2012) and author of *The London Olympics of 2012: Politics, Promises and Legacy* (Palgrave Macmillan, 2015).

Kostas Zervas holds a doctorate from Leeds Metropolitan University in 2012 for a study of anti-Olympic movements. He is a lecturer in sport, health and nutrition in the School of Social and Health Sciences at Leeds Trinity University, UK.

Introduction: Sport and Protest

Jon Dart and Stephen Wagg

On the 30th January 2016, the news agency Reuters carried a story about a Greek second division football match between AEL Larissa and Acharnaikos in the Thessalian city of Larissa. The match was about to begin when all 22 players sat down on the pitch. They, along with the two clubs' coaches and substitute players, remained seated while the following announcement was read out over the public address system:

> 'The administration of AEL, the coaches and the players will observe two minutes of silence just after the start of the match in memory of the hundreds of children who continue to lose their lives every day in the Aegean due to the brutal indifference of the EU and Turkey. The players of AEL will protest by sitting down for two minutes in an effort to drive the authorities

J. Dart (✉) • S. Wagg
Carnegie Faculty, Leeds Beckett University, Beckett Park Campus, Leeds LS6 3QU, UK

© The Editor(s) (if applicable) and The Author(s) 2016
J. Dart, S. Wagg (eds.), *Sport, Protest and Globalisation*,
DOI 10.1057/978-1-137-46492-7_1

to mobilise all those who seem to have been desensitised to the heinous crimes that are being perpetrated in the Aegean[1]" (Reuters 2016).

This incident suggests at least two things: first, that, even in the postmodern twenty-first-century sports, people can recognize that there are more important things in the world than sport and, second, that sport remains—indeed, given the vast swathes of mass media now given over to sport, it is perhaps more than ever—a significant forum for political protest. Some of this protest, as on that day in Greece, uses sport as a platform upon which to highlight an issue outside of the immediate purview of sport. Equally, some protest is levelled at the institutions of sport itself and the governance thereof.

The former, of course, encompasses notable stands against racism, as with the now-iconic clenched-fist salute at the Mexico Olympics of 1968 (see Smith 2011) and the numerous attempts worldwide to disrupt sports encounters which featured representatives of apartheid South Africa (see, e.g., John Minto's chapter in this book); against anti-gay discrimination, as with the demonstration at the Sochi Winter Olympics of 2014 against the so-called 'gay propaganda' legislation passed in the Russian duma the previous year[2] (see BBC 2014, and Helen Lenskyj's chapter in this book); and in favour of the rights and protection of women, as when, in November 2014, members of Sheffield Feminist Network demonstrated outside the city's town hall against the prospect that Sheffield United FC would re-sign striker Ched Evans at the end of his prison sentence for rape (Evans 2014).

Many of the latter kind of protests have been aimed at the institutions of sport and its governance and go to the heart of the capitalist structure of sport and its commercial excesses and abuses. Recent examples here would include the ongoing protests by Newcastle United supporters at their club's sponsorship deal, signed in 2012, with the controversial payday loans company Wonga, which included their marketing a Newcastle shirt without Wonga's logo (Marsh 2015); the walk-out by 10,000 Liverpool supporters chanting 'You greedy bastards...' during a home match in February 2016 in protest at ticket prices (Press Association 2016); and

---

[1] A total of 45 refugees drowned in the Aegean in January 2016 alone. See Connolly et al. (2016).
[2] The law is officially known as *For the Purpose of Protecting Children from Information Advocating for a Denial of Traditional Family Values.*

following growing evidence of corruption in world football's governing body FIFA (Jennings 2015) the showering of its apparently venal president Sepp Blatter with fake money by British comedian-prankster Simon Brodkin in July 2015 (BBC 2015). But the principal protests in this regard appear to have been against the International Olympic Committee (IOC) and have grown in tandem with the massive and escalating public expense generated by staging the Olympics since the mid-1970s. There are signs that these protests are having an effect. At the time writing, three cities have pulled out of bidding process to select host cities for future Olympics: Rome in 2012 (Owen 2012), Boston, Massachusetts in July 2015 (Seelye 2015) and Hamburg the following December (Bohne 2015)—all on grounds of the likely drain on public finances.

This book is an attempt to offer what may be the first thoroughgoing review of the history of sport, protest and activism. It is a contribution to the small but growing literature of which eminent recent examples include Jean Harvey et al.'s *Sport and Social Movements* (Harvey et al. 2015) and Michael Lavalette's *Capitalism and Sport: Politics, Protest, People and Play* (Lavalette 2013), although the latter, while an important book, says comparatively little about protest.

The book is arranged as follows:

In Chapter 2, "'Deeds, Not Words': Emily Wilding Davison and the Epsom Derby 1913 Revisited", British historian Carol Osborne examines male sport as a target for those campaigning for women's suffrage before the First World War. As is quite well known, the suffragette Emily Davison was killed by King George V's horse during the 1913 Epsom Derby while trying to draw attention to the cause of women's suffrage. However, this much publicized and shocking incident was not the only connection between sport and political protest in the early twentieth century. This chapter shows how British suffragettes carried out a series of violent acts against sport and sportsmen that were unparalleled in the English-speaking world.

Chapter 3, "Women's Olympics: Protest, Strategy or Both?" is by sociologist and leading Olympic critic Helen Jefferson Lenskyj and concerns the protests against the exclusion of women from the early modern Olympic Games which resulted in the staging of the first 'Women's Olympics' in Paris in 1922. The Games, first held in 1896, have always privileged some groups of athletes over others. Organized by the IOC, a group of self-elected men (and a small number of women), the Olympic sporting programme is by no means representative of world sporting practices. The eligibility rules, both explicit and implicit, have been variously based on sex, gender, sexuality, social class, nationality, religion, race/ethnicity and/or ability, thereby enabling the participation of some groups while posing barriers to others.

Resistance to the Olympic Games has been in evidence throughout the twentieth century and continues into the twenty-first century. Numerous alternative sporting and cultural festivals organized by members of excluded groups represent the earliest ongoing form of Olympic resistance, such the Workers' Olympics, staged between 1925 and 1937 (see Riordan 1984). Some followed the general Olympic model, while others reflected the distinctive sociocultural values and priorities of the excluded groups. Many alternative games had their beginnings in the early decades of the twentieth century and continue today; others were relatively short-lived, but had important impacts on international sport. The Women's Olympics provide a pertinent example. Many historians treat the establishment of the Women's Olympics as a strategy to force the IOC's hand, at a time when the sporting programme for female athletes was extremely limited. In other words, rather than representing a genuine alternative and a radical political stance, the Women's Olympics are reduced to a mere stepping stone to the 'real' Olympics. This chapter examines these debates and contradictions.

Chapter 4, "A Most Contentious Contest. Politics and Protest at the 1936 Berlin Olympics" by American historians David Large and Joshua Large identifies the 1936 Berlin Summer Olympic Games as a watershed in the emergence of the Modern Olympics and as a lightning rod for political protest. Hosted as they were by the capital city of Nazi Germany, these Games sparked a major international boycott movement in which the USA played a central role. Although the boycott effort eventually failed to materialize, it managed to force some (alas, essentially token) concessions from the German organizers and, more importantly, helped focus world attention on the racist policies of the Hitler government. The 'Boycott Berlin' movement also brought to the fore an issue that would bedevil a number of later Olympiads: the questionable legitimacy of holding a sports festival supposedly celebrating open, internationalist, non-discriminatory values in a largely closed, authoritarian and highly discriminatory venue.

Additionally, albeit on a smaller scale, the 1936 Berlin Games witnessed individual boycotts and protests by athletes against various discriminatory policies of their own nations, as well as objections by ordinary citizens to the staging of an expensive athletic extravaganza at a time of general economic hardship—a recurrent theme in the history of the Olympics. Finally, a Depression-era spectacular that seemed of economic benefit primarily to the rich spurred leftist groups outside the host nation to mount their own 'anti-Olympic' athletic festival, as had also been done in Chicago in 1932 to protest the Los Angeles Games. Due to the outbreak of the Spanish Civil War in 1936, this counter to the 'Nazi Games' was shifted from Barcelona, Spain to New York City. Here was yet another formative moment in the evolution of popular protest against the excesses and abuses of high-level sport in the modern age.

The Games of the New Emerging Forces (GANEFO), held in 1963, are the subject of Chapter 5, "Splitting the World of International Sport: The 1963 Games of the New Emerging Forces and the Politics of Challenging the Global Sport Order", written by Canadian historian Russell Field. In November of that year, approximately 3000 athletes and officials from— but not necessarily officially representing—nearly 50 nations gathered in Jakarta for the inaugural GANEFO, an international multi-sport event that featured 20 sports (predominantly Western sports) as well as cultural festivities. Athletes hailed primarily from recently decolonized countries in Asia and Africa, which were labelled the 'new emerging forces' by Indonesian President Sukarno who created GANEFO as part of his attempt to situate his nation within the non-aligned movement.

GANEFO was an explicit attempt to link sport to the politics of anti-imperialism, anti-colonialism, and the emergence of the Third World. It was organized on the model of the global sporting event made most prominent by the IOC while protesting against the perceived forms of repression symbolized by similar events. IOC President Avery Brundage, who had vigorously resisted campaigns to boycott the 'Nazi' Olympics of 1936, now warned that GANEFO 'might split the world of international sport' along racial lines and in the West, the event was framed in Cold

War terms because of Sukarno's ties to the Indonesian communist party and the Games' sponsorship by China. Indeed, after the hosts, the largest teams in Jakarta represented the powers of the Second World: the Soviet Union and the People's Republic of China, the latter of which financed GANEFO. The chapter explores GANEFO's significance—at a time of heightened geopolitical tensions in Southeast Asia—as a sporting manifestation of the anti-colonial movements of the 1960s, Cold War politics, tensions in the Sino-Soviet split, and a residue of European workers' sport culture. Based upon archival research and oral history interviews with former GANEFO athletes from four countries (including Western Europe and South America), Russell Field expands the analysis to highlight the ways in which GANEFO was an experience that transcended, and often ignored, political rhetoric for the participants, many of whose sporting careers were impacted by sanctions they received for competing.

Celeste Gonzalez Debustamante, a lecturer in journalism studies based in the border state of Arizona, revisits, and extends, what is known about the protest in Mexico City that preceded the Olympic Games in 1968. More specifically, it is about how this event still reverberates in the culture and politics of Mexico and how that reverberation is reflected in contemporary media reportage. On October 2, 2013, thousands of citizens converged on the streets of Mexico City in solemn commemoration of the 45th anniversary of the massacre of hundreds of innocent students and bystanders at the Plaza de Tlatelolco. The 1968 massacre happened just ten days before the inauguration of the XIX Olympiad in Mexico City, the first ever to be held in Latin America.

For the Mexican government and elites, the ability to host the Olympic Games illustrated that the country had reached a high level of modernity, and television executives beamed that vision into the homes of Mexican citizens at home and around the world, and for the first time in colour. For student activists, the Games represented a source of discontent, and that the government had squandered public funds to temporarily beautify the city, while that money could have been used to help achieve the nation's revolutionary goals for those who needed it most, the poor. More than four decades later, the Mexico City Olympics and the Tlatelolco massacre continue to be debated in the public sphere in Mexico. The twenty-first-century activists attempt to keep the memory of the cause and those who died alive, while some of Mexico's old guard would simply like to forget what has become a national stain, but one that remains visible every October 2nd. Through an examination of televised news reports aired in 1968 and 2008, this chapter argues that television reports—and those who participate in its production—actively contribute to a contested public sphere, aiding the construction of collective memories of Mexico's '68. Celeste Gonzalez Debustamante shows how over the years, television coverage of the massacre has changed, along with the country's political opening, mainly during in the 1990s. In 1968, television producers limited images that were broadcast in an attempt to severely downplay the massacre. Yet, four decades later, mainstream television anchors and reporters frequently used the catch phrase, 'The second of October, it is not forgotten', signalling both an acceptance of the government's culpability, and media executives who attempted to appear more independent.

Anti-apartheid protest and its aftermath is the subject of Chapter 7, "Race, Rugby and Political Protest in New Zealand: A Personal Account" which has been written by New Zealand teacher and political activist John Minto, who either led or took part in many of these protests in the 1980s. He suggests that the most notable protests in sport history were those mounted against South Africa's apartheid system of state racism. One of the principal groups involved in the struggle against apartheid was HART (Halt All Racist Tours) formed in New Zealand in 1969. HART campaigned to prevent the South African Springboks rugby team touring New Zealand in 1981. John Minto was chair of HART at the time and in the chapter, he reflects both on the work of HART and on the political progress he and other HART members had hoped to see in South Africa, once apartheid had been dismantled.

Chapter 8, "Fighting Toxic Greens: The Global Anti-Golf Movement (GAG'M) Revisited", written by Anita Pleumarom, an independent researcher, based in Bangkok, is focused on the Global Anti-Golf Movement (GAG'M) and acts as a case study in sport, protest and contemporary eco-politics. It is an axiom of contemporary sociological analysis that, in a postmodern world, dominated by neoliberal doctrines, the gap between the rich (individuals and countries) and poor has widened, leaving a wealthy middle class with ample money to spend on their (sporting) pleasures. This trend has seen a growing demand for golf, a sport with radical implications for land use. The construction of new golf courses in a range of countries across Southeast Asia, more likely for tourism than indigenous use, has driven (invariably poor) people off their land and disturbed local ecosystems. Moreover, the development of golf resorts, frequently touted as creating jobs, has more likely resulted in the export of capital, since the corporations operating the resorts bank their profits elsewhere. One of the crowning ironies here has been in Vietnam, where the defoliant Agent Orange, used by the US military to destroy vegetation in the war of the 1960s, is now being used in golf course construction: in this regard, some have commented that Vietnam won the war, but lost the peace.

In 1996, 'an alliance of groups and individuals concerned with environmental, consumer and human rights issues—will again celebrate their annual No Golf Day aimed at lobbying decision-makers to enforce a worldwide moratorium on golf course construction and stop the promotion of golf tourism' (http://victoriafallsheritage.blogspot.co.uk/2006/08/global-anti-golf-movement-gagm.html. Access 20th February 2014). This movement is still active in this regard and the chapter, written by an independent researcher who monitors world tourism, provides a historical account of the work of the GAGM, identifying the issues, the GAGM campaigns in various countries, and the responses both of the governments concerned and of the golf industry.

Chapter 9, "'Human Rights or Cheap Code Words for Antisemitism?' The Debate over Israel, Palestine and Sport Sanctions", by sociologist Jon Dart, discusses the participation of Israel in contemporary international sport events and how their presence is being contested. The chapter takes as its starting point the Boycott, Disinvestment and Sanctions (BDS) movement's call for sport sanctions to be applied to Israel until it addresses the plight of Palestinian refugees and those living in the occupied territories. The BDS call, first made in 2005, generated increased attention when, in 2013, the Union of European Football Associations (UEFA) awarded the men's Under-21 finals tournament to the country. A commentary is offered on Israel's presence in the sporting world and how the logic of the neoliberalism is the driving factor in the allocation of the UEFA tournament and other major sporting events to countries with records of human rights violations. The reasons why sanctions are seen by some as hypocritical, unfair and inconsistently applied to Israel are assessed. The role of Zionist, antisemitic and anti-racist ideologies in neutralizing criticism of the Israeli state are discussed. The chapter reflects on the often-heard comparison between sanctions applied to apartheid South Africa and those which BDS supporters seek to impose on Israel. Consideration is made of the role of the USA, the legacy of the Holocaust, the legacy of Palestinian identity, and the lack of high profile (sports) spokespersons for 'the Palestinian cause'. The chapter concludes that realpolitik is the basis for understanding the decision by UEFA to award their Under-21 tournament to Israel, despite growing disquiet at the actions of the Israeli state.

Chapter 10, "Chicago 2016 Versus Rio 2016: Olympic 'Winners' and 'Losers'" offers a case study of Olympic protest. Written by UK-based Greek lecturer Kostas Zervas, it is based largely on his own research which examined the campaign of the 'No Games Chicago' (NGC) group which successfully fought against the city's bid to host the Olympics of 2016. Their reasons were explained in their website: 'better hospitals, better housing, better schools, better life for the citizens', in brief, instead of spending millions of dollars and valuable time, seeking an event that has no guaranteed revenue. NGC also accused the Chicago Mayor Richard M. Daley, whose family had dominated the city's politics since the 1950s, of authoritarian and undemocratic behaviour. Most of all, the NGC's campaign sought to start a debate, within Chicago, on the utility of the Olympic Games, and by extension, to challenge Mayor Daley and his practices. They suggested that NGC was representing half of the Chicago population, which had not approved the city's bid and had never been asked about it.

'No Games Chicago' attempted to open up discussion on an issue that concerns every urban centre; its right to take part in important decision-making processes. They countered a well-organized team of politicians, businessmen and PR experts who comprised the official Chicago 2016 bid but, despite the problems, the prohibitions and the closed doors, which they faced throughout their campaign, they apparently succeeded in making their voice heard. And in the legacy of similar notable social movements that were opposed to the Olympic Games in their city, like 'Bread Not Circuses' in Toronto (Lenskyj 2000) and 'No Games 2010' in Vancouver (Shaw 2008), they provided invaluable information about the conflict between the local organizers and the community, where sport mega-events are planned, or hosted. In terms of their campaign, NGC managed in a very short period to make an impact and attract a lot of supporters. This was achieved mainly due to the effective use of 'digital media', which enhanced the campaign with its directness, dynamism and a sense of 'freshness'. These characteristics, along with the immense support, filled the NGC members with confidence and enthusiasm, something which was reflected in their decisions and actions. The anti-Olympic campaign of the Chicago 2016 bid is a fine example of how a society can effectively mobilize and take part in the decision-making process.

Canadian historian Christine O'Bonsawin asks in Chapter 11, "'The Olympics Do Not Understand Canada': Canada and the Rise of Olympic Protests" whether Canada can be seen as the heartland of Olympic protest. The Olympic Games have been hosted in Canada on three occasions, and in the planning and hosting stages of 1976 Montréal, 1988 Calgary, and 2010 Vancouver there have existed omnipresent threats of protest actions directed at Olympic organizing efforts. The chapter explores protest action, as well as explicit boycott campaigns that took place during the preparation and hosting schedules of all three Olympic Games in Canada. In this regard, the chapter has two organizing principles. It examines: (1) protest action that is 'external to the national', which primarily concerns grievances that attempt to coerce and thus modify the practices and policies of the Olympic movement, and (2) protest action that is 'internal to the national', which primarily concerns grievances that attempt to coerce and thus modify practices and policies national in scope (i.e. federal, provincial, municipal, organizational, etc.). However, this is not to say these two realms are mutually exclusive.

A preliminary examination of protest action reveals that in the planning and/or hosting stages of all three Olympic Games, external and internal grievances were expressed, and in some cases, boycott action was successfully executed. Consequently, Olympic organizers in Canada were forced to negotiate external grievances such as the two-China policy, an African embargo, and various equality (i.e. gender and racial) contraventions. Internally, organized political and social units concerned with Indigenous rights, environmentalism, gentrification and fiscal mismanagement, to name but a few, strategically pressured Olympic organizers when held in the country. The 1976 Montréal Olympic Summer Games, undoubtedly, ushered in a new era of protest action, namely the Olympic boycott era. Within the context of Canadian organized and hosted Olympic Games, it is evident that protest actions and boycott efforts have pervaded the organizing and hosting of subsequent Olympic Games in Canada.

Chapter 12, "'The Atos Games': Protest, the Paralympics of 2012 and the New Politics of Disablement" by British sociologist and historian Stephen Wagg, focuses on the Paralympic Games of London 2012 and the protests against one of the sponsors of the Games. On 31st August 2012, a group of variously disabled people and their supporters staged a protest outside the London offices of Atos, a multinational IT services company, based in France, but active in over 50 countries. The protest was one of a series, beginning the previous year, which was provoked by the conjoining of two of Atos' commitments:

1. Following the Welfare Reform Act of 2012, the UK government awarded Atos a lucrative to conduct 'Work Capability Assessments' on people receiving benefits for disability or injury. The conduct of these assessments was rapidly and widely perceived by the people concerned to be an unconcealed assault on their entitlements. (A disabled spokesperson in Leicester two years later told BBC local radio of 'Brown Envelope Syndrome', wherein a disabled person looks with terror at any post contained in a brown envelope marked 'Department of Work and Pensions' and cannot muster the courage to open it without a friend present to offer emotional support—BBC Radio Leicester 19th February, 2014).

2. Atos had been 'IT Partner' to the Olympic Games since 2001 and sponsored the London Paralympics of 2012. This led to a call by disabled groups and others for a boycott of the Paralympics and, following the opening ceremony on 30th August, Olympic officials refuted suggestions that Paralympians had deliberately concealed their accreditation badges, so that the Atos logo could not be seen.

This chapter explores this protest and discusses the complex ideological and political implications of the Games as a media spectacle. It also provides an overview of the changing politics of disablement in Britain between the 1980s and 2012. While some research suggested that public perceptions of disabled athletes, and disabled people generally, became more positive during the Games, the Paralympians were conspicuously promoted as sports celebrities, capable of generating the same public enthusiasm as the able-bodied Olympians, with whom, it was made known, some of them trained. This did not go unacknowledged in government circles and may have strengthened support for their work capability drive: disabled people were now exalted as stars and high athletic achievers —surely it was wrong—and demeaning—for such people to be on state benefits.

Chapter 13, "'*Messing About on the River.*' Trenton Oldfield and the Possibilities of Sports Protest", by Jon Dart, is an analysis of a single act of protest that occurred during the annual elite university Boat Race on the River Thames in 2012. In April of that year, a young Australian activist called Trenton Oldfield disrupted the annual Boat Race between Cambridge and Oxford Universities by going for a swim in the River Thames. For some, Oldfield's swim in a public space was an imaginative and well-executed act of civil disobedience which achieved maximum exposure and caused minimal damage. Live television coverage of the event gave him, with his use of social media, a vast audience and allowed him to promote his manifesto 'Elitism leads to Tyranny'. Oldfield's actions can be seen as emblematic of the individual, autonomous approach to politics in the first decade of the twenty-first century. This chapter considers the opportunities that large sport events offer to individual autonomist protest and the changing dynamic between sport, media and protest created by the emergence of new forms of digital web-based media. Discussion is given to the subsequent treatment of Oldfield by sections of the English media and the UK government who, upset to see their sporting pleasures disrupted, sought to deport him from the UK.

Protest against anti-gay discrimination is the subject of Chapter 14, "Sochi 2014 Olympics: Accommodation and Resistance", by Helen Jefferson Lenskyj. Protests surrounding Sochi's hosting of the 2014 Winter Olympics represent a significant step in Olympic resistance campaigns. For the first time, issues of sexual orientation and Russia's 'anti-gay propaganda' laws—in effect, its anti-gay laws—became the focus of world attention, prompting international demonstrations of outrage and solidarity. The sex–sport association, however, has had a long history. As she has explained elsewhere, 'all the men are straight, all the women are gay' is a sweeping generalization that for decades captured public and media perceptions of sportsmen and women in Western countries (Lenskyj 2013). There are some exceptions: female athletes who embodied grace and aesthetic beauty escaped this stereotyping, while male athletes who did the same—most notably figure skaters—were generally assumed to be gay unless they 'came out' as straight. Lesbian and gay invisibility serves Olympic industry purposes by promoting the myth that sport and politics don't mix. Protests and calls for boycotts in the period leading up to Sochi 2014 amply demonstrate the lasting power of the 'sport-as-special' myth, as well as the impact of the rainbow symbol of global solidarity with sexual minorities.

In the final chapter, Christopher Gaffney, a geographer based in Brazil, examines the huge wave of opposition in the country to the hosting of the FIFA World Cup of 2014 and the Olympic Games of 2016. The 2014 FIFA Confederations Cup (a tournament designed to act as a rehearsal for the nation due to host the FIFA World Cup) was characterized by massive public protests in hundreds of Brazilian cities. Such were the scale and intensity of the social unrest that the successful realization of the tournament was threatened. Violent police reactions exposed a lack of preparation and training while the protests themselves called into question the channelling of public resources into world-class sports facilities, tax exemptions for FIFA and the lack of decent, basic public services in Brazilian cities. The protests that erupted in June of 2013 abated but then gained new impetus with the 2014 FIFA World Cup. As public resources and global attention focused on Rio de Janeiro with the city's preparations for the 2016 Olympic Games, the social movements leading the fight against these massive public outlays again took centre stage.

At the forefront of the protest movement were groups known as Popular Committees of the World Cup (Comités Populares da Copa). These committees formed in every World Cup host city and were the primary drivers of the counter-discourses that framed social resistance to sports mega-events in Brazil. The chapter discusses the formation, social composition and functioning of the first such group in Brazil—the

Comitê Popular da Copa e das Olimpíadas do Rio de Janeiro (The Rio de Janeiro Popular Committee of the World Cup and Olympics—RJPC). Based on the author's personal involvement with the RJPC, interviews with activists and FIFA executives and an examination of the historical trajectory of urban social movements within the Brazilian sociopolitical conjuncture, the chapter examines the increasing role of protest in the production of large-scale sports events.

Naturally, we hope to see greater democratization, both in sport and in the world beyond. Failing that, it seems likely that we will see more protests in sports venues and against sport institutions, such as the Olympics. Certainly, in times of politically imposed 'austerity' in many countries, more and more people will feel confident, indeed impelled, to challenge the hosting of a sport mega-event. Another, more baleful, possibility is that the spectre of terrorism—such as the targeting of the France–Germany match in Paris by Islamic State gunmen in November 2015—will lead to calls for major sporting events to be staged in less democratic countries. In this regard the observation, carried in *The Times* newspaper in 2009, by Formula One boss Bernie Ecclestone that 'If you have a look at democracy it hasn't done a lot of good for many countries—including this one [the UK]… Politicians are too worried about elections.' (Thomson and Sylvester 2009) makes for chilling reading. In all these circumstances, it is to be hoped that greater numbers of sportspeople—players, spectators and others—will recognize and challenge the inequalities and injustices that prevail in sport and in the wider society. We note in this regard the former Formula One champion driver Damon Hill's call for the Bahraini Grand Prix in 2011 to be cancelled rather than rescheduled following the brutal suppression in Bahrain of pro-democracy protests. 'This crisis is an opportunity', said Hill, 'for Formula One to show that it cares about all people and their human rights' (Cary 2011).

References

BBC. (2014). Sochi 2014: Gay rights protests target Russia's games. http://www.bbc.co.uk/news/world-europe-26043872. Posted 5th February. Accessed 2 Mar 2016.

BBC. (2015). UK prankster Simon Brodkin throws money at Fifa's Sepp Blatter. http://www.bbc.co.uk/news/world-europe-33591332. Posted 20th July 2015. Accessed 1 Mar 2016.

Bohne, J. (2015, December 9). 'Fed up of being short-changed': The real reason Hamburg said no to the Olympics'. *The Guardian*. http://www.theguardian. com/cities/2015/dec/09/2024-olympics-why-hamburg-no-hosting. Accessed 1 Mar 2016.

Cary, T. (2011). Damon Hill calls on Bernie Ecclestone and Formula One to abandon Bahrain Grand Prix. http://www.telegraph.co.uk/sport/motorsport/formulaone/8553124/Damon-Hill-calls-on-Bernie-Ecclestone-and-Formula-One-to-abandon-Bahrain-Grand-Prix.html. Posted 2nd June 2011. Accessed 3 Mar 2016.

Connolly, K., Smith, H., & Tran, M. (2016, January 22). Deadliest January for refugees as 45 die when boats capsize in Aegean. *The Guardian*. http://www. theguardian.com/world/2016/jan/22/deadliest-january-45-refugees-die-boats-capsize-aegean. Accessed 1 Mar 2016.

Evans, A. (2014). Feminists protest against Ched Evans outside town hall to 'send message' to Sheffield United. http://www.thestar.co.uk/news/feminists-protest-against-ched-evans-outside-town-hall-to-send-message-to-sheffield-united-1-6955133. Posted 15th November 2014. Accessed 2 Mar 2016.

Harvey, J., Horne, J., Safai, P., Darnell, S., & Courchesne-O'Neill, S. (2015). *Sport and social movements*. London: Bloomsbury.

Jennings, A. (2015). *The dirty game: Uncovering the scandal at FIFA*. London: Cornerstone.

Lavalette, M. (Ed.) (2013). *Capitalism and sport: Politics, protest, people and play*. London: Bookmarks.

Lenskyj, H. J. (2000). *Inside the Olympic industry: Power, politics and activism*. Albany: State University of New York Press.

Lenskyj, H. J. (2013). Reflections on communication and sport: On heteronormativity and gender identities. *Communication & Sport, 1*(1–2), 138–150.

Marsh, M. (2015, July 28). Newcastle United fans sell alternative Toon shirt WITHOUT Wonga label in protest at owner Mike Ashley. *Daily Mirror*. http://www.mirror.co.uk/news/uk-news/newcastle-united-fans-sell-alternative-6155554. Accessed 1 Mar 2016.

Owen, D. (2012). A century after Vesuvius, Rome's Olympic dreams dashed by eruption of Europe's finances. http://www.insidethegames.biz/articles/15877/a-century-after-vesuvius-romes-olympic-dreams-dashed-by-eruption-of-europes-finances. Posted 14th February 2012. Accessed 1 Mar 2016.

Press Association. (2016, February 6). Liverpool fans' walkout protest: Around 10,000 leave in 77th minute over ticket prices. *The Guardian.* http://www. theguardian.com/football/2016/feb/06/liverpool-fans-walkout-thousands-ticket-price-protest. Accessed 1 Mar 2016.

Reuters. (2016). Greek match delayed as players stage sit-down protest over migrant deaths. http://www.theguardian.com/football/2016/jan/30/greek-match-delayed-players-sit-down-protest-migrant-deaths. Accessed 1 Mar 2016.

Riordan, J. (1984). The Workers' Olympics. In A. Tomlinson & G. Whannel (Eds.), *Five ring circus: Money power and politics at the Olympic games* (pp. 98–112). London: Pluto Press.

Seelye, K. A. (2015, July 27). Boston's bid for Summer Olympics is terminated. *New York Times.* http://www.nytimes.com/2015/07/28/sports/olympics/boston-2024-summer-olympics-bid-terminated.html?_r=0. Accessed 1 Mar 2016.

Shaw, C. (2008). *Five Ring Circus, myths and realities of the Olympic Games,* Canada: New Society Publishers.

Smith, M. M. (2011). The "Revolt of the Black Athlete": Tommie Smith and John Carlos's black power salute reconsidered. In S. Wagg (Ed.), *Myths and milestones in the history of sport* (pp. 159–184). Basingstoke: Palgrave Macmillan.

Thomson, A., & Sylvester, R. (2009). Bernie Ecclestone, the Formula One boss, says despots are underrated. http://web.archive.org/web/20090827032853/http://www.timesonline.co.uk/tol/sport/formula_1/article6632991.ece. Posted 4th July 2009. Accessed 3 Mar 2016.

'Deeds, Not Words': Emily Wilding Davison and the Epsom Derby 1913 Revisited

Carol Osborne

In 1908 the newspaper *Votes for Women* published a poem written by Winfred Auld, entitled 'The Fighting Suffragist to the Frightened Politician'.[1] In the first of seven stanzas, Auld pointedly observed that during a period of 60 years suffragists had 'fought a quiet campaign' in which they had 'Petitioned humbly for their rights'—but to no avail. The remaining verses amounted to a statement of intent: women had had enough of the government's empty pledges, there was 'but one thing left to do—To fight.'

Thoughts of the protracted 'quiet' campaign which British suffragists had waged on successive governments since the mid-1860s could not have been far from Emily Wilding Davison's mind when she conspired

---

[1] Winifred Auld, 'The Fighting Suffragist to the Frightened Politician' in *Votes For Women*, 23 July 1908, p. 325. Complete edition available online at: https://news.google.com/newspapers (Accessed March 2016).

C. Osborne (✉)
Leeds Beckett University, 221 Cavendish Hall, Beckett Park Campus, Leeds LS6 3QU, UK

© The Editor(s) (if applicable) and The Author(s) 2016
J. Dart, S. Wagg (eds.), *Sport, Protest and Globalisation*,
DOI 10.1057/978-1-137-46492-7_2

17

to mobilise the 1913 Epsom Derby to draw attention to the still unmet demand of 'Votes for Women'.

Her decision to do so proved fatal. As newsreels of the day show, ducking under the rail at Epsom racecourse and positioning herself in the path of the oncoming horses, whether by design or by default, she appeared tantalisingly close to fulfilling her objective: to attach a silk sash in campaign colours—purple, white and green—to her fancy, Anmer, King George V's horse.

Of the numerous forms of protest adopted by the suffragettes, the name given to the militant campaigners demanding 'Votes for Women', Davison's deed is singularly memorable. As the only woman who died taking direct action for 'the Cause' (as the campaign for The vote was known at the time), her name arguably retains a profile in suffrage history comparable to that of the leading 'Pankhursts'—Emmeline, Christabel and Sylvia—who were instrumental in founding the Women's Social and Political Union (WSPU) in 1903. It was this self-proclaimed militant organisation to which Davison gave total allegiance from 1906 until her death seven years later.

More recently, Davison's prominence in the public consciousness has been further sustained: firstly, due to the centenary commemorations of her death in 2013 when a surge of media interest reflected not only upon the dramatic manner of her demise but also on the associated political cause that occasioned it (Gavron 2015), and, secondly, due to the dramatisation of the event in the acclaimed feature film *Suffragette* (Pathé/Film4/BFI 2015). Whilst the film's narrative turns upon the central figure of a working-class woman and her associated personal struggle as a late recruit to the militant campaign, Davison is acknowledged by the screenwriter, Abi Morgan, as a crucial figure and so was written to 'dip[s] in and out' of the narrative (Pathé/Film4/BFI 2015).[2] In fact, the closing scenes of the film reimagine the circumstances leading up to the Derby Day protest and Davison necessarily becomes more prominent. Her movements are interwoven with those of the working-class heroine, Maud Watts, who accompanies her to Epsom Downs with a view to pinning the sash on the King's horse *before* the race. The attempt is foiled by a vigilant policeman as the two women approach the Royal enclosure.

---

[2] In the audio commentary accompanying the DVD release, Abi Morgan observes that Davison deserves a film in her own right and identifies her as an 'exceptional woman'.

The rest, as they say, is history: the contemporary audience, like the spectators at Epsom racecourse back in 1913, are briskly brought to their senses by the more accurate representation of the event as the figure of Davison becomes central to the impact of the film and race, respectively. In a matter of seconds, she becomes emblematic of the dogged determination of thousands of women who would not bow, either to the government's indifference to their demand for The Vote or the brutal treatment they routinely received from the authorities and members of the general public alike in making it.

Whilst the film *Suffragette* provides a valuable 'factional' entree into the twists and turns of events which informed the strategies of the militant campaign during its highpoint c.1912–1913, a more robust media intervention, focused exclusively on Davison, came in the form of a Channel 4 documentary, *Secrets of a Suffragette* (2013). Presented by the broadcaster, horseracing commentator and feminist Clare Balding, a team of film analysts shed new light onto the final moments of Davison's life in her bid to breach the King's horse as it rounded Tattenham Corner. The analysis of three sets of newsreel footage confirmed that Davison had a much better grasp of the positioning of the field than previously thought. As such, the protest was well calculated—even if the outcome was not. Less surprisingly, the speed of the oncoming horses, estimated as travelling at 35 mph, reveal that in all probability Davison never stood a chance of attaching the silk sash to Anmer. Taken out with considerable force, her 'failed' protest brought down the King's horse and in so doing endangered the life of the jockey, Herbert Jones. He survived, as did the horse, but Davison never regained consciousness and died four days after the race, 8th June 1913, from internal injuries and a fractured skull.

The documentary is useful for having confirmed the more exact details of the circumstances surrounding Davison's protest that day, concluding that her aim was to deliver a publicity coup for the suffrage campaign. It was, however, less emphatic as to whether she truly intended to sacrifice her life for the Cause—the feeling of the current owner of the sash was that Davison 'misjudged it'—in spite of the fact that any considered viewing of the newsreels indicates that a woman of her known intellect and campaign experience could not have been oblivious to the danger of making such a bold protest.

Nevertheless, it is argued that she left no note, carried a return train ticket from Epsom and a postage stamp in her purse. With the benefit of historical distance and for reasons that will be discussed further below, none of these necessarily indicate that Davison believed she would survive; it might equally be argued that she showed a blatant disregard for her own personal safety, simply to seize a rare opportunity to gain maximum publicity in the name of 'Votes for Women'. From that point of view, in her time, Davison was a dangerous woman. The outcome could have been worse: had Herbert Jones, or for that matter the King's horse, also died in the fall that day, it is fair to assume that the relative respect which has accrued to Davison since 1913 for her 'courageous' act would have been if not more muted, then more loudly contested.

Regardless of the harm she might have caused to others then, Davison seemed prepared to become a political martyr. Recently, speaking about the making of the film *Suffragette,* historian Elizabeth Crawford (Pathé/Film4/BFI 2015) endorsed this view, noting that Davison would have been cognisant of the risk she was taking in placing herself on the racecourse. Moreover, the aforementioned incidental evidence cited to inject doubt into this claim—that she carried a return train ticket and postage stamp in her purse that day—is readily dismissed by Crawford: a return train ticket would have been the only type available to purchase; as for the postage stamp, it was customary for suffragettes to carry one to allow the sending of a postcard should they be arrested and held in custody.

This chapter will outline the circumstances which compelled Davison to take such drastic direct action and, in doing so, choose an apparently unrelated sporting event as the platform upon which to stage her political protest. Drawing on Joyce Kay's (2008) research, it will go on to highlight how suffrage protest as associated with sport went well beyond Davison's single act, rendering it more significant in the militant campaign than perhaps hitherto recognised.

Before Derby Day

To understand better why Emily Wilding Davison was prepared to risk and ultimately lose her life in 1913 for the demand of Votes for Women, it is necessary to know something of the prevailing social and political climate in Britain during the long period of protest, generally agreed to have begun in the mid-1860s. It was a demand grounded in a recognition of the inequalities experienced by all women at that time on account of living in an entrenched patriarchal society. Regardless of social class, in Victorian Britain women occupied a position subservient to men, if not entirely in the domestic sphere where women were deemed best suited, then certainly in all spheres of public life where they were denied comparable access to education, occupation, political life and, therefore, all of the social and cultural benefits which followed from these (Perkin 1993; Holloway 2005; Gleadle 2009). Even before the mid-nineteenth century, there is evidence to show that it was a predicament keenly felt by those women belonging to the middle classes who aspired to live an independent life or had been compelled to do so through want of financial (effectively male) support (Vicinus 1985; Perkin 1993). Especially, due to the restrictions females faced in education and employment, the options to pursue independence in a society which formally and informally prioritised male interests over those of females were extremely limited. Whilst women married to financially solvent men were, on the one hand, beneficiaries of the associated security wedlock brought, on the other, they too suffered under the laws of coverture which rendered their person, that of their children and any capital they brought to the marriage the property of their husband (Griffin 2012). It was those with the time to reflect on their personal predicaments, as well as the burdens of their sex more generally, that is, as 'shackled' by laws and gendered social mores who began to publicly voice their dissatisfactions and organise to challenge the disadvantages they encountered throughout life as females.

Relative to other campaigns specifically aligned to the attainment of women's rights (e.g., in education, marriage, sexual health, employment) the campaign for The Vote can be understood as having acquired a special status. This is because it was regarded as the lynchpin in securing women's voice *generally* within the polity (Banks 1981). Possessed of The Vote, campaigners believed that women would be recognised as equal citizens invested with the civil rights commensurate with that equality. As things stood they were, in effect, receptacles for imposed responsibilities and duties on account of the conventional wisdom that they were the weaker and less competent sex. With The Vote in hand, a collective political voice could be mobilised to promote these more specific causes and thereby further women's progress in society overall (Caine 1992). Indeed, still in the throes of leading a decade-long campaign, in 1913, Emmeline Pankhurst asserted: 'We are going through all this to get The Vote so that by means of The Vote we can bring about better conditions, not only for ourselves but for the community as a whole'.[3]

It must not, however, be forgotten that until 1914 when the militant campaign was suspended due to the outbreak of the First World War, the issue of the franchise was as much one of social class as it was of sex: Britain was an emerging democracy, not a fully established one. The majority of working-class men were disenfranchised too and would remain so until after the War's end. This alerts us to the fact that the issue of securing female emancipation via The Vote was a complex and far-reaching one. For example, whilst working-class women were subject to the precedence of male authority in law as well as in custom and practice, different lived experiences conditioned their commitment as working-class suffragists to the Cause as they tussled not only with their own political disempowerment as women but also that of their male kinfolk (Liddington and Norris 2000). It was against this complex backdrop that Emily Wilding Davison, firstly, lived for the Cause and, secondly, died for it.

Embodied Political Protest

As someone whose educational and occupational choices had been restricted on account of her sex, like others in the movement Davison personally felt the impact of what it meant to be a woman living by the standards of patriarchal Victorian society. As her biographers Morley and Stanley (1988) note, it was personal frustration with the inability to move much beyond the position of governess that initially brought Davison to the WSPU in 1906. Coincidentally, in the same year, the *Daily Mail* labelled militant campaigners 'Suffragettes'. The WSPU smartly turned the unflattering label to its own advantage, recognising it as an opportunity

---

[3] 'Freedom or Death', Speech at, Parsons Theatre, Hartford Connecticut, 13 November, 1913.

to bring further distinction between its own strategies of non-violent violent direct action versus the constitutional methods of the National Union of Women's Suffrage Societies (NUWSS), established in 1897. So whilst the latter used what Stanley Holton (1995: 289) refers to as 'the old methods', such as petitions, letters to the press, deputations to Parliament and occasional large public meetings, the WSPU had progressed to tactics associated with heckling at political meetings, obstruction in public places and mass marches on Parliament. From 1909 onwards, there were moves to more aggressive acts, such as window breaking, arson and even occasional targeted attacks on politicians (Rosen 1974). However, no one could have predicted that the stakes associated with militancy would be raised even higher; or that one of its protagonists would become the defining figure of it through an audacious protest at a sporting event and, as a consequence, become a martyr to the Cause.

By 1906 too, the WSPU had relocated from Manchester to headquarters in London. There were also branches all over the country and its advocated campaign of 'Deeds Not Words' was established, justified on account of only gaining illusory traction with the Liberal government, in spite of repeated attempts to engage its representatives to win backing for a women's suffrage measure (Purvis 1995). Suffrage historians tend to regard the Liberal government's back-peddling on the 1910 Conciliation Bill—the first of three parliamentary proposals to give The Vote to around one million wealthy, property-owning women—as one of several tipping points fuelling the militant campaign.[4] But to understand Emily Wilding Davison's decision to choose Derby Day as a platform for political protest, it is important to recognise that the issue of martyrdom was not grounded in repeated acts of government duplicity associated with suffrage bills and successive failed amendments, but in the demand for recognition of women as bona fide political campaigners with legitimate reason to protest.

Hunger strike as a means of protest (and lesser known thirst strikes and sleep strikes) can be understood as the ultimate commitment to the campaign's

---

[4] On suffrage bills see Teele (2014) 'Ordinary Democratization' and Lance (1979) 'Strategy Choices' who alternatively identifies the defeat of the third Conciliation Bill (1912) as a pivotal for the WSPU and, later, Britain's entry into the First World War.

underpinning philosophy of 'Deeds not Words', that is, as opposed to the negotiating (constitutional) approach advocated by NUWSS. The latter organisation was seen by the WSPU's leadership and membership as unable to yield anything meaningful by way of progress, given the failure of peaceable means since the campaign's inception (Stanley Holton 1995). The first campaigner to resort to hunger strike was Marion Wallace Dunlop, a well-to-do artist, illustrator and socialist, who went 91 hours without food in July 1909 on the principle that she should be held under the conditions of the First Division, that is, as a *political* prisoner, not a common criminal.[5] Her 'crime' had been to stamp a quotation from the Bill of Rights on the wall of St Stephen's Hall at the House of Commons: 'It is the right of the subjects to petition the King, and all commitments and prosecutions for such petitioning are illegal' (Rosen 1974: 118). Only a few days into a one-month sentence at Holloway prison, she was released on medical grounds. Wallace Dunlop had taken a unilateral course of action, but this extreme form of embodied protest quickly found endorsement, not only by other suffragettes who followed her initiative but also by the WSPU leadership.

Fearing that it would be only a matter of time before a suffragette death occurred in custody and anticipating negative public reaction, in September 1909, the Liberal government sanctioned forcible feeding to mitigate death by hunger strike as a possible consequence of custodial sentences passed down to protestors. Whilst the WSPU branded this treatment of women as 'torture', the government of the day considered this use of heavy restraint and a crude nasogastric tube inserted into the nose as at best expedient, at worst acceptable on account of the damaging 'terrorist' tactics increasingly deployed by suffragettes. However, rather than cowering the most forthright in the movement, forcible feeding simultaneously hardened their resolve and played into the hands of the well-oiled militant propaganda machine. As Lisa Tickner (1987) has noted, the production of posters, pamphlets and articles in their own periodicals made vivid the degradation endured by women simply for demanding their rights: namely, their entitlement to The Vote as free

---

[5] This event also subject of commemoration by suffrage historian June Purvis: 'Suffragette hunger strikes, 100 years on', *The Guardian*, Opinion Online, 06 July 2009, http://www.theguardian.com/uk/commentisfree Accessed March 2016.

women and due regard for their status as political prisoners when denied that freedom on account of demanding it.

In this battle of wills, the Liberal government has been assessed as having lost public sympathy for what, by any standards, presented as a decidedly *illiberal* approach to the management of albeit unruly women—although that did not prevent continuation of the practice (Geddes 2008). Thus, well before Emily Wilding Davison unilaterally settled upon her own brand of direct action, there was an existing precedent within the militant campaign to engage in high-risk tactics. Firstly, as an effective means of further elevating the demand in the public eye and, secondly, with a view to forcing the hand of Westminster politicians to give in—or otherwise worry about the possible future consequences of not doing so.

'No Surrender!': From Holloway to Derby Day

Davison's Derby Day protest becomes more intelligible when her previous activities as a campaigner are considered. She had worked full-time for the WSPU, written for its paper *Votes for Women,* set fire to pillar boxes and hidden herself with others in the House of Commons.[6] One incident is, however, particularly informative of what we now know was to come. During a term of imprisonment in June 1912, she had not only contemplated, but also had actively endeavoured to take her own life as a means of putting an end to the suffering inflicted on others engaged in the struggle. Having resisted, been restrained and then endured a bout of forcible feeding, on exiting her cell Davison recalled 'I climbed on to the railing and threw myself out on to the wire-netting, a distance of between 20 and 30 feet. The idea in my mind was "one big tragedy may save many others"; but the netting prevented any severe injury.' Refusing to go back to her cell or receive any aid, when the wardresses dropped their guard she turned her attention to an iron staircase:

'... quite deliberately I walked upstairs and threw myself from the top, as I meant, on to the iron staircase. If I had been successful I should undoubt-

---

[6] Entry by Emily Wilding Davison in the *Suffrage Annual & Woman's Who's Who (1913)*, cited in Morley and Stanley (1988), *Life and Death,* pp. 186–87.

edly have been killed, as it was a clear drop of 30 to 40 feet. But I caught once more on the edge of the netting. … I realised that there was only one chance left, and that was to hurl myself with the greatest force I could summon from the netting onto the staircase, a drop of about 10 feet. I heard someone saying, "No surrender!" and threw myself forward on my head with all my might' (Statement, cited in Colmore 1913: 43–48).

By 1913, matters had become dire for the hard-line activists and leading militants when the government not only reneged on its promise to include an amendment to qualify some adult women in a Male Suffrage Bill, but also engaged in a programme of further suppression designed to break the spirit of even the most resilient campaigners. The 1913 Prisoners (Temporary Discharge for Ill-Health) Act, popularly referred to as the 'Cat and Mouse Act', allowed for those on hunger strike to be released when their constitution was on the cusp of breaking down—only to be rearrested and re-imprisoned to finish their sentences once deemed 'fit' to do so (UK Parliament/UNESCO 2012).

In a bid to retain the tactical upper hand, the government had now entered a phase of cynically gambling with women's bodies. When Davison arrived at Epsom racecourse on the 4th June for Derby Day, it would not be an overstatement to say her body was already wrecked as a result of committed physical protests. Even so, in a letter to a friend (January 1913), she reduced her condition to 'bumps', rheumatism and aching in her neck and back (Colmore 1913: 53); this a woman who had sustained injuries to her head, two vertebrae, sacrum and right shoulder blade in the staircase incident. Moreover, since then she had again been on hunger strike in Aberdeen jail, having been charged with assault on a Baptist minister whom she mistook for Lloyd George. She may have, therefore, also privately contemplated the release that would be hers should she not survive her next political endeavour.

Innovation: Sport as a Platform for Political Protest

In spite of the rise of other mass spectator sports from the late 1890s onwards, Davison could not have chosen a better sporting event than the Derby as a platform for what would be her last protest. Although horseracing historian

Mike Huggins identifies the Derby Stakes as having reached its heyday during the Victorian period, in 1913 it still possessed a cross-class appeal. This enabled the event to attract huge attendances and it was, therefore, keenly reported via the press to international audiences.[7] Well into the 1890s, the sweepstake race for 3-year-old horses was 'by far Victorian England's leading sporting, cultural and large-scale *mega-event*' (Huggins 2013: 123). As such, the 'sport of Kings' had not lost its lustre as England's national sport. Thus, Davison could not have made a better choice for her campaigning swansong. As Tickner (1987: 54) observes in relation to the use of [mass] demonstrations to bring further visibility to the demand for The Vote, 'The public demonstration was founded on a politics of "seeing as believing" which, if carefully attuned to the sensibilities of the watching crowds, could be a powerful instrument in winning their sympathy for the cause.' If this is the case, it might be argued that Davison's masterstroke was that she took the notion of 'seeing as believing' to an entirely different level for the spectators on Derby Day that year. The unprecedented sight of a single woman ducking the rails and deliberately placing herself in the path of a field of galloping horses can be projected as having elicited a sense of *disbelief*—why would anyone, let alone a woman, do such a thing?

In the immediate aftermath as racegoers invaded the course to catch sight of the unknown woman lying motionless on the turf, they had their answer: with a WSPU flag wrapped around her body Davison was identifiable as a suffragette and so, by definition, well-understood by the general public at that time as a militant.[8] This perhaps made what she did more comprehensible, although no less shocking. As an example of direct action, it brought sharply into focus the lengths to which a campaigner dispossessed of rights was prepared to go when all other measures to secure them had failed. From this perspective, Davison's extreme gesture continues to be instructive. Drawing on Ralph Turner's determinants of social movement strategies, Lance (1979) maps how frustration

---

[7] For a flavour of the occasion, see 'The Derby 1913: The Race From Start to Finish And Incidents of the Day', *Topical Budget* (Newsreel), available at British Film Institute (BFI) online, www.bfi.org. uk (Accessed March 2016).

[8] Colmore (1913) writes that Davison had told others that she would be going to the Derby and, further, notes that she had collected two 'flags' from WSPU headquarters on the morning of the Epsom race meeting, p. 57.

associated with the diminishing returns from the Liberal government saw the WSPU pass through an early phase of 'persuasion' in keeping with the values of their target group (the Liberal Government), before resorting to increasingly coercive tactics. The difficulty from the outset for the suffrage campaign was not the strength or validity of their arguments for women's rights, but rather that, as an already subjugated group within society, they had nothing of exchangeable value with which to negotiate—except ultimately their bodies. Specifically for the Liberal Party, enfranchising significant numbers of women did not guarantee that the women's votes would be cast in their favour. Thus, in spite of the 'expressive' power of the WSPU as a highly visible mass movement with a sophisticated publicity campaign, ultimately it had no choice but 'to create an impossible situation for the government in which it would be forced to grant women The Vote' (Bearman 2005: 375). In the face of successive Liberal governments' intransigence from 1905 onwards, returning to a 'quiet campaign' was not a viable option. Emily Wilding Davison simply reflected the deeply felt injustice of a woman's place in the world and, for this reason, she intended better notice to be taken of it.

Aftermath

Whether she had lived or died, Davison had generated for herself and the WSPU a 'win-win' outcome in terms of securing publicity for the Cause. Had she attached the sash to Anmer's bridle, the King's horse would have carried the message of 'Votes for Women' across the finishing line at an event which attracted interest 'not just from nearby London, but from throughout Britain, and from Europe, America and the British Empire' (Huggins 2013: 123). As it stood, on the evening of 4th June, the newsreels were already being projected to flabbergasted audiences; by the following morning Davison's death and the motivation behind it had become national and international headline news.

In the aftermath, the publicity generated was born out of both incredulity and the tragic consequences of what has since been described as "a publicity stunt that went horribly wrong" (Channel 4, 2013). With the benefit of hindsight, it is difficult to believe that Davison would have expected—or accepted—her protest to be described as such. Unlike her other failed 'finale' in Holloway prison, this one had apparently gone to plan. As a seasoned campaigner, Davison would have well-understood the mentality of the WSPU and might have anticipated the organisational response to her passing. The funeral procession staged by the WSPU was worthy of an Establishment dignitary and drew a public audience to match (Tickner 1987); renowned suffrage historian June Purvis (2013: 358) has called it 'the last of the great suffragette spectacles'—the occasion secured her in the public psyche as a political martyr.

'It Wasn't Just Emily Davison' Revisited: Sport as a Platform for Political Protest

Whilst it is not clear whether anything in Emily Wilding Davison's decision to target the Derby was informed by it being a site of male privilege, she would have been well aware that women were excluded from the institutionalised sports of flat racing and steeplechase, as they were other male-dominated sports. Indeed, as Joyce Kay's (2008) research reveals, racecourses had been the focus of suffragette ire in England and Scotland before Davison hatched her Derby Day plan. Listing the most significant attacks on sport during 1913, in April alone Ayr, Cardiff and Kelso racecourses saw concerted attempts to wreak material damage to their facilities—the grandstand at Ayr was effectively destroyed after c.£3000 worth of damage was done, and at Kelso too there was 'Attempt to burn down [the] grandstand'. Attacks later that year saw Hurst Park racecourse facing costs of c.£12,000 and Aintree escaping an attempt to burn down the County Stand (Kay 2008: 1343).

Such sites were considered fair game by Suffragettes simply because in Britain sport had been appropriated by men in ways which had benefited their own interests, as consistent with the assumptions and conditions of the prevailing patriarchal society. The codification and institutionalisation of sport in its various forms is well documented. While it did not completely exclude female participation, it severely compromised it (Hargreaves 1994). In attacking the material resources of sport—whether turf, clubhouses, pavilions or grandstands—activists were inflicting a tangible disruption upon male social practice and thereby upon a symbolic pillar of their social status and masculinity. Attacks on sports venues were effectively attacks on property, so were provocative *and* caused real practical inconveniences. Especially affected were well-to-do spectators and players as they incurred losses of seats in the grandstands, the comforts of their pavilions and, perhaps worst of all, a hit to their valuable playing time. Most effective from the latter point of view was the spoiling of bowling and golf greens.

Indeed, during her famous 'Freedom or Death' speech at Hartford, Connecticut, in November 1913, it was golf, 'the great recreation of England', that Emmeline Pankhurst chose as an example of male sporting privilege. Men, she observed, 'so monopolise the golf links that they have made a rule that although the ladies may play golf all week, the golf links are entirely reserved for men on Saturday and Sunday.' At issue then, was not that women were excluded from play, but that they were subject to imposed conditions over which they had no say, even though research has subsequently shown that they often contributed much to the life of a club as members of Ladies Sections (George 2010).

The lack of decision-making power which characterised women's experiences of sport therefore directly mirrored their general predicament within wider society. In Mrs Pankhurst's example of golf, the acid and cutting attacks which temporarily prevented play on 'all the beautiful greens that had taken years to make' amounted to much more than something which caused men inconvenience. In her view, it commanded men to *think* about the direct consequences upon themselves of the extreme methods deployed by the suffragettes. This, Pankhurst hoped, might provoke what she called 'practical sympathy' in men, whereby those who were angry might go to the government and say 'my business is interfered with and I won't submit to its being interfered with any longer because you won't give women The Vote' (Pankhurst 1913). As the tone and content of the speech conveys, by 1913, militants were careless of alienating male sympathy; they had grown tired of their words—the aim was to have men join them in deeds.

Given the relative magnitude of such attacks, as well as Emmeline Pankhurst's explicit acknowledgement of them as a form of effective direct action, it is surprising that suffrage historians have not examined this particular aspect of the militant campaign in any detail. Although published in 1995, a text book which covered key topics about Women's History might still be taken as representative of this over-sight, in that there is neither coverage of sites of sport as a platform for protest (Stanley Holton 1995) nor any mention of the challenges faced by women historically in accessing sport or physical recreation, even though such activities had never been entirely absent from girls and women's lives. Yet, the front cover of the same book depicts two women at the Henley Regatta in 1913, one clutching copies of *The Suffragette*—the same year Emily Wilding Davison died after the Derby and, according to Rosen (1974), Bearman (2005) and Kay (2008) one that saw sites of sport become a more concerted focus of militant protest.

Conclusion

Universal *manhood* suffrage was granted via the Representation of the People Act in 1918. Indeed, the recurring argument that women were 'rewarded' with The Vote as a result of their war work finds a much more convincing parallel in the male example. If men were willing to—and did—lay down their lives for their Country during the conflict then did they not deserve a say in the running of it? The relative rapidity of the bill's passage testifies to there being no meaningful opposition to this single compelling argument made on behalf of men (Blackburn 2011). For women the matter, as ever, remained less clear cut; the point borne out by the fact that not all were granted The Vote without conditions in 1918, unless aged 30 years or over.

Whilst it is true that suffragists and suffragettes held differing views on precisely what type of woman should be granted The Vote (meaning that many amongst their number would have been satisfied with the qualification imposed in 1918), the choice of methods to secure their variable aims was readily discernible as either peaceable or militant, or as Lance (1979) has observed, negotiated or coercive. In the final analysis, the kind of militancy embodied by Emily Wilding Davison, not to mention others like her who targeted sites of sport to progress their demand, seems to have had a more symbolic than tangible impact on the desired outcome of securing Votes for women. Indeed, if Teele's (2014) analysis is accepted, the contribution of the militant campaign was at best marginal, at worst detrimental, to the reaching of a resolution. According to her close analysis, it was the associations and financial arrangements fostered between the constitutional National Union of Women's Suffrage Society (NUWSS) and the Labour Party which sealed the suffrage deal.

The militant campaign and, therefore, the death of Emily Wilding Davison might be considered as less important politically than the prominence afforded to them and the weight of commemorative outpourings suggest. It would seem that the reason historians, feminists and other commentators return so often to Davison as a point of reference is her shrewdness in choosing a great sporting event through which to draw *further* attention to the demand of 'Votes for Women'. Moreover, her incredible demise can still be seen on film—past and present. By 1913, the Cause did not want for promotion—as Tickner (1987) has shown the suffragettes were superior propagandists. Rather, it required something that went beyond public imagination and what had by then become routine outrages contrived by a hard core of women who, in any case, would be put back in their place by men using whatever means necessary to do so. On this count, Davison delivered beyond expectation. Whereas previously the authorities had gone to legislative lengths to deny the movement its martyr, Davison took the power out of their hands with a protest no one could have predicted. Without wishing to detract from the context, reasons and sacrifice Emily Wilding Davison made, it is also worth acknowledging the observation Michael Tanner (2013) made in his well-timed book, *Suffragette Derby:* Davison is important because she was the first to mount a high-profile political protest at a great sporting event and, moreover, be captured on film doing so. Since 1913, others have mobilised sporting events as a platform for protest, when either their own political rights or those of others have been thwarted, thereby drawing down on the suffragette legacy of 'Deeds, Not Words' as a certain strategy to draw attention to their cause.

References

Banks, O. (1981). *Faces of feminism: A study of feminism as a social movement.* Oxford: Martin Robertson.

Bearman, C. J. (2005). An examination of suffragette violence. *English Historical Review, CXX*(486), 365–394.

Blackburn, R. (2011). Laying the foundations of the modern voting system: The Representation of the People Act 1918. *Parliamentary History, 30*(1), 33–52.

Caine, B. (1992). *Victorian feminists.* Oxford: Oxford University Press.

Colmore, G. (1913). *The life of Emily Davison: An outline.* London: The Woman's Press. In Stanley, L., & Morley, A. (1988) *The life and death of Emily Wilding Davison: A biographical detective story.* London: The Women's Press.

Gavron, S. (2015). The making of the feature film Suffragette. *Women's History Review, 24*(6), 985–995.

Gavron, S., & Morgan, A. (DVD, 2015). Special feature: Audio commentary. *Suffragette,* Pathé/Film4/BFI.

Geddes, J. F. (2008). Culpable complicity: The medical profession and the forcible feeding of suffragettes, 1909–1914. *Women's History Review, 17*(1), 79–94.

George, J. (2010). 'Ladies first'?: Establishing a place for women golfers in British golf clubs, 1867–1914. *Sport in History, 30*(2), 288–308.

Gleadle, K. (2009). *Borderline citizens: Women, gender, and political culture in Britain, 1815–1867.* Oxford: Oxford University Press/British Academy.

Griffin, B. (2012). *The politics of gender in Victorian Britain: Masculinity, political culture and the struggle for women's rights.* Cambridge: Cambridge University Press.

Hargreaves, J. (1994). *Sporting females: Critical issues in the history and sociology of women's sports.* London: Routledge.

Holloway, G. (2005). *Women and work in Britain since 1840.* London: Routledge.

Huggins, M. (2013). Art, horse racing and the 'sporting' Gaze in mid-nineteenth century England: William Powell Frith's. *The Derby Day, Sport in History, 33*(2), 121–145.

Kay, J. (2008). It wasn't just Emily Davison! Sport, suffrage and society in Edwardian Britain. *The International Journal of the History of Sport, 25*(10), 1338–1354.

Lance, K. C. (1979). Strategy choices of the British women's social and political union, 1903–18. *Social Sciences Quarterly, 60*(1), 51–61.

Liddington, J. and Norris, J. (2000) (Orig, 1978). *One hand tied behind us: The rise of the women's suffrage movement.* London, Rivers Oram Press.

Morley, A., & Stanley, L. (1988). *The life and death of Emily Wilding Davison with Gertrude Colmore's the life of Emily Wilding Davison*. London: The Women's Press.

Pankhurst, E. (1913, November 13). 'Freedom or death', speech at Parson's Theatre, Hartford, Available at Iowa State University Archives of Women's Political Communication at: http://www.womenspeecharchive.org/. Accessed Mar 2016.

Perkin, J. (1993). *Victorian women*. London: John Murray.

Purvis, J. (1995). "Deeds, not words": The daily lives of militant suffragettes in Edwardian Britain. *Women's Studies International Forum, 18*(2), 91–101.

Purvis, J. (2013). Remembering Emily Wilding Davison (1872–1913). *Women's History Review, 22*(3), 352–362.

Rosen, A. (1974). *Rise up women! The militant campaign of the Women's Social and Political Union, 1903–1914*. London: Routledge & Kegan Paul.

Secrets of a Suffragette. (2013). Channel 4. Broadcast June 2013.

Stanley Holton, S. (1995). Women and the vote. In J. Purvis (Ed.), *Women's history Britain, 1850–1945 an introduction*. London: UCL Press.

Suffragette. (DVD, 2015). Pathé/Film4/BFI.

Tanner, M. (2013). *The Suffragette Derby*. London: Biteback Publishing.

Teele, D. L. (2014). Ordinary democratization: The electoral strategy that won British women the vote. *Politics and Society, 42*(4), 537–561.

Tickner, L. (1987). *The spectacle of women: Imagery of the suffrage campaign 1907–14*. London: Chatto & Windus.

UK Parliament/UNESCO Women's Suffrage Documents. (2012). 'Cat and Mouse' Act, 1913. http://www.parliament.uk/. Accessed Mar 2016.

Vicinus, M. (Ed.) (1985). *Independent women: Work and community for single women, 1850–1920*. London: Virago.

Women's Olympics: Protest, Strategy or Both?

Helen Jefferson Lenskyj

First held in 1896, the Modern Olympic Games have always privileged some groups of athletes over others. Organized by the International Olympic Committee (IOC), a group of self-elected men (and, since 1981, a very small number of women), the Olympic sporting program is by no means representative of world sporting practices. The eligibility rules, both explicit and implicit, have been variously based on sex, gender, sexuality, social class, nationality, religion, race/ethnicity and/or ability, thereby enabling the participation of some groups while posing barriers to others. The founders of the modern Olympics, led by Pierre de Coubertin, prioritized the sporting achievement model, as represented by the faster/higher/stronger Olympic ethos, thereby disadvantaging women as a gender group, as well as rejecting the more inclusive fitness and body experience models of sport that were practiced in many European countries in the nineteenth and early twentieth centuries.[1]

---

[1] For a more detailed explanation of these three models of sport based on Henning Eichberg's work, see Lenskyj, H. (2012) *Gender Politics and the Olympic Industry* (Basingstoke: Palgrave Macmillan), Chapter 2.

H.J. Lenskyj (✉)
University of Toronto, Toronto, ON, Canada

© The Editor(s) (if applicable) and The Author(s) 2016
J. Dart, S. Wagg (eds.), *Sport, Protest and Globalisation*,
DOI 10.1057/978-1-137-46492-7_3

Resistance to the Olympic hegemony has been in evidence from the outset, with numerous alternative sporting and cultural festivals organized by members of excluded groups representing the earliest opposition. Some followed the general Olympic model, while others reflected the distinctive sociocultural values and priorities of the excluded groups. Some of these alternative games, for example, the Workers' Games, had their beginnings in the early twentieth century and continue today; others were relatively short lived, but had important impacts on international sport. One of the most significant examples is the Women's Olympics (from 1926 called the Women's World Games and the Ladies' World Games at the insistence of IOC). They were organized by the Fédération sportive féminine internationale (FSFI) in 1921 under the leadership of Alice Milliat of France and Sophie Eliott-Lynn of Great Britain.

Historians frequently imply that the organizing of the Women's Olympics was a strategy undertaken by FSFI in their campaign to pressure IOC and the International Amateur Athletic Federation (IAAF) to include a full women's track and field (T&F) program in the Olympics.[2] In other words, rather than representing a genuine alternative and a radical political stance, the Women's Olympics were reduced to a mere stepping stone to the 'real' Olympics.[3]

There are at least three possible ways of interpreting the historical facts:

(a) The FSFI took a feminist stance by organizing an international women-only sporting event, controlled by women, which served as a genuine alternative to the Olympics.

---

[2] See, for example, Wamsley, K. and Schultz, G. (2000) Rogues and bedfellows: The IOC and the incorporation of the FSFI, in *Proceedings of Fifth International Symposium for Olympic Research* (London, Ontario: University of Western Ontario), 113–118; Kidd, B. (1996) *The Struggle for Canadian Sport* (Toronto: University of Toronto Press).

[3] The main secondary sources used in this chapter are Guy Schultz's MA thesis, the IAAF and the IOC: Their relationship and its impact on women's participation in track and field at the Olympic Games, 1912–1932 (London Ontario: University of Western Ontario, 2000) and Adams, C. (2002) Fighting for acceptance, in *Proceedings of the Sixth International Symposium for Olympic Studies* (London Ontario: University of Western Ontario), 143–8.

(b) The ultimate goal of the FSFI was to have a full program of women's events in the Olympics, and the Women's Olympics represented a strategy to force the IOC and the IAAF to agree to its demands.

(c) The FSFI adapted its aims and objectives in response to opposition from the IOC and the IAAF, and was eventually forced to compromise and to accept the reduced program offered by IOC.

My focus here is on gender equality in the Summer Olympics in the period 1896–1936. However, readers who are familiar with my critiques of the Olympic industry will be aware that I am not making a liberal feminist case for gender equality in the Olympics today. In my analyses of developments since 1984, I call for the dismantling of the Olympic industry.[4]

Women's Sport in Women's Hands?

Reflecting Western societal norms of the early twentieth century, the FSFI, like England's Women's Amateur Athletic Association, Canada's Women's Amateur Athletic Federation and the American Women's Division of the National Amateur Athletic Federation, upheld the 'women's sport in women's hands' model. Underlying this political position—which contemporary historians have called *maternal feminism*—were some seemingly contradictory rationales. These women tended to take an essentialist view: women, by virtue of their sex, were best suited to control female sport. According to this line of thinking, only women could understand female athletes' physiological and psychological makeup and vulnerabilities, and only women would comprehend the need for protection from sexual harassment and exploitation by male coaches and spectators, the latter a genuine threat but one that was rarely discussed publicly in that era. Maternal feminists also appeared to endorse both the 'female frailty' myth and the 'beauty and grace' rationale favored by conservative men, namely, that women's contributions to the cultural program and opening and closing ceremonies, along with participation

---

[4] See, for example, *Inside the Olympic Industry* (Albany NY: SUNY Press, 2000).

in 'feminine' activities such as gymnastics and tennis, would enhance the sporting spectacle without threatening women's proper place in the social order. And, although this stance appears conservative to contemporary eyes, these women were undoubtedly aware that their equality-seeking initiatives were likely to be either challenged or co-opted by male sport leaders, whereas their quiet efforts at establishing women-only clubs might arouse less male opposition and prove more successful. For their part, female athletes joined women's clubs in large numbers, no doubt because female coaches and administrators bestowed an air of respectability in an era when female T&F aroused widespread censure.

More radical feminists, influenced by suffrage movements in their respective countries, worked toward controlling the organization of women's sport at the national and international levels, and rejected the prospect of taking a subservient position in male-dominated sports organizations. They were not immune, however, to prevailing notions of women's biological uniqueness and responsibilities, sometimes relying on those rationales to gain greater credibility with conservative opponents.

Encompassing some aspects of both political stances, proponents of the liberal/reform position worked toward full female participation in gender-appropriate sports only, while at the same time aiming at meaningful female representation within existing sports organizations at national and international levels. From a twenty-first-century vantage point, one can see that there were limitations to all three positions.

In the broader social context of the early twentieth century, Western medical experts, religious leaders and other guardians of white, middle-class, heterosexual female morality promoted the concept of 'female frailty', an essentialist view of women as solely determined and defined by their reproductive function. Pro-natalist and eugenic propaganda generated idealistic rhetoric concerning the future white middle-class mothers of the nation, and, indeed, the future of the white 'race'. Similarly, vitalism—the theory that human had a fixed supply of 'vital energy' which, in the case of women, must be preserved for future child-bearing duties—served to impose further limits on girls and women. Excessive physical activity was identified as posing the risk of a raft of reproductive problems, ranging from dropped uterus to amenorrhea. Equally important, in the experts' view, was the threat to femininity—a code word for female heterosexual identity when used in the context of sport. An excess of physical stamina and muscle, it was claimed, made women unfeminine and unattractive to men, and the confidence and assertiveness that often accompanied such developments further 'masculinized' sportswomen. These hegemonic views of femininity were as effective as outright prohibitions in the context of sport. Unfortunately, it cannot be assumed that such attitudes became obsolete as the twentieth century progressed. In fact, with the 1960s wave of feminism in most Western countries, the conservative backlash prompted many female sport leaders to revert to the apologetic, 'separate-but-equal' or 'separate-and-different' position of the 1920s and 1930s. In other words, rather than dismissing or challenging sexist and homophobic accusations that female athletes lacked 'femininity', they claimed that they could be both athletic and feminine, and devoted considerable time and energy to proving this.

Track and Field Timelines

In 1896, when Pierre de Coubertin founded the modern Olympic Games, he modeled the event on the Ancient Olympics, an exclusively male sporting competition reported to have been organized to honor Zeus. In doing so, he ignored the ancient Herean Games, a footrace for unmarried women in Ancient Greece, held in honor of the goddess Hera, Zeus' wife. Further following the ancient Greek example of excluding women (on penalty of death, in ancient times), Coubertin famously proclaimed that women's role was on the side lines, applauding their husbands and sons. However, the actual sports programs, including women's events, for the first five Olympics (1896–1912) were determined by the national organizing committees of the host countries, not by the IOC. Women made gradual inroads in these early Olympics, competing in golf and tennis in 1900, tennis, archery and figure skating in 1908, and swimming, diving and tennis in the 1912 Olympics. The situation changed at the 1912 IOC Session, when Coubertin voiced his strong opposition to women's participation, and again at the 1914 Session and Congress.

In his role as president until 1925, Coubertin remained opposed to women's events, especially T&F, although this view was not unanimous within the exclusively male ranks of IOC. This is not to suggest that these men were particularly progressive, but rather that they were politically astute, recognizing that international movements for women's rights had implications for Olympic sport that could not be ignored indefinitely.

Other developments in international sport at this time also influenced Coubertin and IOC. In 1913, after lengthy negotiations, Swedish sports leader Sigrid Edstrom succeeded in persuading Coubertin that a separate international organization, the IAAF, should be established to govern T&F. Since other Olympic sports, including figure skating, cycling, tennis and swimming, had already set up international federations, Coubertin's initial view of a T&F federation as a threat to IOC power may have arisen because of the central place of athletics in the Olympic sports program from the outset. Ironically, as events unfolded, it appeared that the IAAF was more supportive of women's T&F than the IOC, not necessarily motivated by concerns for fairness, or 'conciliation', as one official Olympic source claimed,[5] but rather by awareness of the political implications of women's inclusion or exclusion. Not coincidentally, Edstrom became IAAF president in 1920, a member of IOC in 1920, and an IOC executive member in 1921, thereby facilitating a more cordial relationship between IAAF and IOC.

Some historians view the development of women's sports clubs in France as a response to 'the pronounced hostility' from the male sports establishment and French sport media.[6] Yet French men (and a few women) had established a small number of women's sports clubs, beginning in 1911 with Fémina Sport and, in 1918, the national organization, Fédération des societés féminine sportives de France (FSFSF, later called FSFI). Milliat served as Fémina Sport president in 1918, held the positions of FSFSF treasurer and secretary, 1918–19, and became president in 1919. She did not delay in beginning a campaign to have women's T&F events in upcoming Olympics of 1920 and 1924, making the first unsuccessful request to IOC in 1919. By 1920, FSFSF leadership was completely in women's hands.

---

[5] Athletics and Olympism, ibid., 475.
[6] Marie-Therese Eyquem quoted in Leigh, M. and Bonin, T. (1977) The pioneering role of Madame Alice Milliat and the FSFI in establishing international track and field competition for women, *Journal of Sport History* 4:1, 74.

Western European countries led the way in organizing international women's sports events, beginning in 1921 with the successful Monte Carlo Games, also held in 1922 and 1923. The program offered basketball as well as 11 T&F events, including the 800 m race. Inspired by these games, Milliat and T&F delegates from six countries established FSFI in 1921, with the goal of organizing the Women's Olympics. (Some historians have erroneously credited Milliat with organizing the Monte Carlo Games, which were in fact organized by a local mayor and sports club president.)[7]

Having traveled to the USA, England, Scandinavia and elsewhere, Milliat established a network of sport contacts and maintained a good relationship with the French Ministry of Foreign Affairs. Other French ministries also valued her contributions, inviting her to report on women's physical education and sports. In addition to the growth of women's T&F, Milliat's activism led to the expansion of women's basketball and football in France and some other European countries. However, as well as facing opposition from male sports organizations, FSFI experienced competition from sportswomen who held more conservative positions on women's 'proper place'. While FSFI membership grew to 130 clubs in the years 1920–22, one conservative women's sport federation had attracted more than 500 member-clubs by 1928.[8]

FSFI organized the first Women's Olympics in Paris in 1922, with athletes from England, USA, France, Italy and Czechoslovakia participating in 11 T&F events, including the 1000-m race. Three subsequent Women's Olympics were held in 1926 in Gothenburg, Sweden, 1930 in Prague and 1934 in London, featuring hundreds of athletes and attracting thousands of spectators. The T&F program in the 1930s expanded to 15 events, including the 800-m race, which proved controversial in the 1928 Olympics after false media accounts alleged that most of the runners had collapsed from exhaustion.[9] Women had completed this and

---

[7] Allen Guttman pointed out this common error in his entry, Alice Milliat, in the *Encyclopedia of Women's Sport*, Vol. 2 (New York: Macmillan, 2000), 743.

[8] Terret, T. (2010) From Alice Milliat to Marie-Therese Eyquem, *International Journal of the History of Sport* 27:7, 1157.

[9] Leigh and Bonin, ibid., 78–9.

longer distances in two earlier Women's Olympics without any negative consequences.

In 1923, FSFI again asked IOC to include women's T&F in 1924 Olympics. The IOC ruled that, since it only dealt with international federations, it would not discuss the matter, and it directed the FSFI to direct its request to the IAAF. By 1926, as a result of FSFI's efforts, IAAF had established a special committee on women that comprised three FSFI delegates and three (male) IAAF members, who subsequently controlled technical aspects of women's T&F. The same year, IAAF agreed to include five women's T&F events in the 1928 Olympics on an experimental basis only.

When the FSFI discussed the IAAF's offer, many women were reluctant to accept this diminished program and the precarious future of what IAAF clearly viewed as a trial set of events, but the majority eventually voted in favor. The delegates from Great Britain, however, refused to comply and, in protest, sent no female T&F athletes to the 1928 Amsterdam Olympics, a boycott that dealt a serious blow to the level of competition since British women had dominated the field at the Women's Olympics. Over 100 women from 18 countries participated in five Olympic T&F events in Amsterdam, fewer than half the number of events offered to participants at the two earlier Women's Olympics.

Later in 1928, the FSFI again discussed these issues, voting to send women to future Olympics only if the T&F program were increased to ten events. IAAF voted to retain women's events, but rejected FSFI's demands for ten events; instead they increased the number from five to six, and eliminated the 800-m race. These developments were a major setback for FSFI, whose members continued to demand a full program for the next several years. Men's Olympic T&F had offered 11 events in 1896, and by 1928, had expanded to more than 20.

In 1929, Coubertin was replaced by Belgian aristocrat Louis de Baillet-Latour as IOC president. Although generally portrayed as more progressive on the 'woman question' than Coubertin, one of Baillet-Latour's early initiatives called for women's T&F to be dropped, and for the women's program to be limited to the appropriately 'feminine' sports of gymnastics, swimming, fencing and tennis. When IOC voted to eliminate all women's T&F events, the president of the American Athletic Union and IAAF member, Gustavus Kirby, recommended that IAAF boycott the 1932 Los Angeles Olympics unless women could compete, in some events. With the host country threatening such action, IOC Congress subsequently rejected Baillet-Latour's proposal, and by 1932, a modest program of women's T&F was included in the Olympics on a permanent basis, although the 800-m event would not be reintroduced until 1960. Overall, progress was slow, with only 21 female athletes competing in the 1932 Olympic T&F events, while women comprised less than 10 per cent of all Olympic participants.

FSFI continued its campaign, and in 1934, Milliat wrote to the IOC, in effect threatening a total boycott of female athletes at the upcoming 1936 Olympics:

> The FSFI will accept to give up the World Games if a full program of women's athletics is included in the Olympic Games and under the condition that women are represented within IOC. The FSFI also notes that IOC is less and less willing to open the Olympic Games to women in all sports. For now, it is preferable to keep the Women's World Games which accept all female sports.[10]

The following year, Milliat informed the IOC that FSFI was proposing to 'exclude all participation of women from the Olympic Games since they have their own quadrennial games embracing all feminine sport and controlled by [FSFI]'.[11] Edstrom and other IOC members resorted once more to the position that it would only deal with international federations, not with FSFI. Milliat retired from the position of FSFI president in 1935, citing ill-health, an outcome, according to some historians, of 15 years of struggles with IOC and IAAF, and, to a lesser extent, with the conservative sector of women's sport federations. By 1936, women were admitted to some (male) sports federations and French government funding of FSFI ceased.

---

[10] Milliat cited in Terret, 1163.

[11] Participation of Women in the Olympic Games (May 28, 1935), *Official Bulletin of the International Olympic Committee*, 11

Voting Patterns: IAAF, IOC, FSFI

Given the Euro-centric origins of the modern Olympics, it is not surprising that the future of women's T&F lay largely in the hands of British and European IOC and IAAF members—all of them male. European and British dominance in athletics was also reflected in patterns of female participation in the Monte Carlo Games, the Women's Olympics and the Olympics up to 1936. With two exceptions—1904 in St. Louis and 1932 in Los Angeles—all of these events were held in Europe or Great Britain.

On the question of women's T&F, representatives from some countries can readily be identified as consistent advocates, others as consistent opponents, while a small number wavered on the question. Those generally in favor were France, Austria, Great Britain, Switzerland, Italy and Norway. Finland, Hungary and Ireland were, for the most part, opposed. Of non-European countries, the USA and South Africa were generally positive, while Canada and Australia were inconsistent. From the outset, British delegates took the position that they would not support anything less than a full program of women's T&F events. Moreover, the British Amateur Athletic Association stated that they were not interested in controlling British women's T&F, again not necessarily out of concerns for fairness. In fact, when record-breaking British sprinter Vera Searle requested that women be admitted to this association, she was rebuffed and told to form a women's organization instead. In 1923, she helped to establish the Spartan Ladies, subsequently one of England's leading women's athletics clubs, and became the first secretary of the Women's Amateur Athletics Association (WAAA). According to one obituary, she was 'a traditionalist by nature',[12] the perspective of a twentieth century writer who saw Searle's enthusiasm for women-only clubs as conservative. As late as the 1980s, she opposed moves to integrate women's and men's athletics in one organization.

Most of the male IOC and IAAF delegates who voted against female T&F cited moral and medical grounds for their opposition. The representative from Finland also invoked the classical Greek ideal, claiming

---

[12] Szreter, A. (9 October, 1998) Obituary: Vera Searle, *The Independent*, 7.

that female participation was incompatible with this tradition, thereby conveniently ignoring the (women's) Hera Games of ancient Greece. As well as his reliance on Greek tradition, Coubertin attempted to justify his hostility to gender equality by claiming that 'the French by heredity, by disposition, and by taste were opposed' to gender equality when openly displayed and posing a threat to 'deep-rooted traditions'.[13] As president, he was able to impose his belief system on the rest of IOC members, but some, for example, a later president, Sigrid Edstrom, recognized the strategic value of simply taking control of women's sport rather than continually challenging the fact of its existence.

A comparison of developments in women's suffrage and Olympic sport participation in the early decades of the twentieth century demonstrates some parallels as well as some contradictions. Although Australia granted (non-Aboriginal) women the vote in 1902, Australian sport leaders did not lead the way in promoting women's Olympic participation. By the 1920s and 1930s, when debates were raging over the women's Olympic T&F program, women in most European member countries, as well as in Great Britain, the USA and Canada, had the vote. However, although France (1944), Italy (1946), and Switzerland (1971) were obvious outliers on the suffrage issue, there was significant female participation in T&F in these countries. Indeed, with Milliat leading FSFI, France was a key player. These patterns foreshadowed developments in the second half of the twentieth century, when women's rights movements in some countries took up the issue of equality in sport, while others gave it a low priority.

Alice Milliat, Feminist?

Much can be learned about the FSFI's history by examining what has been written about one of its founders, Alice Milliat.[14] Frequently confronting the all-male IAAF and the all-male IOC as the advocate for women's T&F programs, there is little doubt that, in that historical context, she

---

[13] Coubertin's views, cited by Eyquem, in Leigh and Bonin, ibid., 73.
[14] For sources other than official IOC publications, see, for example, Leigh and Bonin, ibid.; Wamsley and Schultz, ibid.; Terret, ibid.

was viewed as radical and forthright; the 'ladylike' convention that 'you can catch more flies with honey than with vinegar' did not appear to be in Madame Milliat's repertoire. In any event, the men of IOC and IAAF were not noted for their progressive views on any social issue. IOC president (1952–1972) Avery Brundage found her too demanding and a nuisance, and did not hesitate to voice these views to American historian Mary Leigh in a 1973 interview.[15] According to IOC Chancellor Otto Mayer's history of IOC, he and his colleagues lamented the 'abuses and excesses' of the suffrage movement, as they characterized the FSFI's lobbying efforts.[16] Brundage's good friend, IAAF president Edstrom, may have shared these views on suffrage—indeed he wanted FSFI 'to disappear from the surface of the earth'—but, as we have seen, he saw the political expedience of appearing supportive of women's T&F in order to achieve his goal of bringing women's athletics under IAAF control.[17]

A 1927 article in a French sports journal approvingly described Milliat as 'the soul of the women's sport movement... a living example of the modern woman, accustomed to all sports disciplines...'.[18] With a good education and fluency in several languages, she worked as a primary school teacher and later, as a translator. The fact that she was widowed after four years of marriage is often cited as a factor in her subsequent career in sport leadership, on the grounds that she had time and energy as well as organizational skills.[19] It seems likely, however, that she would not have let the responsibilities of marriage and family prevent her advocacy work. Her interest stemmed in part from her own achievements and personal fulfillment in rowing, a sport she took up as a young adult. Like generations of sportswomen since that time, she attributed confidence, resourcefulness and courage to her sporting experiences,[20] while, unlike many of her contemporaries, she generally avoided references to sport as either a threat or a benefit to females' reproductive function.

---

[15] Leigh and Bonin, ibid.

[16] Leigh and Bonin, ibid., 77.

[17] Adams, ibid., 144.

[18] Cited in Quintallan, G. (2000). Alice Milliat and the Women's Games, *Olympic Review* XXVI:31, 27.

[19] See, for example, Leigh and Bonin, ibid., 76; Quintallan, ibid., 27.

[20] Milliat quoted in Quintallan, 27.

Significantly, Milliat understood the relationship between women's political status and the future of women's sport, stating in 1934: '...as we have no vote, we cannot make our needs publicly felt or bring pressure to bear in the right quarters.'[21] However, historians do not agree on the issue of Milliat's feminism, with Terret, for example, claiming that she was 'never a feminist in any formal way and never belonged to any movement... she was a political militant'.[22] Of course, the designation *feminist*, as applied to Western women since the 1960s, could be considered historically inaccurate in the early 1900s context when women like Milliat were advocating for women's rights in sport and in the wider political realm. Furthermore, one could find contemporary liberal 'sport feminists' who 'never belonged to any [feminist] movement' but who have been successful leaders of women's sport organizations. This is not to suggest that the disconnection between 'majority feminism' and 'sport feminism' is a positive trend, but rather that, since sport in Western countries continues to be male dominated, a range of liberal and radical approaches may produce the best outcomes for women.

According to one source, the demise of FSFI demonstrated that it was 'a victim of its own success', having achieved what this author saw as its twin goals of gaining female participation in the Olympics and in the IAAF. Writing in a 2000 issue of the IOC publication, *Olympic Review*, Ghislaine Quintillan also claimed, erroneously, that the 1936 Olympics offered 'an almost complete program of women's athletics',[23] when, in fact, it only offered one sprint event of 100-m, 80-m hurdles, 100-m relay, high jump, discus and javelin, a six-event program that fell well short of FSFI's demands for a 'full program' of 10 events, and yet another example of IOC's broken promises to FSFI. More realistically, Guttman and others have assessed Milliat's victory as only 'partial', given the IOC's failure to expand the program and its decision to eliminate all races longer than 100 m.[24]

---

[21] Milliat quoted in Leigh and Bonin, ibid., 76, from 1934 interview in *Independent Woman*. See also Terret, ibid.

[22] Terret, ibid., 1156.

[23] Quintallan, ibid.

[24] Gutmann, ibid., 744.

Another *Olympic Review* interpretation of these events claimed that female sports leaders' and athletes' 'fear of mixed sport was another stumbling block on the road to unification' between FSFI and IAAF.[25] This alleged fear may well have been motivated by the probability that women's T&F would be subsumed by IAAF, and, as the same item triumphantly concluded, 'in 1936, women's athletics came under the sole authority of IAAF.' The report went on to say, 'Following a vote of thanks to the FSFI for the good work carried out since its foundation, the [IAAF] Congress decides to take over control of women's athletics'[26]—a curt dismissal of 15 years of tireless organizing on the part of FSFI leaders, and an implied message that these men knew what was best for women.

Sophie Eliott-Lynn

Many historical accounts focus on Milliat and tend to neglect the role of Sophie Elliott-Lynn (b. County Limerick, 1896), who seems to have achieved greater fame as an aviator than as an activist. Her first venture in the area of aviation took place during World War I, when she enrolled in the Royal Flying Corps as a dispatch rider (motorcyclist), and her later aviation achievements, including a solo flight from Cape Town to London, reflect her determination to challenge gender-based barriers. The fact that she had been married three times and divorced twice by age 40 further suggests a degree of nonconformity unusual for that era.[27]

In 1923, Elliott-Lynn (then Sophie Pierce-Evans) was elected vice-president of WAAA. Unlike Milliat, she had been interested in sport from a young age and was particularly successful in jumping and throwing events. She probably met Milliat in 1923 at the Monte Carlo Games, where, at the age of 27, she won bronze medals in high jump, javelin and pentathlon. Both women made frequent appearances at IOC and IAAF meetings to press the case for a full T&F program for women, and Eliott-Lynn is

---

[25] Athletics and Olympism (1983) *Olympic Review*, 475.
[26] Athletics and Olympism, ibid., 475–6.
[27] Ware, S. (2005) An Irishwoman's part in admission of women's athletics to the Olympic Games – Sophie Peirce, *Journal of Olympic History* 13, 13–14.

reported to have told the 1926 IAAF Congress that 'women had nothing to gain by participation in the Olympic Games'. One commentator claimed that such a statement 'runs contrary to her other actions and statements'[28]— true, but, as noted earlier, negotiations between FSFI, IOC and IAAF were complex and fluid, and it is quite possible that, in 1926, that statement accurately reflected FSFI's political position. Throughout the negotiations with IOC and IAAF, Eliot-Lynn, along with other British FSFI delegates, displayed their determination to keep women's sport under their control.

Among Eliott-Lynn's other sport-related achievements was the 1925 book, *Athletics for Girls and Women*, published in an era marked by an overabundance of prescriptive, conservative literature on women's sport and physical activity, most of which were written by male doctors and sport leaders. In 1925, she was invited to a Pedagogical Conference held in association with the IOC Congress as a WAAA delegate, and elected to the Medical Sub-Commission. Although British educators generally held a more practical, less overprotective view of girls' and women's physical capacities, particularly on issues relating to reproductive health, than their American and Canadian counterparts, Eliott-Lynn's approach fell squarely in the conservative camp. Admittedly, she was writing primarily as a lay person with a special interest in women's sport administration, and not as a physical educator or doctor. The paper that she delivered to the Medical Sub-Committee reflected a strongly held belief in vitalist theories, as well as an unquestioned assumption that motherhood was the destiny of all girls and young women—all views that no doubt appealed to that male audience. She asserted, for example, that 300 m should be the longest race permitted for women, until it was proven that longer distances posed no threat to reproductive function, particularly the 'ligaments in the pelvic area'—[29] this despite the fact that women ran longer distances in the previous Women's Olympics. However, it is difficult to determine whether Eliott-Lynn's pseudo-scientific statements (based on

---

[28] Ware, ibid. Eliott-Lynn's birth name was Sophie Peirce-Evans, her first husband was Major Eliott-Lynn and her second husband was Sir James Heath. Most sources refer to her as Lady Sophie Eliott-Lynn (also spelled Elliot-Lynn).

[29] Eliott-Lynn, S. (2005, first published 1925) Women's participation in athletics, *Journal of Olympic History* 13, 14.

her own formula of male to female physical capacities) reflected FSFI philosophy, FSFI strategy or her idiosyncratic worldview.

Conclusion

It is easy to analyze the history of FSFI through a contemporary feminist lens, but difficult to assess the political motivations, successes and failures of the federation under the leadership of Alice Milliat and Sophie Eliott-Lynn in that specific sociohistorical context. What is now considered conservative may have been viewed as radical, or politically astute, or pragmatic, in that era. One may conclude, however, that these women and their contemporaries changed the history of women's sport.

References

Adams, C. (2002). Fighting for acceptance. *Proceedings of the sixth international symposium for Olympic studies* (pp. 143–148). London: University of Western Ontario.

Athletics and Olympism. (1983). *Olympic review, 475.*

Eliott-Lynn, S. (2005, first published 1925). Women's participation in athletics. *Journal of Olympic History,* 13, 14.

Leigh, M., & Bonin, T. (1977). The pioneering role of Madame Alice Milliat and the FSFI in establishing international track and field competition for women. *Journal of Sport History,* 4(1), 74.

Lenskyj, H. (2012). *Gender politics and the Olympic industry.* Basingstoke: Palgrave.

Quintallan, G. (2000). Alice Milliat and the women's games. *Olympic Review,* XXVI(31), 27.

Schultz, G. (2000). *The IAAF and the IOC: Their relationship and its impact on women's participation in track and field at the Olympic Games, 1912–1932.* M.A. thesis, University of Western Ontario, London.

Terret, T. (2010). From Alice Milliat to Marie-Therese Eyquem. *International Journal of the History of Sport,* 27(7), 1154–1172.

Wamsley, K., & Schultz, G. (2000). Rogues and bedfellows: The IOC and the incorporation of the FSFI. *Proceedings of fifth international symposium for Olympic research* (pp. 113–118). London: University of Western Ontario.

Ware, S. (2005). An Irishwoman's part in admission of women's athletics to the Olympic Games – Sophie Peirce. *Journal of Olympic History,* 13, 13–14.

A Most Contentious Contest: Politics and Protests at the 1936 Berlin Olympics

David Clay Large and Joshua J.H. Large

Despite claims from Olympic organizers and advocates that the enterprise they cherish is invariably "above politics," the modern Olympic movement has from the outset been rife with nationalistic rivalries, ethnic and religious antagonisms, ideological posturing, and racial prejudice. In other words, for all the talk about "international brotherhood" and "peaceful competition among the youth of the world," the Olympic festivals have always mirrored the contentious political and social realities transpiring in the world outside the athletic arena. Yet no single modern Olympiad has betrayed this state of affairs more prominently than

D.C. Large (✉)
Institute of European Studies, University of California Berkeley, Moses Hall 207, Berkeley, CA, 94720, USA

J.J.H. Large
Departamento de Negocios Internacionales, Carrera 49 N 7 sur 50, Medellín Bl 26, Colombia

© The Editor(s) (if applicable) and The Author(s) 2016   **51**
J. Dart, S. Wagg (eds.), *Sport, Protest and Globalisation*,
DOI 10.1057/978-1-137-46492-7_4

Berlin 1936, those infamous "Nazi Games," which took place against a turbulent backdrop of epic ideological conflict, deep economic depression, and agonizing worries about a possible new military conflagration on the horizon.[1] If we contend, variously, that the International Olympic Committee (IOC) reached bottom, say, when opting for Moscow in 1980, or deciding for Beijing in 2008, or in handing the 2014 Winter Games to Sochi, Russia, it would behoove us to recall what transpired in the Nazi capital and around the world in the summer of 1936. One of the less researched aspects of the 1936 Games is the number of protests and boycott campaigns that they provoked.

Boycott Berlin!

On April 29, 1931, the IOC voted 43 to 16 to hold the Olympic Summer Games of 1936 in Berlin over runner-up Barcelona.[2] At that time, Germany was still a democracy, albeit a beleaguered one, and some of the IOC voters might have hoped that a Berlin-hosted Olympiad might help buttress the so-called Weimar Republic, which was suffering hugely from a surging economic depression and violent assaults from political extremists, most notably the Nazi Party under Adolf Hitler. Less than two years after the IOC venue-vote, Hitler assumed the chancellorship of Germany and the country fell under the control of a government that stood as a mockery of the purported Olympic ideals of fair play, brotherhood, and peaceful competition among all peoples regardless of nationality, race, religion, or ethnicity.

The triumph of Hitler's movement in Germany brought three Olympics-related questions immediately to the fore: first, would the new Nazi government elect to retain an Olympics-hosting obligation taken on by its democratic predecessor? Second, if the German government indeed opted to do this, would the IOC hold to its own choice of Berlin

---

[1] On the 1936 Olympics, see, inter alia, Large, D. C. (2007) *Nazi Games: The Olympics of 1936.* New York: W.W. Norton; Mandell, R. D. (1987) *The Nazi Olympics.* Urbana: University of Illinois Press; Hart-Davis, D. (1986) *Hitler's Games.* London: Century.

[2] Two years later the IOC awarded the 1936 Winter Games to Bavaria's Garmisch-Partenkirchen, but in this essay we are concerned primarily with the Berlin Summer Games.

as the venue for the '36 Summer Games (and also to its later choice of Bavaria's Garmisch-Partenkirchen as the site for the '36 Winter Games)? And finally, if the IOC stuck to its (venue) guns, would all the nations invited to Hitler's elaborate sports parties, above all major democratic countries like America, Britain, France, and Canada, accept the German organizers' invitation? Or, would they try to move the party elsewhere— and/or elect to stay home if Germany remained the host?

With regard to the first question—the Hitler government's stance on hosting the 1936 Olympiad—many people in Germany believed that the Führer would shake off this enterprise like some piece of dreck stuck to his shoe. After all, various National Socialist Party leaders and sports commentators over the years had shown nothing but contempt for the modern Olympic movement, and indeed for most international sporting events, calling instead for purely German competitions and fitness programs based on *Turnen*, or synchronized group gymnastics. In the early 1920s, the Nazi Party had objected to Germans competing with athletes from the Allied countries, which had imposed the "Yoke of Versailles"[3] on the Fatherland. They had also objected to "Aryans" competing with "racial inferiors," such as Slavs, Blacks, and Jews.[4] Hitler himself had called the Olympics "a plot against the Aryan race by Freemasons and Jews."[5]

The Nazi objection to competing with black athletes had special relevance for the American Olympic program. Having had a modest presence in the Summer Olympics in Antwerp (1920) and in Paris (1924), African-American athletes had featured more prominently on the Olympic stage at the Los Angeles Games of 1932. Runners Eddie Tolan and Ralph Metcalfe, labeled the "Sable Cyclones" in the American press, had excelled in the sprints, with Tolan setting a World Record in the 100-m race and an Olympic Record in the 200-m event. For Nazi ideologues, it was a "disgrace" that white athletes, including a German runner named Arthur Jonath, had deigned to compete at all with the likes of

---

[3] A reference to the Treaty of Versailles of 1919, under which Germany had been compelled to disarm and to pay reparations to the victor nations of the First World War.

[4] Haller, G. (1933) "Der Olympische Gedanke," *Nationalsozialistische Monatshefte* 3 (1933): 388–396.

[5] Hart-Davis (1986), p. 45.

Tolan and Metcalfe.[6] *Der Völkische Beobachter*, the Nazis' house-journal, demanded in August 1932 that black athletes be "excluded" from any German-hosted Olympic festival, should one in fact transpire. Given the prospect that the Americans were likely to have black athletes on their team for 1936, probably even more than they had had in 1932, the Nazi paper was dubious about the advisability of a Berlin-hosted Olympiad.[7]

Adolf Hitler, the man in whose purview this Olympics decision clearly fell, did not begin to view the prospect of hosting the quadrennial athletic spectacle more favorably until several months after assuming the chancellorship in January 1933. In March of that year, one of the key members of the German Olympic Organizing Committee, Theodor Lewald, who was half-Jewish, argued in a meeting with the Führer that hosting the Games would provide an invaluable propaganda opportunity for Germany and undoubtedly constitute an economic windfall for the cash-starved treasury. Propaganda Minister Joseph Goebbels argued along similar lines. But what really seems to have erased Hitler's initial opposition to the Olympics project was the prospect of building a grandiose stage for the Games in the Nazi capital, along with the possibility of demonstrating "Master Race" superiority on the athletic field. In May 1933, he therefore let it be known that his government would not only support the Olympic enterprise but would also host the most magnificent Olympic festival ever.[8]

Yet even while belatedly endorsing the Games, the Nazi government, and German sporting associations pursued policies that clashed sharply with Olympic principles of openness and fair play in athletic competition. The Hitler government's program of anti-Jewish discrimination and persecution thoroughly embraced the world of sport, which was forced into conformity with Nazi dogma. In spring 1933, the German Swimming Association banned Jews from its member clubs. Germany's Davis Cup tennis team expelled one of its stars, Dr. Daniel Prenn, because he was Jewish. The Nazi press called for the dismissal of Theodor Lewald on grounds of his part-Jewish ancestry, and he surely would have been

---

[6] Teichler, H.-J. (1991) *Internationale Sportpolitik im Dritten Reich*. Schlondorf: Verlag Karl Hoffmann, pp. 45–47.

[7] "Neger haben auf der Olympiade nichts zu suchen," *Der Völkische Beobachter*, 19.8.32.

[8] Large (2007), pp. 63–64.

pushed out had not the IOC warned that such a move might compromise Germany's chances of holding on to the 1936 Games.[9]

The retention of Lewald, however, was hardly enough to reassure a growing chorus of critics around the world who objected that Nazi Germany was hardly an appropriate site for an athletic spectacle celebrating international brotherhood. Such criticism broached the possibility that the IOC might take the '36 Games away from Germany, or, if that didn't happen, raised the specter of a large-scale boycott of Berlin.

Determined to keep the Games in Germany and to nip in the bud any potential boycott movement, Theodor Lewald convinced the Hitler government to issue a statement promising to respect the Olympic Charter and to welcome to Germany "competitors of all races." The regime added a significant caveat, however: the composition of Germany's own team was nobody's business but Germany's. The Germans hoped the IOC would agree.[10]

And in fact, for the most part the IOC *did* agree. IOC President Henri de Baillet-Latour, a conservative Belgian aristocrat, harbored the view, hallowed among Olympic officials, that the Olympic Committee should avoid taking any "political" positions except in the case of Communist penetration of the Games, which he believed must be thwarted at all costs. Thus, he issued a statement saying the IOC would hold to its 1931 decision for Berlin as long as Germany imposed no racial or religious restrictions on *foreign* participation.[11]

Baillet-Latour's stance essentially guaranteed that the '36 Summer Games would stay in Berlin, but not that they would go un-boycotted (there was little debate or discussion about the Garmisch Games). In fact, an international boycott-Berlin movement now began to crystallize and take on momentum. Interestingly enough, this protest movement had its origins and greatest resonance in the USA—a nation hardly without its own policies of racial discrimination in sport, and elsewhere.[12]

---

[9] Teichler (1991), pp. 45–47.

[10] "Brundage's Views Stir Berlin Press," *New York Times*, 26.4.33.

[11] Lennartz, K. (1994) "Difficult Times: Baillet-Latour and Germany, 1931–1942," *Olympika* 3 (1994): 101.

[12] For example, US Major League Baseball operated an unofficial color bar which stood until 1946. Moreover, when the African American heavyweight Joe Louis fought German Max Schmeling for

Early on, the American push to boycott a Nazi-hosted Olympiad was largely a Jewish affair; American Jews having reacted with alarm and outrage to the Hitler government's anti-Semitic pronouncements and policies. Various Jewish groups asked Avery Brundage, a crusty Chicago-based construction magnate and president of the American Olympic Committee (AOC), to endorse the boycott effort unless the Germans opened their own Olympic program to qualified competitors regardless of religion or ethnicity.[13]

What the Jewish activists did not know was that Brundage was an anti-Semite, as well as a militant anti-Communist and a major advocate of the Nazi-hosted Games and adamantly against any efforts to boycott them. Like Baillet-Latour, Brundage justified his support of the German Games on the grounds that high-level sports and politics occupied independent realms, and that the Olympic movement could survive only if "politics"—that is, left-wing politics—were kept out of it.[14] (Only later, as president of the IOC from 1952 to 1972, would Brundage countenance the participation of the USSR and its Communist allies in the Olympic competitions.)

The German Olympic organizers were understandably pleased by Brundage's stance, but they worried that he might not be able to prevent American Jews from somehow scuttling US participation in the German Games, especially the main pageant scheduled for Berlin. Jewish groups, the Germans knew, were threatening to withhold financial contributions to the American Olympic program should the AOC and the American Athletic Union (whose backing was also needed) vote to send a team to Berlin. Thus, again on the initiative of Theodor Lewald, the Germans decided to concede more ground, in principle at least. In June 1933, they promised not only to observe all Olympic regulations but also that "Jews

---

the world boxing title in June 1936, six weeks before the Berlin Olympics, lynching (murder by hanging) of black people in southern and rural parts of the USA was still common and African Americans were treated as second-class citizens even in areas without official segregation.

[13] Gottlieb, M. (1972) "The American Controversy over the 1936 Olympic Games," *American Jewish Historical Quarterly* 61 (March 1972): 184–185.

[14] The best study on Brundage's views regarding the Olympics and politics is Guttmann, A. (1984) *The Games Must Go On: Avery Brundage and the Olympic Movement.* New York: Columbia University Press.

would not be excluded from membership in German teams" either for Garmisch or Berlin.[15]

The reality, however, was that Jewish athletes, of whom Germany had a sizeable number, were not considered for membership of the German Olympic team. Most notably, a German-Jewish female high-jumper, Gretl Bergmann, was denied the chance to compete in the qualifying rounds for Berlin despite having won the German national championship in 1935. Moreover, anti-Semitic policies in other dimensions of public life in Nazi Germany continued unabated.

Against this backdrop of persistent government-backed racism in Germany, the American boycott movement expanded beyond its original Jewish base to include Catholic and Protestant organizations, labor groups, and the American Civil Liberties Association. On March 7, 1934, a mass rally was held in New York's Madison Square Garden (MSG) to protest Nazi racial policies and to threaten boycotts of German goods along with the Olympics if these policies persisted. Major sponsors of this rally included the Amalgamated Clothing Workers Union, the American Federation of Labor, and the labor-backed National Committee to Aid Victims of Fascism. Also front and center of those calling for action was the National Association for the Advancement of Colored People (NAACP), which insisted that African-American athletes, including that already-famous shoo-in for the American track team, Jesse Owens, must shun "Hitler's Games" on grounds that Nazi Germany was as hostile to blacks as to Jews. The NAACP warned that black athletes might be mistreated if they dared show up in Berlin. Apparently, accepting this argument, Jesse Owens initially declared that he would boycott Berlin should he be selected for the American team.[16]

The diversity of the MSG protest notwithstanding, the prominence of labor's role in it, and in the boycott movement more generally, allowed advocates of an American presence in Berlin to denounce the nay-sayers as one big pack of Communists. (The pro-boycott crusaders provided added fuel for Brundage and his company's liberal use of the red brush

---

[15] Teichler, H.-J. (1989) "Zum Ausschluss der deutschen Juden von den Olympischen Spielen 1936," *Stadion* 5,1 (1989): 47–48.

[16] Large (2007), pp. 77–78, 88–89.

when, at another MSG rally in October 1935, some 20,000 workers cheered a recent storming by leftist longshoremen of the German liner *Bremen* in New York harbor; among other actions, the protestors had pitched a swastika flag into the sea. Predictably, the *Bremen* incident and the supporting rally were denounced by the pro-Berlin faction as Communist provocations.)[17]

Yet writing off the boycott movement as a product of "Jewish-Communist" machinations did not mean it could be ignored. Definitely troubled by the groundswell of anti-Berlin sentiment in the USA, Avery Brundage decided to undertake a "fact-finding" trip to Germany in the fall of 1934, promising to investigate the sporting scene in the Third Reich and to ascertain whether there was any truth to the "rumors" about discrimination. During his trip, he interviewed a few German-Jewish athletic officials, albeit in the presence of uniformed Schutzstaffel (SS) officers. At one point, he put his Nazi hosts at ease by pointing out that his own men's club back in Chicago excluded Jews and blacks. Upon his return to America he unsurprisingly gave the Germans a clean bill of health, saying he saw no evidence of racism and echoed German assurances that there would be no discrimination against any of the foreign athletes competing in Berlin.[18]

Undeterred by Brundage's whitewash of Nazi Germany, the American boycott-Berlin movement continued to expand. In yet another effort to undercut domestic protest sentiment, Charles Sherrill, one of three American members of the IOC and, like Brundage, a strong proponent of keeping the Games free of "politics," traveled to Germany in the summer of 1935 with the goal of persuading Hitler to include at least one Jew in its Olympic program, a gesture he privately equated with the American tradition of the "token Negro." Although he found his task of pressing the Führer somewhat distasteful—he thought America had no business telling the Nazis to eschew racism when it harbored plenty of racism itself, a racism he personally shared[19]—Sherrill warned Hitler that unless

---

[17] "Bremen Rioters Cheered by 20,000," *New York Times*, 9.8.35.

[18] "U.S. Will Compete in 1936 Olympics," *New York Times*, 27.11.34.

[19] Sherrill lamented the "disproportionate representation" of Jews in Washington, who he said were "raising hell" against him and the American pro-Berlin faction in general. See "Mahoney in Clash with General Sherrill," *New York Times*, 24.10.35

he added a Jew to his Olympic team, Berlin-advocates like himself and Brundage might not be able to prevent a US boycott in 1936.[20]

Hitler briskly rebuffed Sherrill, even threatening to call off the Berlin Olympics entirely and to substitute "purely German" games in place of the international festival. Yet this was a bluff. Hitler knew that German-only games would be useless in terms of propaganda, and would not facilitate the desired generation of foreign currency. In the end, he acquiesced to a gesture of compromise tokenism worked out between Sherrill and the German Organizing Committee. The Germans agreed to name a half-Jewish fencer, Helene Mayer, to their team for Berlin. For the Nazis, this decision was made more palatable by the fact that Mayer was an excellent fencer with medal-winning prospects for Germany. Moreover, she looked the part of a perfect Aryan Valkyrie, with long blond hair, blue eyes, and a statuesque physique. Finally, Mayer, who was by then living in exile in Oakland, California, promised not to criticize Hitler's regime in any way.[21]

The Mayer concession weakened the American boycott effort, which was further undermined by a lack of public support from President Franklin Roosevelt, who kept silent on the issue despite calls from the US consular staff in Germany to signal presidential disapproval of Hitler via an American no-show in Berlin. At a crucial meeting of the American Athletic Union in December 1935, Brundage was able to outmaneuver his opponents and, by a very close vote, secure an endorsement of US participation in the German Games. Revealingly, one of the delegates at this meeting who pushed for a negative vote on Berlin felt obliged to reassure his colleagues that "he was neither a Red nor a Communist," while another prominent boycott advocate, Pennsylvania governor George Earle, considered it necessary to couple his denunciations of the Nazis with accusations of Soviet "flagrant breach of trust toward this country and our people through the attempted spread of the Communist doctrine within our borders."[22] This "I'm no Red" plea underscored once again just

---

[20] Aufzeichnung über den Empfang [Sherrill's] am 24.8.34, 4508, Politisches Archiv des Auswärtigen Amts, Berlin.

[21] Large (2007), pp. 86–87. See also Mogulof, M. (2001) *Foiled: Hitler's Jewish Olympian.* New York: RDR Books.

[22] "Governor Earle Urges Ban on Olympics," *New York Times,* 7.9.35.

how vulnerable the entire stop-Hitler campaign was to charges of complicity with Communism. Such concerns would also underlie concerted efforts by the Western democracies to appease Hitler in coming years.

Meanwhile, for its part, the NAACP also proved unsuccessful in its effort to get American blacks, including Jesse Owens, to boycott Berlin. Owens in particular not only hoped to showcase his brilliance on a major international stage but also envisaged parlaying expected Olympic gold into a remunerative post-Games professional career.[23]

America's decision to go to Berlin helped to undercut similar boycott efforts elsewhere in the democratic West, from Great Britain to Switzerland. Crucially, Olympic officials in Europe's leading democracies, Britain and France, opted to wait and see what the US did before making a definitive decision of their own regarding participation at Berlin '36.

Britain's anti-Berlin campaign, spearheaded by the Trades Union Council, chief organ of the country's labor movement, argued vehemently that Britain should have no part in an Olympiad hosted by a regime whose "politicization and militarization" of sport was completely incompatible with the Olympic spirit. Yet the British Olympic Association (BOA), emulating its American counterpart, inclined strongly toward sending teams to Germany. The British Foreign Office, for its part, opposed using sport as a "political weapon" against a German government it was loath to antagonize in any event. America's decision to participate in the German Games constituted the final determinant in the BOA's vote to take the same route.[24]

Canada, which kept a close eye on Britain's Berlin debate, developed a serious boycott movement of its own, led, like America's, by Jewish groups determined to thwart an early inclination on the part of the Canadian Olympic Committee to send teams to Garmisch and Berlin. Canada's anti-German Games faction used public education, organized protests, and political pressure to push their agenda. However, despite numerous rallies and support from organized labor groups, Canada's opponents to the German Games failed to prevent their nation from following Britain

---

[23] On Owens, see Large (2007), pp. 88–90; and Schaap, J. (2008) *Triumph: The Untold Story of Jesse Owens and Hitler's Olympics*. New York: Mariner Books.

[24] Large (2007), pp. 101–106.

into that highly contested Olympic arena.[25] (Canada, we might add, was especially anxious to participate in the Garmisch Winter Games, where it fully expected to win a gold medal in ice hockey. It did not. Britain, deploying some Canadian-born players living in England, managed to secure the gold medal.)

France, directly threatened by the new Nazi Reich, had greater reason to shun Berlin than did Britain or Canada, but its boycott effort ultimately foundered on old divisions within the French sporting community between its bourgeois and working-class components. Bourgeois sports clubs in France had from the outset warmly embraced the modern Olympic movement (which after all was founded by a Frenchman) and even reconciled themselves to Nazi-hosted Games. By contrast, proletarian athletic clubs, having sponsored their own "Workers' Olympics" since 1925, wanted no part of "les jeux Hitlerienne." Yet the workers' clubs and their political allies on the left lacked the clout to determine national policy in this domain. France would send a team to Berlin even though Hitler's Wehrmacht occupied the Rhineland in the interval between the Winter and the Summer Games![26]

In the end, as a matter of fact, *no* nations elected to boycott "Hitler's Games," although Spain failed to send a team to Germany due to the outbreak of the Spanish Civil War in July 1936. (In hindsight, we can see that the collapse of the anti-Berlin crusade pointed up a larger reality about Olympic boycott efforts: they tended to succeed only when backed by national governments, as was the case with the large-scale state-driven boycotts of the 1980s.)

On the other hand, the willingness of all the national Olympic committees to accept the German organizers' invitation to Berlin did not mean that some individual athletes, most of them Jewish, did not choose to carry out personal boycotts of their own. Perhaps the most prominent personal boycotters of Berlin were three Jewish swimmers from Austria: Judith Deutsch, Ruth Langer, and Lucie Goldner. Despite growing anti-Semitism in their country, all three had been named to the Austrian team

---

[25] On the Canadian boycott movement, see Menkis, R. and Harold Troper (2015) *More Than Just Games: Canada and the 1936 Olympics*. Toronto: University of Toronto Press.

[26] Kidd, B. (1980) "The Popular Front and the 1936 Olympics," *Canadian Journal of the History of Sport and Physical Education* 11, no. 1 (1980): 1–8.

by virtue of their stellar performances in the Austrian national championships. They all quit in protest against anti-Jewish persecution in Germany. Such was also the case with Canada's best lightweight boxer, Yisrael "Sammy" Luftspring, who told the *Toronto Globe* that it would have been unconscionable for him to compete in a land "that would exterminate [the Jews] if it could." Three Jewish-American runners considered likely possibilities for Team USA, Milton Green, Norman Cahners, and Herman Neugrass, refused to participate in American Olympic tryouts after learning of Nazi Germany's racist Nuremberg Laws (1935). Yet another boycotter of the trials was the Jewish-American discus–javelin–shotput standout, Lillian Copeland, winner of the female discus completion at the Los Angeles Games.[27]

Alas, personal boycotts by a few individual athletes of conscience did not have much of an impact on the Berlin Games; their absence was barely noticed. Yet, before leaving this topic, we might well ask what might have been the consequence had *entire national teams* from America and other leading democracies decided to stay away. At the very least, a wide-scale sit-out would certainly have ruined Hitler's Olympic party. This in turn would have been a major blow to a regime that had invested huge resources in its Olympic project, and which was still struggling to consolidate its power and prestige at home amid ongoing economic depression, worries regarding Nazism's image around the world, and growing anxieties over the Reich's rearmament program.

Policing, Persecution, and Protest

The German Olympic organizers, along with the Hitler government, were careful to extend a warm welcome to the hundreds of foreign athletes descending on Berlin in the summer of 1936. Orders went out to be gracious and cordial to all visiting Olympians, regardless of how they looked or where they came from. With the exception of a large contingent from Japan, a small team from China, and even smaller ones from India and a few Latin American countries, the foreign squads at Berlin

---

[27] Large (2007), pp. 106–109.

'36 were overwhelmingly Caucasian. America's team for Berlin, however, contained eighteen blacks and six Jews, far more than ever before—and a clear challenge to Nazi doctrines regarding racial homogeneity. Nonetheless, Team USA's Jews and blacks experienced none of the hostility or mistreatment they had been warned to expect from various groups at home. The Jewish athletes would see no signs of open anti-Semitism in the streets, such displays having been banned by the Nazi government for the duration of the Games. Jewish athletes were careful not to draw attention to themselves—unlike the famous Jewish-American swimmer Mark Spitz, who on the occasion of the second German-hosted Summer Olympiad, Munich 1972, observed during a television interview that a lampshade he was sitting next to might have been "made from the skin of one of my aunts."[28] As for the American blacks at Berlin '36, they were greeted quite enthusiastically by the locals. Jesse Owens, his earlier athletic feats well known to German fans, was mobbed by autograph seekers wherever he went. Young women pressed love letters into his hands.

But what Jesse Owens and his black teammates did not know was that they were being closely monitored by the German police, who were determined to prevent any "unsuitable" contacts between the visitors and natives. Fearing possible acts of miscegenation between the black Americans and willing German women, the Gestapo issued 52 warning citations to female citizens "for approaching foreigners, especially colored foreigners, in an unseemly manner."[29]

Moreover, Owens along with all the other visiting Olympians had their incoming and outgoing mail screened by a Gestapo inspection unit stationed at the Charlottenburg Post Office. Among the hundreds of anti-Nazi messages, this unit intercepted was a missive from England addressed to Owens calling on him publicly "to reject with contempt" any medal he might win at the Berlin Games.[30] (Owens, of course, never received this message.)

---

[28] Large, D. C. (2012) *Munich 1972: Tragedy, Terror, and Triumph at the Olympic Games*. Lanham: Rowan & Littlefield, p. 170.

[29] Krüger, A. (1972) *Die Olympischen Spiele 1936 und die Weltmeinung*. Berlin: Bartels & Wernitz, p. 194.

[30] Reichssicherheitshauptamt, Olympische Spiele, R58, 2320, Bundesarchiv Berlin.

Not surprisingly, the Reich security agencies monitoring the visiting Olympians worked in strict secrecy. A Gestapo order decreed: "Under no circumstances must our Olympic visitors be given the impression that they are under police observation, nor must our preemptive measures become a source of irritation to foreign guests."[31] Spies who kept tabs on the comings and goings of the visiting Olympic teams operated under the guise of bilingual "student helpers." The *Kriminalpolizei* station at the Olympic Village was disguised as an "Information Center" and was staffed exclusively by comely female officers.[32]

To the relief of the German authorities, Berlin '36 witnessed no prominent demonstrations of discontent by athletes on the victory podium (unlike at later Olympiads such as Mexico City in 1968 and Munich in 1972). Yet the festival was not without some subtler gestures of courageous opposition to prevailing political realities. The most noteworthy of these involved the winner of the marathon, Kitei Son. Although competing with the Rising Sun emblem on his tricot, Son was actually Korean— latter-day consequence of Japan's 1910 annexation of Korea. On the victory podium, Son duly bowed his head like any patriotic Japanese citizen would, but then proceeded to declare in interviews afterwards that he had made this gesture not out of reverence for Japan but in "silent shame and outrage" over Japan's occupation of the land he considered his own.[33] Son's behavior did not occasion much comment at the time, but the Japanese were so angered by the runner's action that they banned him from competition for life. (Son, however, got the last laugh: 52 years later he carried the Olympic torch into the main stadium on the opening day of the 1988 Seoul Olympics.) In another gesture of courageous defiance, albeit one that did not come to light until much later, a Berlin-born Turkish fencer named Halet Çambel, the first Muslim female competitor in Olympic history, flatly rejected an order from her delegation chief to meet with Hitler as a show of Turkish-German friendship.[34] Unlike many Turks at that time, Berlin-born Çambel was anything but a Nazi

---

[31] Preussische Geheime Polizei, Tätigkeit der Politischen Polizei, 18.7.36, R58, 2320, Bundesarchiv Berlin.

[32] Large (2007), p. 222.

[33] Ibid., pp. 258–259.

[34] "The First Muslim Female Olympian Snubbed Adolf Hitler," *The Daily Beast*, 26.1.14.

sympathizer; she later became an archeologist and in 1937 married the Marxist poet and architect Nail Çakırhan.

Interestingly enough, America's Jesse Owens issued not a word of negative criticism regarding the Nazi Reich despite having initially said he would support the US boycott-Berlin effort. Had Owens received that Gestapo-intercepted appeal to protest Nazi racism on the victory podium in Berlin, he certainly would have ignored it because since arriving in the Nazi capital, he had become something of a Hitler admirer. Far from feeling that he had been "snubbed" by Hitler in the Berlin stadium (as popular lore would have it), he insisted that the Führer had kindly "waved" to him from the stands. Upon his return to America after the Games, he publicly praised Hitler as "the man of the hour," chided the American press for "being unfair" to the German leader, and announced that if anyone had "snubbed" him in Berlin it was not Hitler but *President Roosevelt*, who had neglected to send him a congratulatory telegram.[35]

The Germans' determination to prevent any incidents that might be embarrassing to the Hitler regime or disruptive to the Berlin Games was not of course confined to foreign athletes; vigilant policing applied to Olympic visitors in general, while measures of proactive repression targeted residents of the Nazi capital considered potentially troublesome. Concerned that "foreign communists disguised as harmless Olympic visitors" might try to sabotage the festival in some way, the Gestapo chief in Prussia ordered all his agents to be "on high alert."[36] Fears that underground Communist cells might exploit the Olympic spotlight on Berlin to raise their heads spurred a series of police raids and the mass incarceration of leftist activists in a brand new concentration camp opened on the eve of the Games. Located in the Berlin suburb of Oranienburg, Sachsenhausen *Lager* housed, in addition to political prisoners, various "asocial elements" such as unregistered prostitutes, homosexuals, and beggars. At the time of the Games, this sprawling camp's capacity was already overstretched due to all the "problematical elements" being swept up in the Nazi dragnet. Ad hoc jails like

---

[35] Mandell (1987), pp. 228–229; Large (2007), p. 233; Olympic File, NAACP, Box 384, Library of Congress.

[36] Preussische Geheime Polizei, Tätigkeit der Politischen Polizei, 18.7.36, R58, 2320, Bundesarchiv Berlin.

the Municipal Work and Detention House in Berlin-Rummelsburg thus joined the Nazis' growing repertoire of emergency housing for an imagined army of Olympics disrupters. Finally, to lock away out of sight, a social group considered both "racially inferior" and "a grave moral danger to society"—namely, Gypsies—the German authorities opened yet another new detention facility, the so-called *Zigeunerlager* (Gypsy Camp) in the Berlin district of Marzahn. Although justified as a measure to guarantee Olympic visitors a "clean and safe" environment during their stay, the Marzahn facility remained in use well after the Games, ultimately serving as a way station to later liquidation of this population at camps in the East.[37]

For the Nazis, such proactive repression seems to have paid off, for the Berlin Games went ahead without any major disturbances. During the festival, there were sporadic acts of regime defiance and opposition, but these were few in number and typically involved individuals or small groups. For example, some malcontent set fire to a swastika banner hanging from a balcony, while another painted "Heil Moskau" on a train used by the Italian Olympic team; still another taped Communist slogans to a telephone booth. Slightly more seriously, members of Berlin's Communist underground papered the city with flyers protesting the Olympics as a rip-off of the poor (a sentiment one did not need to be a Communist to share).

The authorities came down hard on any offenders they managed to apprehend, no matter how trivial their offence: they arrested a tailor for yelling "Red Front!" in a tavern; jailed a man for criticizing the Opening Ceremony to his wife; arrested a spectator for failing to give the Hitler salute during the playing of the *Horst Wessel Lied* [*The Horst Wessel Song*—the Nazi Party anthem]. A German journalist received a life prison sentence for informing his foreign colleagues about the repressive reporting guidelines in effect for native scribes during the Games. Long prison terms were handed out to a group of working-class women who complained to visitors about living conditions in

---

[37] Wippermann, W. and U. Brucker-Boroujedi (1987) "Nationsozialistische Zwangslager in Berlin III. Das 'Zigeunerlager' Marzahn," in Ribbe, W. ed., *Berlin Forschungen II*. Berlin: Historische Kommission zu Berlin, p. 191.

Germany, and to a local salesman who made homosexual advances to two visiting Argentineans (although, at the same time, a few previously shuttered gay bars were allowed to reopen during the Games in order to convey the illusion that Nazi-ruled Berlin was a tolerant and cosmopolitan metropolis). In a secret post-Games assessment, Berlin Chief of Police Count Heinrich von Helldorf took pride in a job well done: "The success of the police measures taken on the occasion of the Olympic Games ... is the best demonstration of the rightness of the struggle that we have been waging since the Party came to power against professional criminals, traffic offenders, profiteers and other enemies of the people."[38]

The Nazis also found convincing evidence for the "rightness" of their policies and governmental system in the final outcome of the athletic competitions at Berlin. True, the Americans, and in particular *black* Americans, had managed to dominate the track and field events, but Nazi pundits found a way to "explain" this result in self-serving racist terminology: America had "cheated" by deploying a host of "black auxiliaries," they insisted. Or, as one Nazi official put the matter: "If Germany had had the bad sportsmanship to enter deer or another species of fleet-footed animal, it would have taken the honors from America in the track and field competition."[39] (Actually, to put the German argument in proper perspective, we should note that *Americans* interpreted the black successes in pretty much the same way. American experts, including the track coaches of Team USA, explained the achievements of Jesse Owens and other "race boys" in terms of specific physiological advantages for Negroes derived from a recent "jungle inheritance." Thus it was argued that while blacks might be able to sprint and jump well they would never excel in competitions requiring discipline, intelligence, teamwork, and stamina—sports like long-distance running and basketball!)[40] In any event, despite the humiliations in track, Germany emerged as the

---

[38] Rürup, R., ed. (1997) *1936: Die Olympischen Spiele und der Nationalsozialismus.* Berlin: Argon Verlag, p. 131.
[39] Dodd, M. (1939) *Through Embassy Eyes.* New York: Harcourt, Brace, p. 212.
[40] Large (2007), p. 331.

clear winner in the overall medal count, garnering 89 prizes compared to second-place America's relatively feeble 56. Moreover, because Germany's fascist/militarist soul-mates, Italy and Japan, also did very well, while once-powerful Britain and France lost ground, the Reich could claim that Berlin '36 demonstrated an emerging superiority on the part of tightly disciplined authoritarian nations over effete liberal democracies.

Alternative Games

The most significant public protest against the Berlin Olympics (apart from the international boycott movement) did not take place in Berlin itself, which is hardly surprising given the Nazi regime's instant suppression of opposition gestures of any kind. Having been unable either to shift the locale of the 1936 Summer Games or to keep the world, even the democratic world, from flocking to the festival, a coalition of leftist athletes and their supporters sought to organize a rival "People's Olympics," set to begin on July 19 in Barcelona, Spain. (There was a tradition for this: Along with the above-mentioned "Workers' Olympics" in France, American communists had put together an "Anti-Capitalist Olympiad" in Chicago in 1932 to protest the official "fat-cat" Olympics at Los Angeles in the same year.) As for Barcelona in '36, thousands of athletes indeed showed up for the competition, only to find the Catalan capital paralyzed by the violence attending the outbreak of civil war; the festival had to be called off.

With Barcelona out of the picture, a group of non-Communist labor activists and disgruntled Brundage opponents in the USA decided to pick up the anti-Berlin ball and stage a "progressive and democratic" alternative to the Nazi Games in New York City. With support from New York City Mayor Fiorello La Guardia and New York Governor Herbert Lehman, the organizers secured as a site the same venue that had hosted the US Olympic track and field trials, Randalls Island. Rather grandiloquently, they called their alternative festival "The World Labor Carnival."[41]

---

[41] On the World Labor Carnival, see Shapiro, E. S. (1985) "The World Labor Carnival of 1936: An American Anti-Nazi Protest," *American Jewish History* 74 (1985): 260–72.

This event turned out to be about as global as America's "World Series" in baseball. The vast majority of participants were Americans, the best of them athletes who had narrowly missed the cut for Berlin. Most of these men (there were no women) participated not out of political principle but out of desire for redemption—a need to show that they were just as good as the competitors who had made it to Berlin. A partial exception in this regard was a world-record-holding pole-vaulter named Ben Varoff. Known as "the Jumping Janitor," this custodian from San Francisco held decidedly left-wing views. The small contingent of foreign athletes at Randalls Island also contained some political protestors. Among them was Canadian Henry Cieman, a Jewish race-walker of world-class ability who in all probability could have made the Canadian team had not he decided to boycott the Berlin Games.

Despite the presence of high-quality athletes like Varoff and Cieman, the World Athletic Carnival failed to produce any significant athletic achievements. It also flopped at the box office. A mere 18,000 spectators showed up for the two-day festival even though the Jewish press and union papers had plugged it extensively. The relative paucity of big-name competitors no doubt played a role in this low turnout, as did the painful fact that most Americans, when it came right down to it, would much rather watch baseball or boxing than track and field. The Labor Carnival may indeed have been "an answer to the Nazi Olympics" (as Governor Lehman crowed), but, like similar "alternative Olympics" before and since, it failed to steal much thunder from the official five-ringed circus.[42]

Legacies

As we pointed out earlier, Berlin '36 was not the first Olympiad to occasion hostile opposition or to be blighted by political exploitation and/or racial prejudice, but the sheer *degree* to which that festival violated purported Olympic ideals of international brotherhood and openness made it a negative watchword among critics of the modern Games ever after.

---

[42] Ibid., 271.

(Although not, it must be said, among the IOC itself, which apparently learned nothing from its first experience with Olympics-hosting by a dictatorial, repressive regime: after all, in the immediate wake of Berlin the committee selected militarist Tokyo and Mussolini's Rome to host the—subsequently cancelled—Summer Games of 1940 and 1944.[43] As we know, it would later go on to give hosting privileges to the USSR, authoritarian South Korea, Communist China, and Vladimir Putin's Russia. Not even soccer's FIFA, corrupt and bone-headed as it may be, can match a record like this.)

Having mentioned Rome '44, we should note that Berlin was a negative example for the Italian organizers of the Rome Olympics of 1960. Like Hitler's capital, Rome had a recent fascist past, and hyper-nationalism remained a taboo in some quarters, especially on the Catholic Right and Communist Left. The Italian hosts thus made an effort to downplay patriotic expression, even though the '60 Games coincided with the centenary of Italian national unification.[44] Giulio Andreotti, chairman of the Rome organizing committee (and perennial prime minister), insisted during the opening ceremony that the Olympics were "entirely removed from being national." Instead, he said, they were "fundamentally international and apolitical," their only purpose being to "foster brotherhood across races and nations."[45] In 1960, there were even proposals to "internationalize" Rome by eliminating the cherished afternoon siesta, keeping restaurants open around the clock, and serving spaghetti in the shape of the five Olympic rings. In the end, though, Italians could no more abandon nationalism than had any of the previous hosts, and Italy's credible third-place finish in the medals count produced the predictable orgy of breast-thumping self-congratulation. Only after all the hoopla was over did a

---

[43] As late as the Sochi Winter Games of 2014, Republican presidential candidate Mitt Romney invoked the '36 Berlin Olympics' legacy of "undercutting the Olympic message" in arguing that Vladimir Putin's Russia was a bad choice for any Olympic festival. See "Names and Faces," *San Francisco Chronicle*, 1.6.14.

[44] Bosworth, R.J.B. (2011) *Whispering City: Rome and its Histories*. New Haven: Yale University Press, p. 256. On the Rome Games, see also Maraniss, D. (2008) *Rome 1960: The Olympics that Changed the World*. New York: Simon and Schuster.

[45] Andreotti, G., "L'Anno dei giuochi olimpici," *Studi romani*, 8, January–February 1960, pp. 10–11. Quoted in Bosworth (2011), p. 256

Communist politician remind his countrymen that their Games, like those of Berlin, had in reality been nothing more than a grand party for the "exploiters over the exploited," another victory of the "privileged over the poor."[46]

In no instance, however, did the "anti-model" factor of Berlin become more prominent than in the controversy over Munich 1972, that second German-hosted Olympic festival. The Munich organizers, having won the '72 Games despite their city's heavy historical baggage as the erstwhile "Capital of the [Nazi] Movement," and over loud protests from East Germany that 1972 would be 1936 all over again, or even worse, were determined to put on a show that differed as much as humanly possible from the one staged by the Third Reich. Whereas visitors to Nazi Berlin had found a city awash in swastika flags, Munich in the summer of '72 offered a sea of fluttering Olympic banners in tranquil pastels—the strident reds, blacks, and purples of Nazi and Imperial Germany having been explicitly *verboten*. In marked contrast to the bombastic neoclassical Olympic architecture of 1936, Munich featured modernistic structures designed to remind the world that the Bavarian capital, and indeed West Germany in general, were forward-looking, democratic places, open to the new. The main Olympic Stadium and an adjoining sports complex were covered by a swooping tent-like glass roof, meant to symbolize transparency and inclusiveness.[47] But as strikingly "un-Berlin" as Munich's stylistic regime undoubtedly was, it was in the realm of security, not aesthetics, where the Bavarian festival departed most significantly, and most fatefully, from Germany's previous Olympic experience.

In 1936 Berlin's Reichssportfeld had been rife with uniformed police, soldiers, Gestapo, and SS-men. At Munich's sprawling Olympic Park, by contrast, the primary keepers of order consisted of civilian volunteers dressed in baby-blue leisure suit-like outfits. Rather than pistols or rifles, these pastel-clad guardians of the Olympic peace carried walkie-talkies; they had been trained in anger management, not head busting. Whereas Berlin's Olympic Village had been located on a military base far away from the main stadium, Munich's lay within walking distance of the

---

[46] *L'Unità*, 9.11.60. Quoted in Bosworth (2011), p. 258.
[47] Large (2012) Munich 1972, pp. 22–3.

primary competitive venues. The 1972 Village had a chain-link fence around it, but the barrier in question was only six and one-half feet tall, with no menacing coils of barbed wire running along the top. "We cannot afford to look like a concentration camp," said the chief of security.[48] Yet, horribly, it was just such lax security on the part of the organizers that proved crucially responsible for the greatest tragedy in modern Olympic history: the murder of eleven Israeli Olympians by Palestinian terrorists. The bitter irony of Munich 1972 is that its organizers, in trying so hard to move beyond Berlin '36, ended up awakening memories of the bad old days in the worst possible way.

Eight years after the Munich tragedy, ugly memories of Berlin resurfaced as the Olympic world prepared to celebrate its 22 summer Olympiad in Moscow, USSR. As with Berlin '36, the prospect of an Olympic festival hosted by a single-party dictatorship with imperial ambitions ignited an international boycott movement, led once again by the USA. This time, however, a major boycott actually came off, with some 62 nations (including West Germany, but not Italy, France, or Britain) staying away from the 1980 Moscow Games. Although the Soviet Union's invasion of Afghanistan in 1979 provided the immediate impetus for this boycott, more general objections to the notion of a "tyranny" being allowed to run roughshod over Olympic ideals animated the protest. US Secretary of State Cyrus Vance specifically invoked Berlin 1936 in justifying America's determination to shun Moscow: "I look back on the 1936 Games … and I think in hindsight it was a mistake for us to have attended."[49] Predictably, the IOC also cited Berlin to justify its decision *for* Moscow, insisting that the requirements of Olympics-hosting invariably worked to liberalize and "open up" closed, autocratic societies.[50] Of course, in reality, this happened no more with Brezhnev's USSR than it had with Hitler's Germany—or would work with Communist-controlled China in 2008 or Vladimir Putin's Russia in 2014. As for Moscow 1980, the main consequence of

---

[48] Ibid., pp. 114–115.
[49] Vance is quoted in Smith, G. (1986) *Morality, Reason and Power: American Diplomacy in the Carter Years*. New York: Hill & Wang, 227.
[50] Large (2007), p. 343.

the US-led boycott was a tit-for-tat boycott of the 1984 Los Angeles Games by the USSR and its allies.

If the brown-shirted (or –sheeted) ghosts of Berlin '36 had hovered around Rome 1960, Munich 1972, and Moscow 1980, these ghouls swarmed to center stage when Berlin itself sought to reprise its Olympics-hosting role in 2000. The promoters of Berlin's botched bid proved to be as willfully obtuse regarding the legacy of "Hitler's Games" as was the IOC. Amazingly, they proposed the old Reichssportfeld as the primary venue for recently reunited Germany's coming-out party on the world athletic stage. Even many Germans were appalled by the arrogance and ineptitude of Berlin's Olympic campaign (which included efforts to bribe IOC voters by catering to their every sexual taste), and a boisterous homegrown "NOlympic" movement helped ensure that there would be no second Olympic act for the German capital at the dawn of the new millennium.[51]

Even now, a decade and a half into that new era, following additional examples of Olympic excess and political exploitation (think Beijing and Sochi, above all), Berlin '36 continues to provide a usable case study in modern Olympism at its worst. As Brazil's Rio de Janeiro stumbles uncertainly toward its moment in the Olympic sun, Hitler's bombastic festival still serves as powerful ammunition to those protesting nationalistic hubris, massive public spending, the razing of entire neighborhoods, brutal displacement of long-time residents, and, most grievously, widespread police crackdowns and preventative elimination of "bad elements" in the name of Olympic security.[52] However "successful" Rio may turn out to be, many observers, and not just in Brazil, will undoubtedly wonder whether the traveling Olympic circus, having taken such an errant turn in 1936, has not gone irrevocably astray.

---

[51] Ibid., pp. 338–339.

[52] A good study of Brazil's preparations for the 2014 World Cup and 2016 Summer Olympics is Zirin, D. (2014) *Brazil's Dance with the Devil: The World Cup, the Olympics, and the Fight for Democracy*. Chicago: Haymarket Books. In May 2015 the *New York Times* could note that "killings by police are surging as the authorities clamp down in preparations for the Olympics next year." "Despair, and Grim Acceptance, Over Killings by Brazil's Police," *New York Times*, 22.5.15.

References

Bosworth, R. J. B. (2011). *Whispering city: Rome and its histories*. New Haven: Yale University Press.

Dodd, M. (1939). *Through embassy eyes*. New York: Harcourt Brace.

Gottlieb, M. (1972). The American controversy over the 1936 Olympic games. *American Jewish Historical Quarterly, 62*(March), 181–213.

Guttmann, A. (1984). *The games must go on: Avery Brundage and the Olympic movement*. New York: Columbia University Press.

Haller, G. (1933). Der Olympische Gedanke. *Nationalsozialistische Monatshefte, 3,* 388–396.

Hart-Davis, D. (1986). *Hitler's Games*. London: Century.

Kidd, B. (1980). The popular front and the 1936 Olympics. *Canadian Journal of the History of Sport and Physical Education, 11*(1), 1–18.

Krüger, A. (1972). *Die Olympische Spiele 1936 und die Weltmeinung*. Berlin: Bartels & Wernitz.

Large, D. C. (2007). *Nazi games: The Olympics of 1936*. New York: W.W. Norton.

Large, D. C. (2012). *Munich 1972: Tragedy, terror, and triumph at the Olympic games*. Lanham: Roman & Littlefield.

Lennartz, K. (1994). Difficult times: Baillet-Latour and Germany, 1931–1942. *Olympika, 3*(1994), 99–105.

Mandell, R. D. (1987). *The Nazi Olympics*. Urbana: University of Illinois Press.

Maraniss, D. (2008). *Rome 1960: The Olympics that changed the world*. New York: Simon and Schuster.

Menkis, R., & Troper, H. (2015). *More than just games: Canada and the 1936 Olympics*. Toronto: University of Toronto Press.

Mogulof, M. (2001). *Foiled: Hitler's Jewish Olympian*. New York: RDR Books.

Rürup, R. (1997). *1936: Die Olympischen Spiele und der Nationalsozialismus*. Berlin: Argon Verlag.

Schaap, J. (2008). *Triumph: The untold story of Jesse Owens and Hitler's Olympics*. New York: Mariner Books.

Shapiro, E. S. (1985). The world labor carnival of 1936: An American Anti-Nazi protest. *American Jewish History, 74*(1985), 260–272.

Smith, G. (1986). *Morality, reason and power: American diplomacy in the Carter years*. New York: Hill & Wang.

Teichler, H.-J. (1989). Zum Ausschluss der deutschen Juden von den Olympischen Spielen 1936. *Stadion, 5*(1), 45–64.

Teichler, H.-J. (1991). *Internationale Sportpolitik im Dritten Reich*. Schlondorf: Verlag Karl Hoffmann.

Wippermann, W., & Brucker-Boroujerdi, U. (1987). Nationalsozialistische Zwangslager in Berlin III. Das 'Zigeunerlager' Marzahn. In W. Ribbe (Ed.), *Berlin Forschungen II* (pp. 189–201). Berlin: Historische Kommission zu Berlin.

Zirin, D. (2014). *Brazil's dance with the devil: The World Cup, the Olympics, and the fight for democracy*. Chicago: Haymarket Books.

Splitting the World of International Sport: The 1963 Games of the New Emerging Forces and the Politics of Challenging the Global Sport Order

Russell Field

'The Third World was not a place,' Vijay Prashad (2007, p. xv) notes at the outset of *The Darker Nations*. 'It was a project.' A visible, if now little remembered, sporting manifestation of this project was the Games of the New Emerging Forces (or GANEFO), an explicit attempt to link sport to the politics of anti-colonialism. This international multi-sport event, which took place in Jakarta, Indonesia, on 10–22 November 1963, attracted approximately 3000 athletes and officials from—but not necessarily officially representing—48 nations. They met in the Indonesian capital and competed in 20 athletic events (virtually all of them Olympic and Western sports) as well as cultural festivities. Athletes hailed primarily from recently decolonized countries in Asia and Africa (as well as former colonies in Latin America), which were labelled the 'new emerging forces' by Indonesian President Sukarno who created GANEFO as part of his attempt to situate his nation as a regional power.

R. Field (✉)
University of Manitoba, 118 Frank Kennedy Centre, Winnipeg, MB, R3T 2N2, Canada

© The Editor(s) (if applicable) and The Author(s) 2016
J. Dart, S. Wagg (eds.), *Sport, Protest and Globalisation*, DOI 10.1057/978-1-137-46492-7_5

77

In such an environment, GANEFO was an explicit attempt to link sport to the politics of anti-imperialism, anti-colonialism, and the emergence of the Third World or non-aligned movement following the 1955 Bandung conference. The event's slogan—Onward! No Retreat! — spoke to these ideological origins. However, the immediate genesis of GANEFO was a controversy that arose from Indonesia's hosting of the IVth Asian Games in August 1962. When the hosts refused entry to the teams from the Republic of China (Taiwan) and Israel, the Asian Games became embroiled in diplomatic tension and the host's relations with India turned sour as Asian Games official, G.D. Sondhi from India, bore the public brunt of local criticism (American Embassy Jakarta 1962, p. 1). The sporting consequence of the Asian Games troubles was that the International Olympic Committee (IOC) suspended Indonesia, whose National Olympic Committee (NOC) promptly withdrew from the IOC and set about planning GANEFO.

The nascent event took shape at a preparatory conference held in Jakarta in April 1963. There, the hosts were joined by representatives from Cambodia, the People's Republic of China (hereafter China), Guinea, Iraq, Mali, Pakistan, North Vietnam, the United Arab Republic, and the USSR, with Ceylon and Yugoslavia attending as observers. This meeting formulated a plan for GANEFO and a subsequent conference intended to solidify the political unity of the new emerging forces, and concluded by noting that the 'countries invited to take part in the First GANEFO in Djakarta are those accepting the Spirit of the Asian-African Conference in Bandung and the Olympic Ideals' (GANEFO 1963a). Such an invocation of 'Olympic ideals' is intriguing given that Sukarno publicly promoted GANEFO by taking a critical stance towards the IOC and the colonial role played by the Olympic Games (though this can be read as a reaction to the suspension of the Indonesian NOC following the IV Asian Games[1]).

---

[1] An argument can be made that, if the IOC had not suspended the Indonesian NOC over the issues at the Asian Games, then Sukarno would not have started GANEFO. Given the speed with which GANEFO came to fruition the event was consistent with the Indonesian leader's approach to geopolitical issues, but there was no hint of any sort of dissent with international sport prior to the incidents at the Asian Games (in fact, hosting the Asian Games in the first place would suggest some interest in being part of mainstream sport).

As athletes gathered for the opening ceremonies on 10 November, they were greeted by a crowd of approximately 100,000 who inaugurated an international multi-sport event that very much resembled the Olympics, complete with such symbols and rituals as a torch relay, flame lighting, a parade of athletes in front of the head of state, and the raising of the 'movement's' flag. The event concluded with a closing ceremony and in between these two celebrations, spectators watched competitive athletic events adjudicated according to accepted international standards where the winners were awarded gold, silver, and bronze medals. All international athletes stayed in an athletes' village of bungalows while the sporting competitions were supplemented by a variety of cultural exhibitions and performances.

Although framed as a gathering of the Afro-Asian nations of the non-aligned movement, or Third World, the largest teams (after the hosts) represented the Second World: China and the Soviet Union. In addition, individuals from workers' sport clubs and student groups in Europe and the Americas also made their way to Jakarta. Of the 48 'nations' in attendance, 17 competed in only one or two sports and two others (Belgium and Bolivia) arrived too late to compete, likely by design (GANEFO 1963b). The sporting–political implications of competition at GANEFO for athletes from around the world are a significant part of the analysis that follows. At a post-Games meeting of international representatives, Sukarno sought to enshrine GANEFO as the sporting arm of a larger but eventually unrealized political movement. GANEFO II as it was known was initially scheduled for Cairo, before being shifted to Beijing, and subsequently disappearing from the athletic calendar after the onset of the Cultural Revolution. Except for an Asian GANEFO hosted by Cambodia in 1966, the Games themselves were never held again.

With GANEFO having faded from historical view, it can be too easily dismissed as having taken place only once in geographic obscurity, as far as Western observers are concerned. Contesting this conclusion is reason enough to reconsider and re-centre GANEFO, but this chapter argues that the event had repercussions throughout the world and was taken extremely seriously in November 1963 in halls of power both sporting and diplomatic. GANEFO was conceived as a sporting outcome of the non-aligned movement that emerged out of the

1955 Bandung conference (Chakrabarty 2010; Prashad 2007; Wested 2005), but the event also had repercussions in the First and Second Worlds. Despite its contemporary significance, GANEFO is largely absent from histories of Indonesia and the period of Sukarno's presidency (e.g., Brown 2003; Gelman Taylor 2003; Legge 1972 makes brief mention of the event). The event, however, can be understood within the Indonesian leader's anti-colonial, nationalist, and Third Worldist positions, as well as his efforts to promote a homogenized sense of Indonesian identity in the face of ethnic and regional diversity, and criticisms of the Java-centrism of his government (Hadiz 2006; Roosa 2008).

GANEFO has also received relatively little attention from historians of sport (exceptions include Connolly 2012; Gitersos 2011), obscured by higher profile (in the West) civil rights struggles connected to sport in the 1960s, including the suspension of South Africa's apartheid regime by the IOC (e.g., Booth 2003) and the American 'black power' demonstration by US sprinters Tommie Smith and John Carlos at the 1968 Mexico City Olympics (e.g., Hartmann 2003). The few considerations of GANEFO (e.g., Lutan and Hong 2005; Sie 1978) have framed the event as an expression of Indonesia's independence and Sukarno's efforts to link politics and sport, while GANEFO was also important in attempting to position Indonesia within the non-aligned world (Adams 2002; Majumdar and Mehta 2008). Kidd (2005, p. 153), albeit briefly, considers GANEFO from the perspective of newly decolonized countries using a global sporting event as a forum for the expression of their redefined international status, by enabling 'third-world participation, in terms that third-world leaders could shape.' But accounts that consider the geopolitical implications of GANEFO focus on Chinese financial support as evidence of GANEFO's implication within Cold War politics. For Guttmann (1992, p. 109), the 'impetus for GANEFO may, in fact, have come from Beijing, where the Chinese Communists were eager to embarrass the IOC.' Torres and Dyreson's (2005, p. 73) characterization of Sukarno, 'with support from Beijing and Moscow,' founding GANEFO 'to counter the IOC's "bourgeois" control of world sport' also suggests considering GANEFO in the terms of 'the bipolar Cold War world' (Chen 2009, p. 427). Little attention has been paid to Third World, North American, or European responses, or the ways in which GANEFO reflected Sino-Soviet relations in Southeast Asia.

GANEFO is a valuable lens through which to examine the politicized nature of sport in the 1960s, the politics of decolonization and development, and the ideological and diplomatic history of the Cold War, the Sino-Soviet split, and the emergence of the Third World or non-aligned movement. This chapter begins with a consideration of Western responses to GANEFO. In anticipation of the event, State Department officials in the USA and the diplomatic legation in Jakarta debated how to counsel allied nations interested in participating in GANEFO. Beyond the political repercussions, in the aftermath of the Games, the governing bodies of international sport meted out the competitive consequences for those athletes who did participate. When turning to the interests tied to GANEFO in Southeast Asia—both in the host country, Indonesia, and the Games' primary sponsor, China—this distinction between politics and sport melts away. Finally, consideration is given to the non-emerging forces who found themselves competing in Jakarta—from members of workers' sport clubs in Europe to university students from South America—to highlight GANEFO as a lived experience for thousands of athletes that is not defined solely by the event's political posturing. A consideration of the broader interests wrapped up in GANEFO reveals that an event generally thought of as either Third World resistance to the dominant sport model or a communist event dominated by China actually has multiple meanings.

The First World Anticipates GANEFO

In September 1963, two months in advance of the event, a US State Department communiqué, over the signature of Secretary of State, Dean Rusk, was sent to US embassies around the world instructing diplomatic officials that 'On balance, the disadvantages of GANEFO to the Free World far outweigh any discernible advantages' (Department of State 1963a). In considering how US diplomats both framed and recommended dealing with GANEFO, it is worth noting the ways in which anti-communist Cold War politics were conflated with a racialized, postcolonial discourse.

The US analysis of the events surrounding the 1962 IVth Asian Games, which led to Sukarno creating GANEFO, focused on 'communist' elements. During the riots that accompanied the Asian Games, 'Efforts were made by Communist elements in the mob, and subsequently, by the Communist press, to picture the United States as the driving force behind Sondhi,' the Indian IOC member and putative head of the Asian Games committee, who was spearheading efforts to sanction Indonesia for excluding Taiwan and Israel (American Embassy Jakarta 1962, p. 3). Furthermore, one report noted that although 'the United States came through the Asian Games controversy relatively unscathed ... there is little doubt that had the Games been cancelled, the weight of Indonesia opinion would have held the United States responsible' (Ibid, p. 12).

Generally, the Asian Games incident was seen as a reflection of the influence of China and the United Arab Republic (a short-lived union of Egypt and Syria, 1958–61) over Indonesian affairs, and consequently a sign of Sukarno's leftist leanings (Ibid, p. 5). In the nascent Third World/non-aligned movement, Prashad (2007, p. 9) argues, 'Communism as an idea and the USSR as an inspiration held an important place in the imagination of the anticolonial movements from Indonesia to Cuba.' In the 1960s, this was of particular concern in the West, so much so that the State Department concluded that 'GANEFO, if held, will be heavily exploited by the communists' (Department of State 1963a).

Unease over GANEFO was couched in diplomatic reports in a dismissive colonial language. In general, Indonesian foreign policy was framed by US embassy officials in Jakarta as based on 'confrontation,' 'emotional nationalism,' and 'a chip on the shoulder attitude resulting from the belief that Indonesia was kicked around for three hundred years and, consequently, is now entitled to get whatever it wants' (American Embassy Jakarta 1962, p. 15). Indeed, the Asian Games were organized by a 'prestige-hungry and internationally insecure Indonesian leadership' (Ibid, p. 14), which led to, what the State Department called, 'Sukarno's half-baked concept of "new emerging forces"' (Department of State 1963a).

But Cold War concerns, especially given US interests in the region, were preeminent. The US Embassy in Jakarta was worried about a 'significant political consequence' in the aftermath of the Asian Games controversy: 'a chorus of voices from the extreme left has been raised here in favor of Indonesia taking the lead in organizing a new Afro-Asian-Latin American sports federation' (American Embassy Jakarta 1962, p. 15). This would become GANEFO, and nearly a year later, the State Department still had 'serious doubt' about 'Indonesia's willingness and ability to keep the communists from making a propaganda circus of GANEFO' (Department of State 1963a).

Embassy officials in Jakarta framed the issues raised by GANEFO as germane to US interests in these geopolitical terms. The USA and USSR had much to lose if GANEFO were to succeed, especially at the expense of the Olympic Games, as a flourishing 'colored Olympics' meant the prospect of disempowering the West. If GANEFO was successful, especially if African nations joined the cause, then the emerging forces could offer a particularly racialized alternative to the IOC Games. But beyond sporting dominance, there were regional concerns raised by GANEFO; 'its very existence as a forum for Communist propaganda can exacerbate Indonesia-U.S. relations' (American Embassy Jakarta 1963, p. 9). There was concern for the potential impact upon 'neutralist' nations in the region, for example, Cambodia, and for the strain placed upon US allies in the region (Japan, the Philippines, and Pakistan), who might feel compelled to send athletes to Jakarta.

It was in this context that US officials worked out a response to GANEFO. It was the opinion of the US Embassy in Jakarta that 'the U.S. should ignore GANEFO as much as is possible on the quite legitimate grounds that international sport is a private and not a governmental affair in the U.S.' (Ibid, p. 2). The Rusk communiqué cited earlier (Department of State 1963a) noted, 'Our relations with Indonesia could suffer substantially, however, if we were actively to oppose GANEFO or to be found lobbying against it.' But if not lobbying, then influence and opposition were certainly applied, with the Embassy arguing 'it also might not be out of place to discretely make known to our Asian and African friends the part to be played by Communist China in supporting GANEFO.' International missions were advised to 'take no initiative in raising the question of participation in GANEFO with the countries to which you are accredited.' However, if the question of GANEFO participation was raised with US diplomatic officials, then the State Department prescribed responses, one each for the First and

Third World. There were few concerns that its First World allies would compete in Jakarta—although it is worth noting that Japan was included in this group and the Japanese had considerable internal struggles over participating in GANEFO. So the State Department, framing GANEFO as 'a relatively minor matter,' did not 'intend to take any overt steps to prevent holding the Games or to discourage participation by our friends.' Those same friends were told that they could discourage their own allies in the Third World, but US policy was limited to expressing 'reservations' and to directing nations to consider the consequences with international sporting bodies.

Sporting Consequences: The IOC Responds to GANEFO

Officials from international sporting bodies were certainly concerned over GANEFO's explicit mixing of politics and sport (Field 2011). According to IOC President Avery Brundage, 'the Indonesia Government has thrown down a challenge to all international amateur sport organizations, which cannot very well be ignored' (Brundage 1963a). While Brundage wanted sport to be 'independent' of political influence, Indonesian Foreign Minister, Dr. Subandrio, contended that 'sport cannot be separated from politics, and Indonesia uses sports as a political tool to foster solidarity and understanding between nations' (IOC Circular Letter 1965). Otto Schantz (1997, p. 130) notes that for Sukarno 'sport was inextricably linked with politics, whereas Brundage regarded this standpoint as "an abnegation of one of the most fundamental and important principles of the Olympic Movement."' The Marquess of Exeter, David Burghley, head of the International Amateur Athletics Federation (IAAF: the governing body of international track and field), a member of the IOC executive, and a key Brundage ally on the issue of GANEFO, noted that his organization was fighting the biggest battle that had ever been fought to save amateur sport. 'If we lose, the politicians will then have the gate open to destroy the authority of our sport' (Burghley 1964).

There were other crises crossing Brundage's desk in the late 1950s and early 1960s, many of which were a result of Cold War politics and diplomacy. NATO's Allied Travel Commission was refusing travel visas to the West in the early 1960s for many athletes from the Soviet bloc, which especially hampered East German participation and was part of a larger debate about how and whether to recognize the two Germanys in international sport. The IOC and individual International Sport Federations (IFs) were not only working out how to deal with two Germanys, but also two Chinas, two Koreas, and subsequently two Vietnams. There was also the inherent racism of apartheid-based South African Olympic teams and the teams from countries with state-funded athletes, whose amateur status was debated in some corners. To this already bubbling cauldron was added the challenge of integrating newly decolonized countries of Asia and Africa into international sport.

Brundage feared that the 'GANEFO Games might split the world of international sport asunder,' and, on the day before the opening ceremonies, he asserted that GANEFO 'is unquestionably the first move in a campaign to take over international sport in one way or the other' (Brundage 1963b). As GANEFO would reveal, the post-war era of decolonization was creating a series of suddenly independent former colonies with their own national sport organizations that wished to be recognized within the world of international sport. Too often for Brundage's liking, government officials were intimately involved in sporting organizations in countries that often did not have sufficient administrative architecture to separate the two institutions in the ways that the IOC insisted upon. Brundage observed that 'Sport in most of these countries is apparently quite firmly in the hands of the politicians,' who 'threaten the very foundations of amateur sport' (IOC Circular Letter 1964). It was in Southeast Asia and Africa, indeed in many of the countries labelled by Sukarno as the new emerging forces, that Brundage hoped the IOC could inculcate Olympic values in the face of the concerning influence that governments had over the organization of sport.

With the Olympic Games scheduled for Tokyo in 1964, as Senn observes (1999, pp. 130–1), GANEFO was a rival attempt 'to reach out to the Asian sport world.' An immediate concern for the IOC and international sport leaders was the potential for unrest that GANEFO represented only ten months before the first IOC Games in Asia. The response was to threaten competitors with a ban from the Tokyo

Games if they competed in Jakarta. The rationale for this lay in the rumoured participation of athletes from China and North Korea (and to a lesser extent North Vietnam and Palestine, neither of which was recognized by the IOC). 'Red' China, in the parlance of the Cold War, had seceded from the Olympic Movement in the late 1950s over the issue of the recognition of the Republic of China (Taiwan). Similar withdrawals of the individual IFs whose sports comprised the Olympic programme had followed. Athletes from mainland China then did not belong to any IF and, according to international rules, athletes from IF-member countries risked suspension if they competed against athletes from non-member countries. A similar penalty faced sportspeople who competed against athletes from Indonesia, due to the suspension of the Indonesian NOC, and Indonesian competitors were a certainty in Jakarta.

Determining eligibility for Olympic competition, the IOC argued, rested with the IFs, which allowed the IOC to distance itself from overtly policing GANEFO participation, insisting instead that any discipline to be meted out would come at the discretion of each IF. Nevertheless, the documentary record makes clear that IOC officials took a very hands-on approach to participation in GANEFO, warning their constituent NOCs of the potential for sanctions in advance of GANEFO and then attempting to gather information on who did participate in Jakarta—details that were passed on to the IFs. IOC concern over GANEFO extended, for example, all the way to the Andes.

On 23 December 1963, the president of the Comite Olimpico Boliviano, Dr. Roberto Staszeski Lucero, wrote to the IOC Chancellor, Otto Mayer. He was responding to a letter that Mayer had sent to all NOCs requesting that they confirm whether or not athletes from their country competed at the recently ended GANEFO (IOC Circular Letter 1963). In his response—copied to both Brundage and the IAAF—Staszeski emphasized for IOC leaders that 'No bolivian [sic] sportsmen took part at the GANEFO GAMES at Djakarta last month' (American Embassy La Paz 1963). Yet Bolivians had participated in GANEFO. Nevertheless, the distinction made by the Bolivian NOC, and one that would become important as the IOC and IFs sought to police international sport, was that the Ministers of Foreign Affairs and Education had 'sent a symbolic group of university sportsmen[2] with the meaning of an international courtesy.' Similar remonstrations—making the distinction between unofficial athletes and recognized 'sportsmen'—were received at the offices of the IOC throughout December 1963 and January 1964, both from NOCs whose countries did and did not have competitors at GANEFO.

The presence of 'unofficial' athletes, second-rate sportspeople, provided Western officials, both sporting and diplomatic, with a narrative around which to dismiss GANEFO. 'According to the Press,' the delegates to the 61st IOC Session were told in January 1964, 'the Games were on the whole badly organized' (IOC Session 1964). Diplomats questioned the presence of competitors from Western nations such as France and the Netherlands, while the fact that only China sent its best athletes was taken to suggest the mediocre nature of GANEFO as an athletic competition. According to the Canadian embassy (Canadian Embassy Jakarta 1963, p. 1), 'The French community here is quite embarrassed by "their" athletes. They frankly say that they are worse than third-rate and are communists. The Dutch contingent is drawn from a social-cultural-labour club and similarly the athletes from South America all belong, as near as we can gather, to leftist organizations.'

No mention was made of the threatened IF/IOC bans, a year before the Tokyo Olympics, which had much to do with the athletic composition in Jakarta. Similarly, an Australian diplomatic report (Department of Foreign Affairs [Australia] 1964, p. 3) concluded that 'the success of GANEFO is in many ways quite spurious,' especially since many of the African and Latin American teams had been fully funded by China—as though gatherings of the non-aligned world were less genuine if nations could not overcome on their own their post-colonial material circumstances. British diplomat A.G. Gilchrist (Gilchrist 1963, p. 4) came to the conclusion—remarkable given the efforts by the IFs and IOC to discourage

---

[2] Because there are no detailed records of participation, it is not clear whether the Bolivian contingent included female athletes.

There were certainly female sports at GANEFO and women competed in a number of events. But while GANEFO was framed rhetorically as a progressive alternative to the IOC Games, that progressive vision did not extend to the gendered (exclusively male) organization of the Games themselves. In many ways, GANEFO is interesting because it contested IOC hegemony but, in this regard, reproduced the very event it was protesting.

participation in Jakarta—that 'GANEFO showed that with remarkably few exceptions the sport nations of the world were unwilling to interfere in any important way with the present organisation of world sport.' He was left to ask, 'What then, if anything, did the Indonesians achieve with GANEFO?' before answering his own question: 'I am strongly inclined to believe that at the international level they achieved remarkably little … there is little evidence to show that GANEFO made any real impact on world opinion or on more than a handful of people who matter.'

Southeast Asia and the Second World

The irony of such a colonial stance in regard to an event celebrating a decolonizing world is clear. For GANEFO mattered a lot in cities such as Jakarta and Beijing. With China sympathetic in the early 1960s to nationalist governments in Southeast Asia as it attempted to build its geopolitical sphere of influence (e.g., Guanhua 2003; Harding 1981; Zhai 2006), and challenging the Olympic Movement over the recognition of Taiwan, it saw a benefit to supporting Sukarno's efforts at hosting an alternative international sporting event. Taomo Zhou (2013, p. 7) argues that 'Sukarno's conceptualization of "new emerging forces" of the formerly colonized world echoed Beijing's strategic thinking.' After the host nation, China sent the largest team to GANEFO with 229 athletes having been selected at a national trial held in Beijing on 14 September 1963. GANEFO would represent China's return to international sporting competition and Chinese sport officials saw much to gain from the event's success (e.g., Field 2014; Shuman 2013; Xu 2008).

If the 1955 'Bandung [Conference] provided the terrain to end China's isolation from world opinion and support' (Prashad 2007, p. 37), GANEFO was its sporting equivalent. As Zhou (2013, p. 6) notes, 'China's proclaimed solidarity with the Third World also served as a propaganda tool for winning the hearts and minds of the developing world, where the competition for influence among the Western powers and the Soviet Union intensified in the 1960s.' While Sukarno was sympathetic to China during the Sino-Soviet rift in the late 1950s and early

1960s, African nations were thought to be in the Soviet camp (Lüthi 2008). China saw an opportunity in GANEFO and, according to an Associated Press report (1963), 'offered to pay expenses of invited African states to the GANEFO Games.'

The use of GANEFO to curry favour with non-aligned nations went beyond Cold War boundaries and highlighted Second World tensions between China and the USSR, as the event was implicated in Sino-Soviet struggles for influence in Southeast Asia. While GANEFO presented China with an opportunity to assert its influence through sport, the USSR faced trickier diplomatic options. Its return to the Olympic Movement in 1951 and the position of Konstantin Andrianov on the IOC executive suggest the importance of sport to Soviet foreign policy (Riordan 1993). GANEFO required balancing a 'stance as the champion of the Newly Emerging Forces ... without compromising their position within the IOC' (Senn 1999, p. 132). As Parks (2009, p. 180) notes, 'Soviet leaders felt compelled to participate in this endeavor in order to maintain the Soviet presence in Asia, but rather than risk sanctions from the IOC or IFs, the Sports Committee recommended to send only teams "not connected with the IOC or IFs."'

Although reports from the US embassy in Jakarta (American Embassy Jakarta 1963) that suggested GANEFO would be 'heavily exploited by the communists' risked portraying GANEFO in a Cold War binary, the nuance of the Sino-Soviet split was not lost on diplomats. In observing that to win influence with the nations of the emerging forces, Soviet officials would have 'to overcome the disadvantage of their Caucasian origin,' the US embassy also noted that: 'The U.S. and the U.S.S.R. can only lose by the establishment of GANEFO as a true "colored Olympics" rather than simply another A-A [Afro-Asian] propaganda show. Conversely, the Chinese Communists can only gain by this situation' (American Embassy Jakarta 1963, p. 9). Even an Associated Press wire story (1963) framed GANEFO in terms of the Sino-Soviet split: 'Russia's dilemma is Communist China's delight. The more trouble Red China can stir up in the Communistic world, some observers believe, the more status Russia will lose and the more China will gain in this athletic facade for political throat cutting.'

GANEFO was most often portrayed in the Western media as a 'red' event because of Sukarno's ties to the Indonesian communist party (PKI) and the Games' sponsorship by China. But there was also a significant nationalist development subtext to GANEFO. As Tania Li (2007, p. 1) notes, 'Programs that set out to improve the conditions of the population in a deliberate manner have shaped Indonesian landscapes, livelihoods and identities for almost two centuries.' Sukarno, in the late 1950s and early 1960s, was more of an anti-imperial nationalist than the Cold War communist he would eventually be portrayed as. He toured the globe in the late 1950s meeting world leaders and at the time was often characterized by the Western press as charismatic and modern. In seeking to apply these adjectives to a post-Dutch Indonesia, Sukarno accepted assistance from both sides of the Cold War divide, which was reflected during GANEFO. International athletes and guests arrived in Jakarta and travelled along a highway paid for by the USA, stayed in the International Hotel constructed by Japanese investment, and competed or spectated at the massive Bung Karno sports complex originally built by the Soviet Union for the IVth Asian Games in 1962.

Allies, Warm Bodies, and Contested Meanings

While Indonesian and Chinese officials hoped that GANEFO would inaugurate a quadrennial rival to the IOC Games, as the event approached there were concerns that the threatened sanctions by the IFs would have a negative effect on participation. Despite the rhetoric of the emerging forces, fears that the numbers of athletes would be smaller than hoped for compelled the organizers to consider a variety of entrants. This led to what the US Embassy in Jakarta, characterized in September 1963 as 'Indonesian willingness to accept any type of sports delegation, whether sent by a national sports association or merely by a "progressive" group in the country of origin' (American Embassy Jakarta 1963). But not all GANEFO competitors fit so easily into these new categories or were comfortable with the politics of the organizers.

Somewhat surprisingly, on the surface, given the event's framing around the emerging forces, there was Western European participation at GANEFO. Athletes from Finland, France, Italy, and the Netherlands competed in a variety of events. These athletes represented leftist student unions, but were predominantly the legacy of the European workers' sport movement, such as the six track-and-field athletes who represented TUL, the Finish workers' athletics club (e.g., Laine 1996; Riordan 1999.) Most compelling was the participation of Dutch athletes, primarily in the pool. One swimmer, 16-year-old Gude Heyke, won a gold medal and received a warm reception from the Indonesian crowd (as did the Dutch team at the opening ceremonies). Given the Netherlands' colonial history in the region, this surprising public support was noted in 1963 in both the Dutch and Indonesian press (Applaus voor Nederlandse vlag 1963).

But to appreciate the complexity of GANEFO, consider the variety of rationales by which competitors from South and Central America found themselves in Jakarta. First, the event attracted sportsmen and women sympathetic to the aggressive anti-colonial rhetoric of the Games. This was certainly true in the case of Cuba, where support for GANEFO became a rhetorical battleground between officials of the still-nascent revolutionary government and the sporting organization of the anti-Castro forces in Miami (USA), the Union Deportiva de Cuba Libre. Also, in interviews some Argentinean participants in GANEFO recall that athletes from Dominica and some from Chile and elsewhere in South America were more overtly 'leftist' than they were.

Second, invitations to GANEFO were extended through governmental and diplomatic channels rather than through sporting ones, and as a consequence, officials in many nations were wary of the potential consequences of participating in the event. Hence the 'symbolic gesture' of countries such as Bolivia, noted earlier, to send competitors while hoping to protect the Olympic eligibility of their more accomplished sportspeople. Indeed, at least 16 members of the Bolivian delegation left for Jakarta two days after GANEFO had begun, ensuring that their 'unofficial and strictly symbolic' participation would not include being visible at the event's opening ceremonies (American Embassy La Paz 1963).

General Jose de Jesus Clark, head of the Comite Olimpico Mexicano and mindful of the fact that his country was bidding to host the 1968 summer Olympics, was especially uncomfortable with the pressure that the Indonesian government was applying. They offered him the position of GANEFO regional vice-president, while at the same time 'laying stress on the commercial engagements between both countries' (Clark 1964). He wrote to Brundage that Mexico's president was supportive of the NOC position that Mexico would not participate in GANEFO, but 'Mexico was forced to be present in Indonesia' because of 'business – mainly in the textile field – with Indonesia' (Clark 1963). In the end, the Mexican government (and not the NOC) sent badminton and table tennis players to GANEFO, aware that these were non-Olympic sports. But it was Mexico's cultural delegation to GANEFO—a well-known mariachi band—that became one of Latin America's most visible contributions in Jakarta.

Finally, athletes especially those from Latin America represented the sporting arms of university student associations. For example, Brazilian athletes at GANEFO included a basketball team comprised of male university students from Rio de Janeiro associated with the Confederação Brasileira de Desportos Universitários. The participation in GANEFO by university student groups from South America points to the nuance required in examining the rhetoric surrounding the Games and degree to which this was embraced by all competitors. Indonesia subsidized travel costs for many of these students and, as one US diplomat noted, 'students rarely, if ever, decline free invitations to travel' (American Embassy Montevideo 1963). But not all student sporting associations shared the same politics. In Uruguay, according to the characterization of the US Embassy in Montevideo, 'the radical pro-Castro Federation of Uruguayan University Students (FEUU) has responded favorably to an invitation by the Indonesians,' in part because 'the Games probably have a political appealing theme' (Ibid). The Uruguayans had been approached with an informal invitation to GANEFO earlier in 1963 at the Fédération internationale du sport universitaire (FISU) world university games (Universiade) in Porto Alegre—and FISU was another organization whose leaders' politics troubled the mainstream of international sport.

Student footballers from Argentina were also invited to GANEFO following a university competition, this one a match in Chile. The official invitation to the Argentine government was initially passed along to the NOC, which checked with the IOC and declined to participate. But with the invitation framed primarily as a university sporting competition—Argentine participants in GANEFO insist that they were unaware of the event's political overtones until they arrived in Jakarta—the decision to participate and select a team was left to sport officials at the University of Buenos Aires. It was student swimmers and water polo players who made the trip to Indonesia, initially with the support of their institution. Even as the risk of suspension loomed, the team's organizers were called by the nation's president, Arturo Illia, who encouraged them to travel to Jakarta to represent Argentina, telling them he would do what he could to protect them from suspension. It turned out that there was little he could do.

The IFs, led by athletics (IAAF) and aquatics (FINA), moved after GANEFO to impose the penalties for participation that had been threatened in the lead-up to the Games. Both organizations debated lifetime bans and Argentine swimmers and water polo players were initially suspended for 99 years by their domestic aquatics federation. Three track-and-field athletes from Chile were suspended for life. A swimmer from Buenos Aries decided not to risk his chances of making it to the 1964 Olympic team and chose instead to play football at GANEFO with his university colleagues; nevertheless, the Argentine aquatics federation suspended him. Even members of the Olympic family opposed to GANEFO were troubled by such a response. As General Clark of Mexico wrote to Avery Brundage, 'It is a negative and drastic attitude to disavow a Nation because some unknown swimmers who are not related to the Olympic Committee in any way, go to Indonesia to compete' (Clark 1963). Most suspensions were eventually lifted, but the majority of them were imposed long enough to prevent the penalized athletes from competing at the 1964 Tokyo Olympics.

Conclusion

It is worth reconsidering GANEFO because it has been obscured by higher profile struggles in the 1960s around sport and human rights in the USA and the South African anti-apartheid movement, and because GANEFO took place in Southeast Asia, hidden by the more prominent glare of Western sport which has dominated understandings and histories of international sport. Yet the dismissal of GANEFO as a second-rate Asian event is reason enough to reconsider it, both in an attempt to acknowledge the 'other' within international sport and because in 1963 GANEFO was viewed as anything but a sideshow in the offices of Western institutions, both sporting and governmental. As historian Laine (2007, p. 12) has encouraged, Western Olympic and sport scholars need to re-evaluate how they 'approach otherness … The same recognition might be necessary when one investigates the general effects of the International Olympic movement.'

Take the case of the Philippines. An avowedly anti-communist US ally in the region, the Philippines was also signatory in the summer of 1963 to the Maphilindo Accord, which sought to unite the pan-Malay peoples of Malaysia, the Philippines, and Indonesia (Hamanaka 2010, pp. 34–5). As such, Philippine president Macapagal had offered assurances to Sukarno that the Philippines would send athletes to GANEFO later that year, much to the consternation of Filipino sport authorities who feared the repercussions their athletes would face. It was the position of the US Department of State, communicated to the Embassy in Manila, that 'the Philippines should be discouraged from such action' and 'discreet measures undertaken to dissuade Philippine participation in GANEFO' (Department of State 1963b). These discreet measures involved mobilizing 'American assets' to apply 'pressure' be it through the IOC (specifically the American members of the IOC) or sympathetic Philippine media.

Whether such influences made a difference is unclear. In the end, the Philippine athletes who went to GANEFO represented military and government clubs, so any eventual punitive actions they faced from international sport bodies would not imperil the athletes that the Philippine Olympic Committee planned to send to Tokyo a year later. Overall, it is unclear whether any of the manoeuvring by the West to mitigate GANEFO's influence mattered—although it would be naïve to think that it had no effect. But while Sukarno imagined GANEFO as a quadrennial rival to the IOC Games, by 1965 he had been ousted, and the 1967 GANEFO originally planned for Cairo was subsequently moved to Beijing, before disappearing from the sporting calendar altogether with the Cultural Revolution.

But in September 1963, GANEFO's star was still waxing, and, according to the US Embassy, in Jakarta, 'It is still too early to tell to what extent GANEFO will prove to be a new rallying point in the anti-white campaign'[3] (American Embassy Jakarta 1963, p. 2). And it is in this

---

[3] With the USA 'combating' Chinese influence in Southeast Asia and Soviet influence in Africa, this racialized language could be read in Cold War terms—so that 'white' was code for 'anti-communist.'

context that GANEFO reveals that while, as Odd Arne Wested (2005, p. 74) argues, 'the processes of decolonization and of superpower conflict may be seen as having separate origins, the history of the twentieth century cannot be understood without exploring the ties that bind them together.' Through this lens, GANEFO answers Chen's (2009, p. 423) call to 'think more about the intertwined histories and multiple legacies for contemporary political projects of Third Worldism and the Third World.' When the First World did think of the Third, Prashad (2007, p. 8) argues, it rarely left behind the stereotypes of the colonial era, adding to those the belief that the 'darker peoples' were now even more likely, in his words, to be 'poor, overly fecund, profligate, and worthless' in the absence of their colonial masters. GANEFO is a useful lens through which to 'consider the ways in which the Third World operated simultaneously as a political category of radical promise and of Euro-American discipline in the 1960s' (Chen 2009, p. 425). It was on the sporting fields of Jakarta for two weeks in November of 1963 that the geopolitical struggles of the Cold War, Second World tensions, and Third World aspirations played out.

References

Archival Sources[4]

American Embassy Jakarta. (1962). 'The Asian Games Controversy', airgram A-593, 28 November, RG 59, EDU 15-INDON 1963, NARA.

American Embassy Jakarta. (1963). 'Indonesian Sports, Politics and GANEFO', airgram A-216, 4 September, RG 59, EDU 15-INDON 1963, NARA.

American Embassy La Paz. (1963). 'Bolivian Participation in GANEFO (Games of the New Emerging Forces)', airgram A-396, November 14, RG 59, EDU 15-INDON 1963, NARA.

American Embassy Montevideo. (1963). 'Uruguayan Student Participation in Indonesian Games (GANEFO)', 19 October, RG 59, EDU 15-INDON 1963, NARA.

---

[4] The National Archives and Records Administration, College Park, Maryland is referred to as NARA; the Olympic Studies Centre, Lausanne, Switzerland is referred to as OSC.

Applaus voor Nederlandse vlag. (1963). *De Telegraaf* (Amsterdam), 22 November, Folio 2, Archief Harry Stapel, International Institute of Social History, Amsterdam, Netherlands.

Associated Press. (1963). Undated/untitled newswire report, box 201, Avery Brundage Collection, University of Illinois Archives.

Brundage, A. (1963a). Letter to Syed Wajid Ali, 11 September, D-RM01-PAKIS/002, sous-dossier 4: Correspondance, 1960–1969, OSC.

Brundage, A. (1963b). Letter to Otto Mayer, 9 November, CIO-PT-BRUND-CORR/7063, sous-dossier 3: Correspondence, septembre-decembre 1963, OSC.

Burghley, B. (1964). Letter to Otto Mayer, 18 September, Athletisme, Correspondance, 1915–1966, sous-dossier Correspondance, 1959–1963, OSC.

Canadian Embassy Jakarta. (1963). 'GANEFO', numbered letter 467, 16 November, File 12671-CH-40, RG 5554, Library and Archives Canada, Ottawa.

Clark, J. (1963). Letter to Avery Brundage, 16 December, box 201, Avery Brundage Collection, University of Illinois Archives.

Clark, J. (1964). Letter to Otto Mayer, 17 January, box 201, Avery Brundage Collection, University of Illinois Archives.

Department of Foreign Affairs [Australia]. (1964). Despatch to Australian Embassy Jakarta, 16 January, File: GANEFO, RG 25 – 10919, Library and Archives Canada, Ottawa.

Department of State. (1963a). 'Indonesian Sponsorship of GANEFO', Department of State communiqué to all embassies, CA-2996, 16 September, RG 59, EDU 15-INDON 1963, NARA.

Department of State. (1963b). 'Philippines and GANEFO', Department of State communiqué to American Embassy Manila, A-97, October 11, RG 59, EDU 15-INDON 1963, NARA.

GANEFO. (1963a). 'Rules and Regulations of the 1st GANEFO (Games of the New Emerging Forces) Djakarta 1963' *GANEFO Bulletin*, No. 1, July: 8, RG 59, EDU 15-INDON 1963, NARA.

GANEFO. (1963b). 'Medals of GANEFO I,' GANEFO Press Release No. 118/E, 4 December, Jeux du GANEFO, Correspondence, 1963, OSC.

Gilchrist, A.G. (1963). 'The Games of the New Emerging Forces', British Foreign Office report, 20 December, File: GANEFO, RG 25 – 10919, Library and Archives Canada, Ottawa.

IOC Circular Letter. (1963). #252, 15 December, OSC.

IOC Circular Letter. (1964). #255, 24 February, OSC.

IOC Circular Letter. (1965). #279, 12 April, OSC.

IOC Session. (1964). Minutes of the 61st Session of the International Olympic Committee, Innsbruck, Austria, 27–28 January, OSC.

Secondary Literature

Adams, I. (2002). Pancasila: Sport and the building of Indonesia – Ambitions and obstacles. *International Journal of the History of Sport, 19*(2/3), 295–318.

Booth, D. (2003). Hitting apartheid for six? The politics of the South African sports boycott. *Journal of Contemporary History, 38*(3), 477–493.

Brown, C. (2003). *A short history of Indonesia: The unlikely nation?* Crow's Nest: Allen & Unwin.

Chakrabarty, D. (2010). The legacies of Bandung: Decolonization and the politics of culture. In C. J. Lee (Ed.), *Making a world after empire: The Bandung moment and its political afterlives.* Athens: Ohio University Press.

Chen, T. (2009). Third World possibilities and problematic: Historical connections and critical frameworks. In K. Dubinsky, C. Krull, S. Lord, S. Mills, & S. Rutherford (Eds.), *New world coming: The sixties and the shaping of global consciousness.* Toronto: Between the Lines.

Connolly, C. (2012). A politics of the Games of the New Emerging Forces (GANEFO). *International Journal of the History of Sport, 29*(9), 1311–1324.

Field, R. (2011). *The Olympic movement's response to the challenge of emerging nationalism in sport: An historical reconsideration of GANEFO.* Lausanne : IOC Library. http://doc.rero.ch/record/24926

Field, R. (2014). Re-entering the Sporting World: China's Sponsorship of the 1963 Games of the New Emerging Forces (GANEFO). *International Journal of the History of Sport, 31*(15), 1852–1867.

Gelman Taylor, J. (2003). *Indonesia: People and histories.* New Haven: Yale University Press.

Gitersos, T. (2011). The sporting scramble for Africa: GANEFO, the IOC and the 1965 African Games. *Sport in Society, 14*(5), 645–659.

Guanhua, W. (2003). "Friendship First": China's Sport Diplomacy during the Cold War. *Journal of American-East Asian Relations, 12*(3–4), 133–153.

Guttmann, A. (1992). *The Olympics: A history of the modern Olympic Games.* Urbana/Chicago: University of Illinois Press.

Hadiz, V. (2006). The left and Indonesia's 1960s: The politics of remembering and forgetting. *Inter-Asia Cultural Studies, 7*(4), 554–569.

Hamanaka, S. (2010). *Asian regionalism and Japan: The politics of membership in regional diplomatic, financial and trade groups.* Abingdon: Routledge.

Harding, H. (1981). China and the third world: From revolution to containment. In R. H. Solomon (Ed.), *The China factor: Sino-American relations and the global scene.* Englewood Cliffs: Prentice-Hall.

Hartmann, D. (2003). *Race, culture, and the revolt of the Black athlete: The 1968 Olympic protests and their aftermath.* Chicago/London: University of Chicago Press.

Kidd, B. (2005). Recapturing alternative Olympic histories. In K. Young & K. B. Walmsley (Eds.), *Global Olympics: Historical and sociological studies of the modern games.* Amsterdam/New York: Elsevier.

Laine, L. (1996). TUL: The Finnish worker sport movement. In A. Kruger & J. Riordan (Eds.), *The story of worker sport.* Champaign: Human Kinetics.

Laine, L. (2007). 'The IOC, the globalisation of sport, and gender in the revolutionary 1960s: A Nordic perspective' Unpublished keynote address, 'Sport in a global world – Past, present and future,' joint World Congress of the International Society for the History of Physical Education and Sport (ISHPES) and the International Sociology of Sport Association (ISSA), Copenhagen.

Legge, J. D. (1972). *Sukarno: A political biography.* Sydney/London/Boston: Allen & Unwin.

Li, T. (2007). *The will to improve: Governmentality, development, and the practice of politics.* Durham/London: Duke University Press.

Lutan, R., & Hong, F. (2005). The politicization of sport: GANEFO – A case study. *Sport in Society, 8*(3), 425–439.

Lüthi, L. (2008). *The Sino-Soviet split: Cold War in the communist world.* Princeton/Oxford: Princeton University Press.

Majumdar, B., & Mehta, N. (2008). *Olympics: The India story.* New Delhi: HarperCollins.

Parks, J. (2009). *Red sport, red tape: The Olympic Games, the Soviet sports bureaucracy, and the Cold War, 1952–1980.* Unpublished PhD thesis, University of North Carolina.

Prashad, V. (2007). *The darker nations: A people's history of the third world.* New York/London: The New Press.

Riordan, J. (1993). Rewriting soviet sports history. *Journal of Sport History, 20*(3), 247–258.

Riordan, J. (1999). The worker sport movement. In J. Riordan & A. Kruger (Eds.), *The international politics of sport in the 20th century.* London/New York: E & FN Spon.

Roosa, J. (2008). President Sukarno and the September 30th movement. *Critical Asian Studies, 40*(1), 143–159.

Schantz, O. (1997). The presidency of Avery Brundage (1952–1972). In *The International Olympic Committee – One hundred year: The idea, the presidents, the achievements* (Vol. II). Lausanne: International Olympic Committee.

Senn, A. E. (1999). *Power, politics, and the Olympic Games.* Champaign: Human Kinetics.

Shuman, A. (2013). Elite competitive sport in the People's Republic of China, 1958–1966: The Games of the New Emerging Forces. *Journal of Sport History,* 40(2), 258–283.

Sie, S. (1978). Sports and politics: The case of the Asian games and the Ganefo. In B. Lowe, D. B. Kanin, & A. Strenk (Eds.), *Sport and international relations.* Champaign: Stipes Publishing.

Torres, C., & Dyreson, M. (2005). The Cold War games. In K. Young & K. B. Walmsley (Eds.), *Global Olympics: Historical and sociological studies of the modern games.* Amsterdam/New York: Elsevier.

Wested, O. A. (2005). *The global Cold War: Third World interventions and the making of our times.* Cambridge: Cambridge University Press.

Xu, G. (2008). *Olympic dreams: China and sports, 1895–2008.* Cambridge/London: Harvard University Press.

Zhai, Q. (2006). China and the Cambodian conflict, 1970–1975. In P. Roberts (Ed.), *Behind the bamboo curtain: China, Vietnam, and the world beyond Asia.* Washington, DC: Woodrow Wilson Center Press.

Zhou, T. (2013). Ambivalent alliance: Chinese policy towards Indonesia, 1960–1965, Working Paper # 67. Washington, DC: Woodrow Wilson International Center for Scholars.

"Memorias del '68: Media, Massacre, and the Construction of Collective Memories"

Celeste González de Bustamante

On October 2, 2013, thousands of citizens converged on the streets of Mexico City in solemn observation of the 45th anniversary of the massacre of hundreds of innocent students and bystanders at the Plaza de Tlatelolco. The 1968 massacre had happened just ten days before the inauguration of the XIX Olympics in Mexico City, the first ever to be held in Latin America.

For Mexican officials, the ability to host the Olympic Games illustrated that the country had reached a high level of modernity, and television executives beamed that vision into the homes of Mexican citizens at home and around the world, and for the first time in color. For student activists, the Games represented a source of growing discontent, as they claimed the government had squandered public funds simply

Parts of this chapter were first presented as a paper at the annual conference of the Rocky Mountain Council for Latin American Studies March 4–7, 2009, Santa Fe, New Mexico.

C.G. de Bustamante (✉)
School of Journalism, University of Arizona, Marshall Building, 336, 845 North Park Avenue, Tucson, AZ 85721, USA

© The Editor(s) (if applicable) and The Author(s) 2016
J. Dart, S. Wagg (eds.), *Sport, Protest and Globalisation,*
DOI 10.1057/978-1-137-46492-7_6

to beautify the city, while that money could have been used to help achieve the nation's Revolutionary goals for those who needed it most, the poor. Students claimed, "We weren't against the Olympics as a sports event, but we were against what the Games represented economically."[1] Witherspoon summed it up well: the XIX Olympiad "united politics, culture, diplomacy, and athletes as no Olympics before or since."[2]

More than four decades later, the Mexico City Olympics and the Tlatelolco massacre continue to be debated in the public sphere in Mexico. Twenty-first-century activists attempt to keep the memory of the massacre alive, and while some of Mexico's old guard would simply like to forget what has become a national stain, the media make the tragic event visible at least once a year on October 2nd. Through an examination of televised news reports aired in 1968 and 2008, this chapter argues that television coverage—and those who participate in its production—contributes to a contested public sphere, aiding the construction and reconstruction of collective memories of Mexico's '68. Over the years, television coverage of the massacre has changed, along with the country's political opening, mainly during the 1990s. In 1968, television producers (under pressure from the Mexican/the International Olympic Committee/their station bosses?) limited the images that were broadcast and attempted to severely downplay the tragedy. Yet, four decades later, mainstream television anchors and reporters frequently used the catch phrase, "The second of October, it is not forgotten," signaling both an acceptance of the government's culpability, and media executives who perhaps attempted to appear more independent.

It is important to note that in the USA, and in much of Europe, the 1968 Olympics are remembered as a time of racial conflict and protest. It was during these Games that two US Olympic team runners, Tommy Smith and John Carlos stood on a platform during the medal ceremony with black armbands and their fists raised in protest against racial inequality and discrimination in the USA.

---

[1] Cited in Witherspoon (2008).

[2] Witherspoon, *Before the Eyes of the World*, 5.

Adela Micha, one of Mexico's preeminent news anchors, sat at a slick news desk at Televisa's Chapultepec studios on October 2, 2008. Moments before her news program "The News by Adela" went on the air at eight in the evening, a floor director gave her a countdown "three, two, ___,":

Hello, how are you? Good evening. In today's news: several marches were underway in Mexico City to commemorate the 40th anniversary of the '68 massacre.[3]

Natural sound and images of a crowd aired for an additional 30 seconds after Micha's voice-over. The anchor continued, stating that throughout the city, marches had taken place, starting in various parts of the capital, en route to the main square, and on the corner of Isabel La Católica and Francisco Madero streets, marchers threw rocks and damaged buildings in the historic district and that:

In a moment I am going to give you all the details, a complete chronicle of what happened this afternoon.[4]

Images and sounds of a chaotic crowd moving through the streets appeared during the anchor's voice-over, including a shot of two young masked men kicking in a door of a downtown business.[5]

News coverage ensued on Mexico City's Channel Three—the newest national commercial network, as anchor Yuriria Sierra announced:

Today is the second of October, and of course, every year this calls us to remember what happened forty years ago with the mass slaying of students at Tlatelolco, one of the darkest pages in the contemporary history of our country.[6]

---

[3] Adela Micha, *Las noticias por Adela*, Televisa, October 2, 2008.
[4] Ibid.
[5] Ibid.
[6] Yuriria Sierra, *Cadena Tres Noticias, Segunda Emisión*, Cadena Tres, 2 October 2008.

She continued, "Four decades have past, and '68 is not forgotten."[7]

Television, print, and online journalists extensively covered the 40th anniversary of the massacre at the Plaza de las Tres Culturas on October 2 and 3, 2008.[8] On Mexico City commercial and public television, at least 89 reports focused on October 2.[9] The online versions of major Mexico City dailies *La Jornada*, *El Universal*, and *Reforma* all devoted special sections to the subject. Journalists reported on events commemorating the massacre, and also produced audiovisual reconstructions of what happened on October 2, through the use of historic still photographs and moving images, and broadcasting interviews with the activists who had participated in the student movements of 1968.

On Francisco Zea's newscast on Channel Three, ex-leader of the student movement Angel Verdugo, appeared in an eight-and-a-half minute special report in which he told his story of survival on the night of October 2, four decades ago.[10]

Clearly, *el '68*—the somewhat euphemistic term that refers to the more than 300 students, local residents, and bystanders who were killed by government troops in the Nonoalco-Tlatelolco neighborhood on October 2, 1968—was on the minds of journalists and citizens throughout the capital 40 years later.

---

[7] Ibid.

[8] Over the past 40 years, scores of Mexican scholars have generated a sizable body of literature (in Spanish) focusing on the student movements of 1968 and the massacre at the Plaza de la Tres Culturas from a variety of perspectives. Some of the most prominent include: Aguayo (1998), Garín (1998), Poniatowska (1971), Revueltas (1978), García and Monsiváis (1999). Aside from Poniatowska (1992), few works in the English language have emerged. Carey (2005) is the only scholarly book written in English that analyzes the student movements and subsequent massacre at Tlatelolco from a gendered perspective; primarily interested in the Olympics of 1968, Witherspoon (2008) analyzes the myriad of controversies surrounding the XIX Olympiad, including the government crackdown on students; Diana Sorensen examines the work of literary figures such as Octavio Paz and their relationship to politics, modernity, and post-modernity in Sorensen (2007).

[9] A graduate student at El Instituto Tecnológico y Estudios Superiores de Monterrey (Mexico City Campus) used the university's digital database of all Mexico City television programming to locate items related to the 40th anniversary of the massacre. A search of the database using the phrase "2 de octubre" generated 89 news items on the subject.

[10] Francisco Zea, *Cadena Tres Noticias*, Cadena Tres, 2 October 2008.

In Mexico, this particularly dark moment in its history unfolded, as noted, in stark contrast to one of the country's moments of crowning glory—the hosting of the XIX Olympiad, which began just ten days later on October 12, 1968. The close temporal proximity in which the two events occurred made it difficult, then and now, to separate the two. Both tragedy and triumph must be considered in concert if a deeper understanding of these historic events is to be gained. Despite the importance of the country hosting the first Olympics in Latin America, and the international esteem that came along with that, what is collectively remembered first in Mexico about the autumn of 1968 is the massacre that happened just days prior to the inauguration of the Games.

Yet, the news coverage of the 1968 Tlatelolco massacre 40 years later represented a marked difference from the limited and what scholars have defined as "official" coverage of the massacre when the events were actually unfolding. What happened between 1968 and 2008 that can explain the stark difference in media coverage? Why is that important? And, what is the relationship between sport and protest in this case? These are the questions that this chapter aims to answer.

By juxtaposing the news coverage of the student movements, Olympics, and Tlateloco massacre in 1968, with the news coverage of the events in 2008, this chapter examines how media portrayals of the events of 1968 shifted over time. First, the chapter offers some historical context in which to understand the medium of television in mid-twentieth-century Mexico as well as the events of the summer of 1968, paying specific attention to the preparations for the XIX Olympiad and student protests which criticized the government for hosting the Olympics in Mexico City. Next, the chapter provides an explanation and synopsis of the 1968 news coverage of the student movements and Olympics. Then, the chapter fast-forwards 40 years to 2008 to present an analysis and discussion of contemporary news coverage of the historic events of 1968. Finally, I will explain what the shifting collective memory portrayed on television signifies for the nation, and how this case advances the theoretical framework of hybrid framing as a way to understand news media and history.

Hybrid Framing Theory and the Importance of Collective Memory

Collective memories compose part of a "master narrative" that individuals and groups use to create a sense of national identity.[11] Since independence and certainly after the Mexican Revolution (1910–1920), national officials sought to create a sense of *lo mexicano* or Mexican consciousness. While Mexico is exceptional in this regard, attempts to define and construct national identity have perhaps been more fervent in Mexico compared with other Latin American countries.[12]

As one of the most important forms of mass communication, television and its programming contribute—in no small part—to the formation of collective memory. As Thomas Benjamin states, "poets, journalists, politicians, and writers are often more influential in composing the master narrative than are professional historians."[13]

The plural form "collective memories" is useful because temporal and geographical contingencies can play a significant role in influencing what and how so-called "hot" moments are remembered.[14] Hot moments refer to those times when societies access their own significance. Further, with respect to television, news coverage may offer interpretations of daily social reality that can illuminate "*a* world, not *the* world."[15] And, while it may be true that, as Patricia Fournier argues, on an individual level "we all have our own '68," analysis of cultural products such as newscasts, which reflect and shape national memory, reveals a coherent sense of meaning regarding *el '68*. With the goal of isolating the significance of *el '68*, this chapter aims to answer the question: What can television news coverage, 40 years after the *matanza* (slaughter) tell us about popular memory, sport, and protest?

---

[11] Benjamin (2000).
[12] Chorba (2007). Chorba argues that Mexico's proximity to the USA and its revolutionary past represent are two factors that distinguish Mexican "national discourse," 8.
[13] Benjamin, *La Revolución*, 14.
[14] Lévi-Strauss (1966), Schwartz (1982).
[15] Becker and Celeste González de Bustamante (2009).

I have elected to examine what news journalists, as "cultural authorities," chose to remember and how they choose to remember it.[16] Zelizer uses the term "cultural authority" in her study of television journalists who covered the assassination of John F. Kennedy. She argues that it was at this pivotal time that television journalists established themselves as cultural authorities, meaning that from the 1960s onward, they played a crucial role in determining collective memory.

In contrast to their North American counterparts, Mexican television journalists held less in the way of independent cultural authority in the 1960s.[17] For example, in the 1960s, officials from the Secretariat of Communications would review news scripts and censor items that they deemed unfit for air.[18] Government spies kept a close watch on the activities of journalists. On-camera presenters who deviated from the government's official story found themselves out of work.[19] As the country moved slowly toward democracy during the second half of the twentieth century, especially since 1968, and as a result of the events of *el '68*, members of the media began to assume more control over cultural production.

Throughout the twentieth century, despite state efforts to control the media and civil society under a semi-authoritarian regime, there were limits to what scholars refer to as cultural hegemony. As this chapter will reveal, the cultural products of 2008 provide evidence of those limits.[20]

It is through television news that one can begin to see how relations between the media and the state have altered over the past 40 years, which in turn helps to explain the stark difference in news coverage between 1968 and 2008. The second argument posited here is that news coverage has changed dramatically, reflecting a shift in the country's political, social, and cultural *milieux*. The focus of news coverage has changed somewhat from the use of official voices to discuss *el '68* to the inclusion

---

[16] Zelizer (1992).

[17] Cole (1996), Christlieb (1982), Fernández et al. (2000), Ferreira (2006), Lawson (2006), Hughes (2006), Medellín (1992), Miller and Darling (1997), Gómez (2002), Saragoza (forthcoming), Sinclair (1999), Skidmore (1993), Delabre (1988) and Delabre (1985).

[18] Camp (1985).

[19] One of the few to offer alternative voices during 1968, Jorge Saldaña has been fired from Telesistema Mexicano (and later Televisa) and Televisión Independiente de México more than a dozen times. Interview with the author, 5 August 2008.

[20] Celeste Gonzalez de Bustamante (2012), Martín-Barbero (1993).

of former student leaders and witnesses as experts on the subject. Further, in contrast to the media landscape in the 1960s and 1970s, women now constitute major players in the realm of cultural authority. In 1968, few women worked on the air and those who did were relegated to covering "cultural events."

The third argument related to changing media practices posits that technological advances have contributed to the shifts in power between the state and the media, not just in Mexico, but throughout the world.[21] Satellite communications coupled with the Internet and digital technologies have changed, though not necessarily completely diminished, the role of the state and its ability to control the "hearts and minds of the people." At relatively low cost, everyday citizens can contribute to the global media landscape by posting their stories and other forms of cultural expression on the Internet, which has created new "channels of resistance."[22]

A useful concept in thinking about visual media and shifts in cultural hegemony is a theoretical framework I call *hybridity of framing*. The concept draws upon *cultural hybridity* and *framing*, which emerged from communication studies. Cultural hybridity holds that when two or more cultures converge, the social practices and beliefs of each group influence one another to the extent that a new distinct culture emerges.[23] Framing analysis concentrates on the manner in which a news producer or writer emphasizes some elements of an event or issue over others, with the goal of making a news report meaningful.[24] For example, television news producers frame events in particular ways, but viewers may re-frame the same events differently, in ways that they deem more just, useful, and meaningful. Hybridity of framing helps explain how news producers in 1968 may have attempted to downplay student movements, and how young people might have interpreted the same issues/events in a distinctly different way, a manner that in some cases fostered the ability to build domestic and international solidarity. The framework is of

---

[21] Fox et al. (2002).
[22] Duwmunt (1992), Jackson Lears (1985).
[23] Canclini (2001), González de Bustamante (1993).
[24] Gamson and Modigliani (1989).

particular value to historians who analyze change over time. In this case, hybrid frames help to explain the marked contrast between how television news producers portrayed students in 1968 and how they depicted the group of individuals four decades later.

As Steve Anderson notes, numerous theorists have criticized television as a medium "unsuitable for the construction of history," because of its ability to be both in the present and the past simultaneously, and that television creates "cultural amnesia."[25] Yet, as I would argue with Anderson, instead of perpetrating collective forgetfulness, "TV has modeled highly stylized and creative modes of interaction with the past."[26] Further, it is the medium's ability to interact with the past in ways that are creative and entertaining for viewers, compounded by its ability to reach a mass audience that adds to television's power to influence the construction of collective memory.

Mexican Television in Historical Context

Although engineers and businessmen had been developing television since the 1930s, the first official public broadcast on television in Mexico did not occur until September 1, 1950, with the fourth state of nation address by then president Miguel Alemán Valdez. From their inception, private interests have driven the development of broadcast media—both in radio and television. That is not to say the state showed no interest in either medium. The very early days of radio were marked by government support, as generals in the Revolution used radio to circumvent the Revolutions enemies, and then later Mexican presidents such as Lázaro Cárdenas used the medium as a way to disseminate the tenets of the Revolution.[27] State officials sought to influence the development of television as well. The government subsidized the building of the infrastructure necessary for long-distance television, including the 1968 completion

---

[25] Anderson (2001).

[26] Anderson, "History, TV and Popular Memory," 20.

[27] Castro (forthcoming).

of a national network of microwave towers and a satellite center, which helped the country broadcast the Olympic Games.[28]

Nevertheless, it was businessmen, mainly those based in Mexico City, such as Romulo O'Farril Sr., Emilio Azcárraga Vidaurretta, and Guillermo González Camarena, who developed the industry, albeit with some intervention from the state.

The moguls of early television competed against each other for the first five years, trying to develop the level of advertising necessary to sustain three separate networks, but not all advertisers were ready to climb aboard. According to Azcárraga Vidaurretta, during the early days of Mexican television, he lost millions of pesos in the competition between the three networks. Reconciled to the realities of the fledgling industry, the three networks combined in 1955 to create Telesistema Mexicano, which resulted in a television monopoly that would continue almost until the end of the twentieth century. By 1968, there were two main commercial networks on the air, with Telesistema Mexicano, based in Mexico City, being the most important. The other, Television Independiente de Mexico, was run by the so-called Monterrey elite of entrepreneurs based in Nuevo León.

As on the radio, sporting events quickly became a mainstay of programming on television. In the 1950s, Mexican television executives and producers participated in several "experiments," which gave them the experience necessary to take on the broadcasting of the Olympic Games in 1968. The Pan American Games of 1955, which were held in Mexico City, represented one such opportunity.

One of the first things many of the country's viewers might remember about television in the 1960s is the formal nature of reporting that was broadcast on a daily basis. Frequently, newscasts were filled with Presidential visits to various places around the country and abroad. News reports reflected the version of the country's affairs favored by the political and economic elite. Tumult outside of the country was highlighted over internal conflict. News coverage of the 1968 Olympics can be understood in this light and a discussion of coverage of the Games and the Tlatelolco massacre is presented below.

---

[28] González de Bustamante, "*Muy buenas noches*," 19.

After the 1968 Olympics, sporting events continued to receive top billing on Mexican television programming lists. The country's hosting of the 1970 football World Cup provided media moguls and politicians with another opportunity to construct and present an image of a modern Mexico. Throughout the tournament, television anchors highlighted the day's sporting events in their newscasts. News coverage of politics paled in contrast to the amount of airtime devoted to sports. For example, the number of stories about the 1970 Presidential election campaign, which unfolded at the same time as the World Cup, took up less airtime than news about the soccer matches of the day.[29]

Several technological advances in the 1960s changed the television landscape. The advent of satellite communications made it possible for the country to transmit domestic news internationally. Of course, the 1968 Olympics represented one of the most important events of the decade. Videotape represented another crucial technological development; in contrast to film, it did not have to be processed. This dramatically sped up news and entertainment production routines and flows. Telesistema Mexicano, a commercial monopoly at the time, could produce telenovelas and newscasts much more quickly, and distribute the company's programming in a more efficient and cost-effective way.

One of the most significant industry developments during the 1970s was the creation of Televisa in 1973, which resulted from the merging of Telesistema Mexicano and Television Independiente de Mexico. This merger forged an alliance between two powerful business families: the Monterrey, Nuevo León, based Garza-Sadas, and of course, the Azcárragas who had dominated the Mexico City-based commercial media landscape since the 1940s, beginning with radio and then television.

Mexican television in the 1980s can be characterized as an era of exportation. With the improvement of satellite technology, programming could be beamed to faraway countries. Televisa became a dominant player in media markets throughout Spanish-speaking Latin America and in parts of the USA. The most popular genre of television programming was telenovelas (soap operas), but news was also exported. Entertainment

---

[29] González de Bustamante, "Muy buenas noches," 177–204.

programs were exported to more than 100 countries around the world, and anecdotes abounded from viewers from places as far away from Mexico as the former Yugoslavia, who claimed to have learned how to speak Spanish by watching Mexican telenovelas.[30]

The network took advantage of satellite communications and followed Ted Turner's Cable News Network (CNN). In 1988, Azcárraga formed his own news network ECO (Empresa de Comunicaciones Orbitales/ Orbital Communications Enterprise). The network transmitted news throughout the Latin America and in Spain.[31]

Since 1968, Televisa's credibility had been in question, and trust in the network has continued to plummet over time. One blow to the network's integrity came after distorted news coverage of Chihuahua state elections in 1986 aired on Televisa newcasts. Claims of election fraud came from well-respected intellectuals, as well as the Catholic Church and ordinary citizens, but the network ignored and failed to report evidence of irregularities.[32]

As in other parts of the world, neoliberal politics took hold in Mexico during the 1990s, which had consequences for the media. For example, as part of his privatization program, President Carlos Salinas de Gortari offered a state-owned television station to commercial interests. Ricardo B. Salinas Pliego, no relation to the president, eventually purchased the station, and turned it into Televisa's first real competitor of the twentieth century. In 1995, TV Azteca moved into the sports market when it landed a contract to televise the US NBA (National Basketball Association) games in Mexico. TV Azteca also began to compete with Televisa in entertainment programming.

With the beginnings of a political opening, news media also began to increase the diversity of political coverage they offered to viewers and readers.[33] Nevertheless, in general, the mainstream media outlets steered away from criticizing the traditional "untouchables": the President, the

---

[30] Richard Cole, foreword in Celeste González de Bustamante, "Muy buenas noches."

[31] Fernández and Paxman, *El Tigre*, 328.

[32] Fernández and Paxman, *El Tigre*, 315.

[33] Hughes (2006), Morris (1999).

church, and the military; and two new untouchables entered the picture, the business elite and large-scale drug kingpins.[34]

In the twenty-first century, television sports programming continues to be one of the most lucrative genres in over-the-air broadcasting. One indication of the convergence or vertical integration of the sports industry is Grupo Televisa's ownership of Azteca Stadium where the soccer team America, which is also owned by the same media company, plays. The stadium functions as the locale for major sporting events, as a production warehouse, as well as a news and television sports archive. Indeed, the company's news and sports archive are located beneath the stadium in tunnel 27.

In the new millennium, Televisa's monopoly of the last century has been replaced by a duopoly between TV Azteca and Televisa. Communications reforms that were introduced by President Enrique Peña Nieto in 2013, that theoretically would allow new media entrepreneurs to enter the market, have yet to alter dramatically the political economy of the country's media system.

The Student Movements and the Olympics of 1968

The student uprisings of 1968 began in the summer leading up to the Olympics as street fights among students from two rival high schools, which had affiliations with two of the national universities, the UNAM, National Autonomous University and the Instituto Politécnico, Polytecnic University. On July 22, police cracked down using brutal force on the students, thereby inspiring university students to join in and to begin protesting police abuse. The news media's general and contentious characterization of the students involved in the street fights as "thugs," drew a growing number of youth to the movement.[35]

As autumn drew near, students felt that the government was ignoring their demands, and continued to criticize the state officials and the

---

[34] Hughes, *Newsrooms in Conflict*, 84.
[35] Carey (2005).

amount of money the federal government was spending to host the Games. They drew attention to the fact that the state was neglecting a large group of people who lived in poverty, and who had not benefitted from the so-called "Mexican economic miracle."[36]

By September 18th, with the Olympic events due to begin in less than a month, the military moved onto the national university campus and occupied the UNAM, where some of the sporting events were scheduled to take place, forcing student protestors to relocate. Student organizers who belonged to the National Strike Committee chose to move their base to the Plaza de Tlatelolco. The plaza was located in a middle-class neighborhood, in the midst of a modern housing development with a large esplanade in the middle of several high-rise apartment buildings. The forced relocation of the students' base only seemed to strengthen their convictions and ire against the state.[37]

The government's preoccupation with, and its use of force against, student dissidents fit within the politics of dual containment, which included the state's fight against wider domestic unrest, and its concern about possible foreign agitators who might have communist ties, and who could influence young and politically minded activists.

On the evening of October 2, just ten days before the inauguration of the Olympic Games, *granaderos* (riot police) moved into the Plaza de Tlatelolco during a protest that included thousands of people marching in the esplanade. At around six o'clock in the evening, gunfire erupted as thousands scrambled to find shelter. Many did not. The shooting continued for several hours. In the end, several hundred students, local residents, and bystanders were killed, and many others wounded. Thousands of students were arrested and taken to jail.

Dozens of international journalists, who were there to cover the Olympics, also covered the massacre. Italian journalist Oriana Fallaci was at the scene on October 2 and was wounded in the crossfire. She survived to write about the ordeal.[38]

---

[36] González de Bustamante, "*Muy buenas noches,*": Mexico, Television and the Cold War, 149.

[37] Carey, *Plaza of Sacrifices: Gender, Power and Terror in 1968 Mexico*, 128–130.

[38] González de Bustamante, "*Muy buenas noches,*" 151.

News Coverage of the Tlatelolco Massacre and the Olympics

In national newspapers and in television news reporting, writers tended to downplay the state violence against the students. Given the consolidated and centralized nature of Telesistema's monopoly, and the network owners' close relationship with high-level governmental officials, television news reporting of the events of October 2 was perhaps the most limited in the overall media landscape of the time. Forty-seven years after the massacre, at the time of his death, Jacobo Zabludovsky, a leading news figure during the twentieth century, continued to be criticized for downplaying the government's role in the violent events of 1968.

Throughout the summer of 1968, television news producers depicted student activists as a threat to the nation, and in some cases called them "terrorists" and "gang members." These types of portrayal continued after the crackdown at Tlatelolco.[39]

News coverage of the massacre included police allegations that the movement was instigated by outside agitators. The attempt to link foreign interests to domestic unrest represented a common tactic among government officials throughout the twentieth century, as a way to delegitimize and disrupt social movements.[40]

In stark contrast to the news media's negative portrayals of dissident youth, Olympic athletes were hailed as positive symbols of national promise and development. Female hurdler Norma Enriqueta Basilio was framed as the embodiment of Mexican modernity and national pride. Basilio, who was from the country's north, became the first woman in modern times to carry the Olympic torch and light the Olympic cauldron on the opening day of the Games.

One of the ways that news producers sought to downplay domestic unrest was to increase coverage of sporting events. The Olympic Games provided Telesistema Mexicano with the ideal event to which it could turn its news focus, and which enabled the network to reduce the number

---

[39] González de Bustamante, *"Muy buenas noches,"*: *Mexico, Television and the Cold War*, 150–163.
[40] González de Bustamante, *"Muy buenas noches,"*: *Mexico, Television and the Cold War*, 161.

of reports devoted to the events and fallout of the Plaza de Tlatelolco massacre.

Twenty-five years after Tlatelolco and the XIX Olympiad, Basilio was featured on a special television program that reflected on the importance of the Olympiad for the country and for Televisa. She remarked that she had felt a great responsibility and that the experience "was unforgettable."[41] That year Televisa did not produce a program that commemorated the massacre at Tlatelolco. Partido Revolucionario Institucional / Institutional Revolutionary Party (PRI) rule was still firmly in place at the federal level. Despite apparent fissures in the party, the PRI would remain in power at the level of the presidency until 2000.

Coverage of 1968 Four Decades Later

By 2008, for the first time in seven decades, Mexico had seen two opposition party candidates accede to the presidency, first with the election of National Action Party (PAN) candidate Vicente Fox Quesada in 2000, then again in 2006 with the election of Panista Felipe Calderón Hinojosa. The move toward political pluralism, and a more robust and inspired civil society, coincided with a new generation of more critical and independent news media.[42] Independent voices appeared more often in print than on television, given that a virtual duopoly remained firmly in place, as a result of the so-called Televisa Law, which favored the concessions (licenses) of the two most prominent television networks of the time, Grupo Televisa and TV Azteca.

Nevertheless, by 2008, television news media producers and anchors did reflect an overall sentiment in favor of the student protesters of 1968. The collective memories that were aired presented a stark contrast to the government "affiliated" news media of the 1960s.

Three examples from the large amount of news coverage related to commemorative ceremonies on October 2, 2008 illustrate the myriad of ways in which producers of television contributed to the *memorias del '68*. Rather than focus on how news reporters covered the marches

---

[41] González de Bustamante, "*Muy buenas noches,*": *Mexico, Television and the Cold War*, 165.
[42] Hughes (2006).

on October 2, 2008, this chapter examines those reports whose focus is more historical and included historical images and sounds, in contrast to those reports, which focused on the 2008 marches.

Yuriria Sierra, Cadena Tres Noticias, Segunda Emisión

On the afternoon of October 2 2008, *Cadena Tres Noticias* aired a 2:33-minute report on commemorative events. News anchor Yuriria Sierra read from a script stating that "the Governor of the capital committed himself to find justice for the fallen."[43] As she continued her introduction to the upcoming report, an over-the-shoulder graphic appeared with a still image of Marcelo Ebrard, Mexico City governor, with the text, "the unhealed wound." The report filed by Enrique Sanchez was the fourth story in the program related to October 2, including the show's lead report.

Sanchez's report began with dramatic music and a montage of historic still photographs.[44] The report focused on a ceremony held at the Plaza de las Tres Culturas[45] headed by Ebrard who vowed to stand by el Comité 68, a group of survivors of the movement who are still pressing the government for answers regarding the massacre.[46] Standing at a podium, with ex-leaders of the movement at his side, Ebrard affirmed:

> The Mexico City government will stand with the Committee in its struggle, so that there is justice and the truth be clarified, and so that the Mexican state assume the responsibility and respect for those events. The second of October will not be forgotten.[47]

---

[43] Sierra, *Cadena Tres Noticias, Segunda Emisión*, October 2, 2008.

[44] Ibid.

[45] The esplanade at the Plaza Tlatelolco is also known as the Plaza de las Tres Culturas, in reference to the three cultures, European, Indigenous and Mestizo, which have influenced Mexican history and culture.

[46] Elaine Carey and José Augustín Roman Gaspar, "Carrying on the Struggle: *El Comité 68*," NACLA Report, (May 2008), online at http://nacla.org/node/4631 (accessed 28 February 2009).

[47] Sierra, *Cadena Tres Noticias, Segunda Emisión*, October 2, 2008.

In the report, Ebrard also mentioned that the problems that the students fought to overcome in 1968, such as inequality and abuse, remain four decades later.[48]

As the sound of a trumpet played a solemn rendition of the national anthem, Ebrard lowered the Mexican flag at half-staff in the Plaza de Tlatelolco, while a crowd of people stood by.[49]

A split screen revealed simultaneously moments of the past and present, with images of Ebrard dedicating a white floral wreath in memory of the fallen students on the left side of the screen, and black-and-white images of students running through the plaza on the right.[50] The next set of split-screen images included a juxtaposition of present-day soldiers standing at attention in the plaza on the left side of the screen, and a 1968 image of a gun-toting soldier in the plaza on the right of the screen.[51]

In addition to the Mexico City governor, reporter Sánchez included contributions from ex-student leader and member of el Comité 68 Felix Hernández Gamundi, who pledged to seek international aid in an effort to find and prosecute those responsible for the massacre.[52]

Sierra's introduction to Sanchez's report and the report itself illustrates how television journalists tended to emphasize certain elements over others regarding *el 2 de octubre*. First, the 1968 students and their cause remained the focus of the report. Throughout the piece, in the narrative, through historic still images, and dramatic sound, Sanchez characterized student protestors as victims of tragedy, yet at the same time Sanchez suggests that they represent victors of history because they helped move the country from authoritarianism to democracy.

Second, as with other reports on the subject, Sanchez's report emphasized continuity regarding what has become a dual struggle: of fighting for justice in the names of the victims of the massacre, as well as present-day efforts to end abuse and inequality throughout the country. Sound bites with Comité 68 member Hernández, who vowed to seek international

---

[48] Ibid.
[49] Ibid.
[50] Ibid.
[51] Ibid.
[52] Ibid.

assistance, and Ebrard's statement about current "abuse" in the country, offer evidence that both of these struggles live on. In addition, text used in graphics such as "an unhealed wound" served to further the notion of continued problems.

The goal to keep the memory of Tlatelolco alive represents a third element that journalists tended to emphasize in their reports on October 2, 2008. The use of historic images at the beginning of the report, along with dramatic sound, split-screen images of 2008 images on the left and 1968 images on the right, shown midway through the report, literally connect the past with the present. Finally, Ebrard's statement at the podium that October 2 "is not forgotten" served as another reminder to viewers that the memory of Tlatelolco should live on.

The inclusion of key national symbols provided visual cues for viewers, adding significance to the event. National symbols of mourning such as the flag at half-staff, the trumpet playing the national anthem, and the laying of a floral wreath all worked together to strengthen the magnitude and solemnity of the ceremony.

Television journalists learned to use these types of symbols in the early days of the medium.[53] Throughout the 1950s, on Independence Day, photographers shot images of the president laying floral wreaths at the *Columna de Independencia*, which provided visual proof for the stories they told.[54]

Francisco Zea, Cadena Tres Noticias, Cadena Tres

One of the ways the news media contributed to keeping alive the memory of Tlatelolco was through testimonial-style reports and reconstructions of events which used the voices of those who had participated in the events. Ana Ignacia Rodriguez, a member of the Comité '68, was featured in a 4:48 minute report on Channel 22's newscast hosted by Laura Barrera. In the report, Rodriguez stood in the Plaza de las Tres Culturas

---

[53] President Miguel Alemán Valdes brought Mexico and Latin America into the electronic visual age with the broadcast of his fourth *informe* on September 1, 1950.
[54] González de Bustamante, "Muy buenas noches."

and recounted the night of October 2 when she and her friend ran from police while bullets flew overhead.[55]

On Cadena Tres's early morning show, news anchor Francisco Zea introduced an 8:30 minute special report that focused on the "survival story" of ex-leader of the movement, Angel Verdugo.[56] In his introduction to the report, Zea began his newscast asking the question, "October second is not forgotten, how many times have we heard that phrase?" As he continued with an introduction to the first report of the newscast, he characterized October 2, 1968 as "a watershed moment in the country's history, a dark chapter that left a wound that has not yet healed."

Like the first example mentioned in this chapter (i.e., Sanchez's report), Zea's report (mini-doc) began with dramatic sound, but in contrast to the other report, the piece started with simple black-and-white text that read, "For 40 years, the former young polytechnical student, Angel Verdugo, did not visit Tlatelolco."[57]

For the next eight minutes, Angel Verdugo, in his own words, told his harrowing story about how he arrived at the Plaza de Tlatelolco, found himself in the midst of gunfire, sought refuge in an apartment in the Chihuahua building, and how he was sheltered by a helpful family the morning after the violence.

Historical images from the Mexico City daily *El Excelsior* that had "never been published before" were inserted into the documentary-style special report. In the piece, Verdugo led the reporter and viewers to the Chihuahua building and apartment number 615, whose inhabitants offered him and his friend refuge throughout the night of the massacre.

It is difficult to overstate the power of telling the story through eyewitnesses. Elena Poniatowska's chronicler approach in *La Noche de Tlatelolco* gave its readers the sense of "being there" on that violent night and provided some of the initial firsthand accounts to be published. Forty years later, the use of firsthand accounts, although framed by news producers, provided viewers with a sense of realism and legitimacy. Personal testimonies from survivors such as Angel Verdugo, "La Nacha" Rodriguez,

[55] Laura Barrera, *Canal 22 Noticias*, Canal 22, October 2, 2008.
[56] Francisco Zea, *Cadena Tres Noticias*, Cadena Tres, October 2, 2008.
[57] Ibid.

and Félix Hernández Gamundi, coupled with the inclusion of historic images, worked to achieve the reporter's goal: to keep the memory of Tlatelolco alive.

Oscar Estrada and YouTube.com

The reports mentioned so far in this chapter represent cultural products that stemmed from large media outlets—Cadena Tres, Televisa, Canal 22; however, digital media and technological advances have enabled everyday citizens to contribute to the shaping of collective memory. As a result of relatively low-cost video and editing equipment, coupled with the speed and global reach of the Internet, citizen journalism has reached an unprecedented level.

Producers and reporters no longer must rely on the infrastructure of large-scale media conglomerates such as Televisa and TV Azteca to get their stories out. A 2008 YouTube.com search for video clips related to *el '68* using the key phrase "2 de octubre" yielded 1420 items. One of those clips included a report produced by self-proclaimed citizen journalist Oscar Estrada. Estrada originally filed his report titled, "Mexicans observe 1968 massacre," for the global citizen journalist website Instablogs.[58] This India-based media organization touts its virtual newsroom as a place where "everyone has a viewpoint."

As Internet anchor Sukhmani introduced Estrada's piece, she called the massacre, "contemporary Mexico's most traumatic atrocity and one that remains unresolved."[59] The anchor was positioned on the left side of the screen and a 1968 still photograph of Mexican troops crouching down behind a concrete wall, with rifles pointing in the air played on the right side.

Like his media counterparts on Cadena Tres, Estrada began his report with 1968 black-and-white images at Tlatelolco square. He reported that the events of *el '68* had "left a deep scar in Mexican society. That scar could only be healed by full disclosure of the facts, by bringing the

---

[58] Oscar Estrada, "Mexicans observe 1968 massacre," October 6, 2008, accessed December 1, 2008, http://tv.instablogs.com/entry/global-report-06-october-2008/jacqui/
[59] Oscar Estrada, "Mexicans observe 1968 massacre."

perpetrators to justice and providing reparations to the victims' families." A somber music track played during Estrada's voice-over.[60] The 1:17 minute report did not include interviews or video of the numerous marches held in Mexico City on October 2, 2008.

Estrada's report mirrored those produced by his counterparts in larger news operations in three intriguing ways. First, although his report appeared via an international platform, he nevertheless explained the events through a national lens. The news anchor's, as well as the reporter's language, using phrases such as the country's "most traumatic atrocity," and a deep wound in "Mexican society," stressed the importance of the 1968 events in national rather than global terms. Because Estrada's report was hosted on a website with an international focus, he could have put the massacre into a wider context by mentioning that 1968 represented a tumultuous year in many parts of the world. His decision to describe the violence solely in national terms remains consistent with coverage aired on Mexico City television.

Second, although Estrada did not include interviews with survivors of the massacre, like his Mexico City counterparts, his voice-overs and images emphasized the students'/victims' roles in the tragedy. The use of historic black-and-white images of students in the plaza made the students the clear focus of the report.

Finally, Estrada's portrayal of the Tlatelolco massacre as something that has left "a deep-wound that can only be healed through the unveiling of the truth," coincides with other reports that emphasize the continuity of the story—in other words, *el '68* remains an on-going social problem, and will continue to be so until justice has been (seen to be) served.

Olympics Remembered

While news coverage on daily broadcasts in October of 2008 did not reach the same level of reporting as the 40th anniversary of the massacre, some news documentaries about the Olympics in Mexico

---

[60] Ibid.

were produced. One of them aired on Channel 11, the state-funded arts and culture station Channel 11 (XEIPN), which is based in Mexico City.

Produced by Gustavo Sanciprián Marroquín, the documentary opened with a montage of black-and-white still images of athletes participating in a variety of sporting events, alongside a dramatic soundtrack underneath the video. The narrated documentary described the economic context of the country, which at the time had an annual economic growth rate of 6 percent. The voice-over continued stating that Mexico was a post-World War II society in which young people began to listen to a new rhythm, rock and roll music, as music and video of the Beatles played in the background.[61] In contrast to the somber moments remembered in the news stories that reconstructed the Tlatelolco massacre, the Olympic documentary highlighted the heroic moments of Mexican athletes such as Felipe Muñoz, who won a gold medal in the men's 200 meters breast stroke event, and José Pedraza who won a silver medal in speed walking. According to the narration, the Olympics were a memory "worth reliving." And, from the words of an Olympic athlete trainer, it was "something unforgettable."

Similar to the coverage of the 40th anniversary of the massacre, Sanciprián Marroquín's documentary included iconography and symbols commonly attached to the Olympics. For example, the set of graphics used in the documentary featured some of the same iconographic logos of the 1968 Olympics, including the well-known "Mexico '68." Another similarity between the Olympic documentary and the news coverage of the anniversary of the massacre was the inclusion of the voices of the participants. Hurdler Norma Enriqueta Basilio, who played an important image-making role in 1968, also was featured numerous times in the documentary. According to Basilio, the 1968 Olympic Games "opened a place for women" in sports.

Felipe Muñoz talked about the importance of the Games for the country, and suggested that after 1968, "Mexico was seen differently in the eyes of the world." The documentary program ended with archival

---

[61] Gustavo Sanciprián Marroquín, "México '68: 40 aniversário," Canal 11, Mexico City, October 2008.

footage of the closing ceremony of the Olympic Games, and shots of a mariachi group composed of dozens of musicians dressed in the archetypical *charro* costume.[62]

What was different about this documentary compared to one that had been produced and aired on Televisa 15 years earlier in commemoration of the 25-year anniversary of the Olympics was the recognition of the social turmoil that the country experienced in the 1960s. In an interview with the head of the national Olympic organizing committee, Pedro Ramírez Vásquez, he remarked that:

> When the social anxieties among the youth began in Berkeley, Stanford, Berlin, in Paris, it was in May of 1968. The young people who worked with us would go down and participate in the marches, and when the protest was over they would come back and continue working on the Olympics… They were the same ones.

Although Ramírez Vásquez made no mention of the massacre just ten days before the Olympics began, the inclusion of his statement about social unrest among youth did represent a break from earlier documentary works about the 1968 Olympics, when no acknowledgment of the student movements and their connection to the Olympics was included in documentary work.

Conclusion

The overwhelming number of news reports about the Tlatelolco massacre aired in 2008 demonstrated that television producers perceived October 2 to be an important date in Mexican history; hence their decision to cover the events surrounding the 40th anniversary. However, beyond the media's acknowledgment of October 2 as a significant historical date, a cohesive sense of the who, what, and how, *el 68* should be remembered surfaced in news coverage.

---

[62] Gustavo Sanciprián Marroquín, "México '68: 40 aniversário," Canal 11, Mexico City, October 2008.

First, reporters and anchors chose to privilege the voices of former student leaders over official accounts. Not one of the reports in the analysis included an interview with former military personnel, or public officials who were in office 40 years before. Media opportunities for individuals such as General Marcelino Barragán, the Minister of Defense in 1968, to comment on the events of 1968 have been eclipsed and replaced by individuals such as Ana Ignacia Rodriguez and Felix Hernández Garamundi. Through their decisions to focus on members of the Comité 68, television journalists have silenced those who traditionally provided the so-called "expert" testimonies of 1968.[63] This represents a dramatic change from four decades earlier when journalists tended to exclude students' voices from television.[64] In other words, the framing of the 1968 events in 2008, while they include still and moving images from the past, have been reconstructed or re-framed in a hybrid way that privileges the testimony of social activists over that of government officials.

We can begin to understand the magnitude of this media shift if we consider what Telesistema Mexicano (what would become Televisa in 1973) aired on the second anniversary of *el 2 de octubre* in 1970. The lead story on *24 horas*—to date, the country's longest running news program[65]—focused on President Gustavo Díaz Ordaz and his naming of Manuel Bernal Aguirre as the new Secretary of Agriculture and Livestock. Throughout the entire newscast, anchor Jacobo Zabludovsky did not once mention the events at Tlatelolco.[66] Zabludovsky ignored the anniversary of the Olympics as well.[67]

By 2008, television journalists were highlighting the continuing saga of *el 68*—meaning that survivors and the country "should not" rest until those responsible for the violence have been brought to justice. From the words of international anchor, Sukhmani, who stated that the atrocity remained unresolved, to Mexico City anchor Francisco Zea's assertion that the massacre left "a wound that has not been healed," viewers were

---

[63] Celeste González de Bustamante (2005), Witherspoon (2008) and Zolov (2005).
[64] Ibid.
[65] *24 horas* aired from 1970–1997.
[66] *24 horas*, Telesistema Mexicano, 2 October 1970.
[67] Ibid.

given to understand that the ordeal continues, figuratively speaking, and that the final chapter has yet to be written.

In conclusion, news coverage in 2008 sought to ensure that the memory of 1968 remained fresh in the minds of viewers. Through the use of historic photos, testimony from survivors, and even statements from anchors, it was clear that journalists worked to ensure that "*El dos de octubre no se olvida*," and Mexico itself has not forgotten the events.

The manner in which journalists in 2008 covered the events which surrounded the student uprisings and subsequent crackdown of 1968 (that is, what was included as well as excluded) signaled profound changes on the airwaves and reflected an adjustment in power relations between the media and state. In essence, journalists have replaced government officials as cultural authorities of the twenty-first century.

There has also been a shift in those who deliver the news. In contrast to the 1960s and 1970s, women now anchor news programs on all of the major and minor networks, as well as the Internet, and function as arbiters of cultural authority.

Technological advances in low-cost video recording and editing equipment, as well as the Internet, have opened the field of audiovisual media. While new media sites such as Instablogs.com and YouTube.com make the cultural playing field more inclusive, allowing everyday citizens to contribute to the shaping and reflecting of collective memories, the field is by no means completely level. Although, on average, 91.5 percent of the nation's population now has television sets at home, and almost 98 citizens out of 100 living in the capital do, a digital divide remains.[68] In 2015, just over 40 percent of the population used the Internet.[69] That figure pales in comparison to the figure for usage in the USA, where around 86 percent of the population uses the Internet.[70]

Further, the average citizen who might produce audiovisual expressions of collective memories on the Internet cannot compete with the "cultural capital" and resources available to the media powerhouses

---

[68] Comisión Federal de Telecomunicaciónes (COFETEL), with figures from INEGI.

[69] Internet Live Stats, accessed August 1, 2015 http://www.internetlivestats.com/internet-users/mexico/

[70] Internet Live Stats, accessed August 1, 2015, http://www.internetlivestats.com/internet-users/united-states/

such as Televisa and TV Azteca. As with cultural hegemony, the cultural authority that television journalists wield has its limits. On December 2, 2008, the national senate passed a law proclaiming "*el 2 de octubre*" as a national day of mourning.[71] Article 18 dictates that on this date, the flag shall be raised at half-staff in memory of "the victims who fought for democracy." Clearly, cultural authority has not been completely usurped by the media and popular groups.

The widespread news coverage of events surrounding the 40th anniversary of the Tlatelolco massacre demonstrates that rather than fostering "cultural amnesia," television constructs, as well as reflects, a collective memory about the massacre.

Missing from the cultural memory in 2008 was an obvious reflection or recognition about the actual Olympic event. In general, the reports focused on the events of Tlatelolco and did not make reference to the temporal proximity of the Games. Indeed, none of the reports analyzed in this chapter mentioned the international context in which the Tlatelolco massacre unfolded. Several historians have treated the connection between the Olympics and the massacre, but journalists chose to not make that connection and the relationship remained off-screen on October 2, 2008.

References

Aguayo, S. (1998). *1968: Los archivos de la violencia*. Mexico City: Grijalbo/ Reforma.

Anderson, S. (2001). History, TV and popular memory. In G. R. Edgerton & P. C. Rollins (Eds.), *Shaping collective memory in the Media Age* (p. 19). Lexington: University of Kentucky Press.

Becker, B., & Celeste González de Bustamante (2009). The past and the future of Brazilian television news. *Journalism: Theory, Practice and Criticism, 10*, 45–68.

---

[71] Notimex, "Declaran luto nacional el 2 de octubre," Excelsior Online, 2 December 2008. http:// www.exonline.com.mx/diario/noticia/primera/pulsonacional/declaran_luto_nacional_el_2_de_ octubre/432433 (accessed on 28-2-2009)

Benjamin, T. (2000). *La Revolución: Mexico's great revolution as memory, myth, and history* (p. 14). Austin: University of Texas Press.

Camp, R. A. (1985). *Intellectuals and the state in twentieth century Mexico* (p. 189). Austin: University of Austin.

Canclini, N. G. (2001). *Culturas híbridas: Estratégias para entrar y salir de la modernidad*. Buenos Aires: Paidós.

Carey, E. (2005). *Plaza of sacrifices: Gender, power and terror in 1968 Mexico* (p. 40). Albuquerque: University of New Mexico.

Castro, J. (forthcoming). *Radio in revolution: Wireless technology and state power in Mexico, 1897–1938*. Lincoln: University of Nebraska Press.

Celeste González de Bustamante. (2005). *"Muy buenas noches": Mexico, television and the Cold War; Elaine Carey, Plaza of Sacrifices*. Albuquerque: University of New Mexico Press.

Celeste Gonzalez de Bustamante. (2012). *"Muy buenas noches: Mexico, television and the Cold War*. Lincoln: University of Nebraska Press.

Chorba, C. (2007). *Mexico, from Mestizo to multicultural: National identity and recent representations of the conquest* (p. 8). Nashville: Vanderbilt University Press.

Christlieb, F. F. (1982). *Los medios de difusión masiva en México*. Mexico City: J. Pablos.

Cole, R. (Ed.) (1996). *Communication in Latin America*. Wilmington: Scholarly Resources.

de Bustamante, G. (1993). "Muy buenas noches," xxix; Jesús Martin-Barbero. *Communication, culture, and hegemony: From the media to mediations* (p. 151, trans: Fox, E., & White, R. A.). London: Sage.

Delabre, T. (Ed.) (1985). *El quinto poder*. Mexico City: Claves Latinoamericanas.

Delabre, R. T. (Ed.) (1988). *Las redes de Televisa*. México: Claves Latinoamericanas.

Duwmunt, T. (1992). *Channels of resistance: Global television and local empowerment*. London: British Film Institute.

Fernández, C., Paxman, A., & Tigre, E. (2000). *Emilio Azcárraga y su imperio Televisa*. Mexico City: Grijalbo.

Ferreira, L. (2006). *Centuries of silence: The story of Latin American Journalism*. Westport: Praeger.

Fox, E., Waisbord, S., & Politics, L. (2002). *Global media*. Austin: University of Texas.

Gamson, W., & Modigliani, A. (1989). Media discourse and public opinion on nuclear power: A constructionist approach. *The American Journal of Sociology, 95*(1), 1–37.

García, J. S., & Monsiváis, C. (1999). *Parte de Guerra, Tlatelolco 1968: Documentos del General Marcelino García Barragán: los hechos y la historia.* Mexico City: Siglo/Aguilar.

Garín, R. A. (1998). *La estela de Tlatelolco: Una reconstrucción histórica del movimiento estudiantíl del 68.* Mexico City: Grijalbo.

Gómez, G. O. (2002). "La televisión en México,". In G. O. Gómez & N. Maziotti (Eds.), *Histórias de la televisión en América Latina: Argentina, Brasil, Colombia, Chile, México, Venezuela* (pp. 59–70). Barcelona: Gedisa.

Hughes, S. (2006). *Newsrooms in conflict: Journalism and the democratization of Mexico* (p. 83). Pittsburgh: University of Pittsburgh.

Jackson Lears, T. J. (1985). The concept of cultural hegemony: Problems and possibilities. *The American Historical Review, 9*(3), 567–593.

Lawson, C. (2006). *Building the fourth estate: Democratization and the rise of the free press in Mexico.* Berkeley: University of California Press.

Lévi-Strauss, C. (1966). *The savage mind* (p. 259). Chicago: University of Chicago Press.

Martín-Barbero, J. (1993). *Communication, culture, and hegemony: From the media to mediations* (trans: Fox, E., & White, R. A.). London: Sage.

Medellín, F. M. (1992). *Televisa: Siga la huella.* Mexico City: Claves Latinoamericanas.

Miller, M., & Darling, J. (1997). Emilio Azcárraga and the Televisa Empire. In W. A. Orme Jr. (Ed.), *A culture of collusion: An inside look at the Mexican press.* Miami: North-South Center.

Morris, S. D. (1999). Corruption and the Mexican political system. *Third World Quarterly, 20*(3), 632.

Poniatowska, E. (1971). *La noche de Tlatelolco: Testimonios de historia oral.* Mexico City: Ediciones Era.

Poniatowska, E. (1992). *La noche, Massacre in Mexico* (trans: Lane, H. R.) Columbia: University of Missouri Press.

Revueltas, J. (1978). *México 68: juventud y revolución.* Mexico City: Ediciones Era.

Saragoza, A. (forthcoming). *The media and the state: The Origins of Televisa*

Schwartz, B. (1982). The social context of commemoration: A study in collective memory. *Social Forces, 61*(2), 374–402.

Sinclair, J. (1999). *Latin American television: A global view.* New York: Oxford University Press.

Skidmore, T. (1993). *Television, politics, and the transition to democracy in Latin America.* Washington, DC: Woodrow Wilson Center Press.

Sorensen, D. (2007). *A turbulent decade remembered: Scenes from the Latin American sixties*. Palo Alto: Stanford University Press.

Witherspoon, K. B. (2008). *Before the eyes of the world: Mexico and the 1968 Olympic games* (p. 120). DeKalb: Northern Illinois University Press.

Zelizer, B. (1992). *Covering the body: The Kennedy assassination, the media, and shaping of collective memory*. Chicago: University of Chicago Press.

Zolov, E. (2005). Showcasing the land of tomorrow: Mexico and the 1968 Olympics. *The Americas, 61*(2), 159–188.

Race, Rugby and Political Protest in New Zealand: A Personal Account

John Minto

Introduction

This chapter describes the effects of racism on the New Zealand Rugby Union (NZRU) and on rugby relations between New Zealand and apartheid South Africa. It then charts the growing opposition to this racism in New Zealand. It concludes with an appraisal of political developments in South Africa since the dismantling of apartheid in the early 1990s—a period of uninterrupted government by the African National Congress (ANC).

J. Minto (✉)
Former National Organiser and National Chairperson of HART Aotearoa,
21 York Street, Waltham, Christchurch, 8023, New Zealand

© The Editor(s) (if applicable) and The Author(s) 2016   **131**
J. Dart, S. Wagg (eds.), *Sport, Protest and Globalisation*,
DOI 10.1057/978-1-137-46492-7_7

Rugby and Race in New Zealand Society: Some History

In the three decades from 1960, New Zealand experienced turbulent times with the heady, visceral mix of race, rugby and politics changing dramatically the political landscape of the country. The politics of race relations in New Zealand became intermingled with the politics of the struggle against apartheid in South Africa. Māori nationalist aspirations and freedom for black South Africans collided with a conservative rugby union (hereafter, rugby) establishment and cynical political strategies. Campaigns were organised, boycotts planned and disruptive protests undertaken. Māori and black South Africans were excluded from rugby on the basis of race. Fierce protests ensued, during which grandstands were torched; glass and nails strewn across rugby fields; goalposts levelled; thousands arrested; elections won and lost; an Olympic boycott became an Olympic walkout with the issues coming to a head in pitched battles on New Zealand streets during the infamous 1981 Springbok tour which bitterly divided the country.

In New Zealand, the outcome for Māori was positive but in South Africa, it became a story of betrayal.

From its earliest days, rugby culture has been deeply embedded in New Zealand's national life. When it was introduced in the 1870s, it spread rapidly through the country and was taken up eagerly by Māori and Pakeha (New Zealanders of European descent) alike. In a colonial society where there was plenty to divide Māori and Pakeha, rugby was one area where the cultures came together in shared experiences.

Unlike most British colonies, New Zealand was settled on the basis of an agreement—Te Tiriti o Waitangi (The Treaty of Waitangi), signed in 1840 between the British crown and the chiefs of most Māori tribes.

In the decades which followed, the treaty was largely ignored by European authorities as dishonest dealings to separate Maori from their land became commonplace. This pressure on Maori and their land increased once European settlers increased in number and had the backing of British colonial troops. Māori tribes fought a series of wars from the 1840s to the 1880s to defend their lands from theft and to assert their right to "tino rangatiratanga" (Māori self-determination) guaranteed by the Treaty. However, despite many brilliant military victories Māori tribes were eventually overcome by a standing army of colonial troops and "friendly Māori" who had joined the British in various battles initially to settle inter-tribal grievances. Vast tracts of Māori land were confiscated, much of it the most productive land in the country. Having lost their economic base many Māori communities went into rapid decline which included decimation by disease. Māori were politically marginalised.

Despite this, a love of rugby provided common ground and a point of positive contact for the cultures through clubs and associations around the country. Rugby became the national sport and several tours to Britain and Australia were undertaken by Maori rugby sides and New Zealand national sides towards the end of the nineteenth century. Māori rugby evolved through Maori participation in the "All Blacks" (the national team, named after their team kit), but also developed its own tradition of selecting Māori All Black teams to play visiting teams and to go on their own tours as a separate side.

The first New Zealand national side was selected in 1884 to tour Australia and included two Māori players. In 1888, an all Māori national team, the New Zealand Natives, predecessor of the Māori All Blacks, toured Australia and Britain.

In South Africa, a similar settler culture adopted rugby and the Springbok emblem to represent their national sport. Deep and pervasive racism on the part of South African Europeans prevented rugby becoming a positive point of contact between the races as it had become in New Zealand.

Rugby links with South Africa developed after the First World War when a combined New Zealand Services Team played in South Africa but only after South African rugby officials sent the following telegram to the South African Embassy in London from where the New Zealand team was leaving:

Confidential. If visitors include Maoris tour would be wrecked and immense harm politically and otherwise would follow. Please explain the situation fully and try arrange expulsion

As a result, Rangi Wilson, an All Black who had played in 10 tests, was excluded from the squad. He was forced to remain on the boat (which was carrying troops home to New Zealand from the First World War) when it berthed in Durban, South Africa while the team played ashore.

The first Springbok tour to New Zealand occurred two years later in 1921 and a similar incident occurred which highlighted the blatant racism at the heart of South African sport and society. The Springboks played a New Zealand Māori team in Napier after which Charles Blackett, a South African reporter following the team, sent this telegram home:

> Most unfortunate match ever played. Bad enough having play team officially designated New Zealand natives but spectacle thousands Europeans frantically cheering on band of coloured men to defeat members of own race was too much for Springboks, who frankly disgusted

The telegram was leaked to a New Zealand newspaper and caused a furore with the local rugby union saying "whoever was responsible for the telegram does not know or understand how highly the Māori race is regarded by his Pakeha fellow citizens".

Strong rugby links developed between the two countries but because of South African race-based laws (institutionalised in 1948 as "apartheid" by the Afrikaner Nationalist Party) Māori players were prevented from playing for the All Blacks on trips to South Africa. From 1919 to 1970 race-based All Black teams, with Maori excluded, were selected to tour South Africa. In 1928, for example, the All Black fullback George Nepia—often cited as the greatest All Black fullback in New Zealand history—was not selected. The NZRU justified this policy on the grounds that South Africa operated a colour bar and in any case they did not want to subject Māori players to racism. The players and general public were never consulted.

Opposition to de-selecting Māori players on tours to South Africa gained momentum, however, and by 1960 when another such tour was planned, marches were held to oppose it. The group CABTA (Citizens All Black Tour Association) was formed to fight the issue under the slogan "No Māoris – No tour!" with many prominent Māori and Pakeha leaders speaking out. At the forefront were Māori All Black legend George Nepia and leaders of the Māori battalion which had fought with great distinction as a separate unit in the Second World War.

The questions being debated were compelling: Hadn't Māori fought for New Zealand in the First and Second World Wars? Why were they being denied the right to represent their country? Why was the NZRU discriminating against Māori players? Why was South African racism dictating racist practices in New Zealand sport?

Several Māori players were likely to be selected in the All Black team although most were pressured to make themselves unavailable to save the NZRU embarrassment. CABTA was successful in mobilising public opinion and large street marches were held in the lead-up to the tour. However, the racially selected All Black team left and played in South Africa just a few months after the Sharpeville massacre had seen 69 black South Africans shot dead (most shot in the back as they ran from police machine gun fire) after a protest against the hated "pass laws" of the apartheid state.

The Springboks toured New Zealand in 1965 and another All Black tour to South Africa was planned for 1967. This time public opinion was much more firmly against the tour and under pressure from the government, the NZRU declined the invitation on the ground that Māori players were unable to be selected to tour. South Africa rugby officials and politicians changed tack and in November 1967 confidentially agreed to issue an invite to New Zealand to send its best team regardless of race. South African Prime Minister Vorster agreed to the invitation on three conditions: that there would not be too many Māori players, that those Māori players selected would not be "too black" and that no controversy should attend their selection and dispatch. In effect, the Māori players were to be afforded the status "honorary white" so they would be able to travel together, stay in the same hotels as the All Blacks and avoid the apartheid laws applying to black South Africans. It was a breakthrough of sorts and the rugby union prepared to tour in 1970 but the focus of debate was changing.

From the late 1950s, the liberation movements, the ANC and PAC (Pan-Africanist Congress), had been calling for an international boycott of South Africa to force the regime to abandon its apartheid policies. This call was gathering steam internationally and was taken up within the United Nations and the Commonwealth Secretariat. Across Africa and in India, for example, solidarity with black South Africans and support for the boycott was strong but in the white British Commonwealth countries conservative governments and sports bodies ignored the call.

H.A.R.T

HART (Halt All Racist Tours) was formed in New Zealand in 1969, with Trevor Richards as leader, to campaign against the 1970 tour. The other high-profile group protesting the tour was CARE (Citizens Association for Racial Equality) under Tom Newnham's leadership. One reason HART was formed was that CARE had a bad reputation in conservative circles and it was felt another group should be formed without the same baggage. Ironically, HART quickly became "public enemy number one" amongst the rugby-following public as it had a strong base in the country's universities and took a more activist approach.

HART was unusual amongst anti-apartheid groups internationally in that it recognised both the ANC and PAC as liberation movements on the basis both groups were recognised by the United Nations and the Organisation of African Unity. Most other anti-apartheid groups recognised either one or the other, but not both.

The public debate was fierce with the NZRU saying South Africa's internal policies were their business while HART and CARE activists took an internationalist stance in support of oppressed black South Africans. All Black great Ken Gray shocked the rugby fraternity by making himself unavailable for selection for the tour and revealed later this was because of his opposition to apartheid. In 1970 the All Blacks finally left New Zealand with three Māori players and one Samoan player, Bryan Williams, included in the touring team. In South Africa, the black and coloured sections in the grounds, wired off from the white sections, cheered madly for the All Blacks. Bryan Williams, nicknamed "the golden boy", was a particular crowd favourite. The All Blacks returned with some soul-searching by players like Chris Laidlaw who had taken time out to see conditions in South Africa first-hand. Laidlaw said the tour should not have proceeded and later worked in the Commonwealth Secretariat and helped strengthen the boycott of apartheid South Africa.

Meanwhile, the early 1970s saw the Māori nationalist struggle being taken up by a new generation of young Māori activists in groups such as Nga Tamatoa (the young warriors) and the Polynesian Panthers (modelled on the US Black Panthers). They rightly accused the state of dishonouring the Treaty of Waitangi and pointed to the institutionalised racism which had led to appalling statistics for Māori communities in areas such as health, education and employment. In many high-profile protests, they put race relations high on the national agenda. This was a profound shock for Pakeha New Zealand which had prided itself on good race relations and in a patronising way assumed Māori saw things the same way. Pakeha saw the country as unlike the USA, where race riots were regularly seen on our black and white TVs, or Australia, where the situation for aboriginal people had always been a racist disgrace.

The year 1975 saw the first land march by Māori from the top of the North Island to parliament in Wellington under the slogan "Not one more acre" referring to the ongoing, government-supported appropriation of Māori land. Land protests followed in several areas including Raglan where the golf course was occupied and Takaparawha (Bastion Point)—the last remaining area of Māori land in Auckland which the government was trying to subdivide for housing. Māori activists and supporters occupied Takaparawha in 1977 and stayed for 506 days before a massive police operation cleared the site.

Māori activists identified with the struggle in South Africa and welcomed South African representatives such as Denis Brutus from SANROC (the South African Non-Racial Olympic Committee) who toured New Zealand seeking international support for the sports boycott. In 1973, the Springboks had been planning to tour but anti-apartheid sentiment was growing and a strong campaign by HART and CARE saw the tour called off by Norman Kirk's Labour government on the basis of a police report which said there would be widespread disorder and violence if it proceeded. Kirk had promised not to intervene in the tour during the 1972 election campaign but as Prime Minister said that in the best interests of the country he could not grant visas to the Springboks. There was a furious public backlash at the tour's cancellation and this was a significant factor in Labour losing the following election in 1975 when the pugnacious conservative politician Robert Muldoon took the National Party to victory.

The year 1976 saw the All Blacks again plan to tour South Africa and this time there was no question of any government pressure to stop it. Nationwide demonstrations were held against the tour and the All Black trials were disrupted but the tour went ahead and when the team left it was seen off by a senior National MP who said it went with the "goodwill and blessing" of the New Zealand government.

The year 1976 was also the year of the Soweto uprising when black school children protesting at having to learn their lessons in the Afrikaans language were fired on by the South African police. The riots which followed were brutally suppressed with more than 600 children shot dead in the six months following the June 16 protests. Caught up in the riots were some of the touring All Blacks players and a tear-gassed Ian Kirkpatrick (loose forward) stumbled to a South African police vehicle pleading "Help me – I'm an All Black". Never have so few words better summed up the contradiction between rugby and the political context in which the game was being played.

So as the world watched black children dying on the streets, they also saw the All Blacks entertaining supporters of the regime in stadia across South Africa. Africa was incensed as were most Commonwealth nations and the United Nations Special Committee Against Apartheid. Before the tour, appeals had come from all over the world calling on the Muldoon government to act and stop the tour, but to no avail. Muldoon had issued a very public snub to the head of the Supreme Council for Sport in Africa, Abraham Ordia, who had come to New Zealand to speak directly with the government. Muldoon refused to see him telling the media that Ordia could "stew in his own juice". These comments were welcomed in redneck, rural New Zealand but played poorly internationally. The world was appalled at New Zealand's insensitivity and Prime Minister Muldoon's intransigence. New Zealand became an international pariah.

African sporting organisations attempted to have New Zealand expelled from the Olympic Games in Montreal later that year and when this failed, 29 African and Caribbean counties walked out of the games in protest at New Zealand's sporting links with South Africa. One of the glamour events of the games, the 1500 metres, was to have pitted New Zealand's great runner John Walker against Tanzanian champion Filbert Bayi. Bayi was the rising star of African, and international, running. Both men had impressive performances in the lead-up to the games. It was time for the showdown. However, Tanzania was one of the 29 nations who pulled out of the Games which left Walker to win the race largely unchallenged.

A year later, as a direct result of the 1976 All Black tour, the Commonwealth Heads of Government met at Gleneagles in Scotland and drew up the Gleneagles Agreement which required each Commonwealth country to take "all practical steps" to discourage sporting links with South Africa. With political activity by black South Africans all but disallowed (e.g., the ANC and PAC had been banned) sport was one area where people could organise and use the sports structures to help build the anti-apartheid struggle. To this end, SACOS (the South African Council on Sport) was formed in the early 1970s to bring together sporting codes to play non-racial sport. Rugby is a good example of how sport was organised under the regime's apartheid structures. There was SARB (South African Rugby Board for white players only) SARF (South African Rugby Federation for coloured players only) and SAARB (South African African Rugby Board for black players).

SACOS rejected these divisions and SARU (South African Rugby Union) organising non-racial rugby was affiliated to SACOS. SACOS-linked teams and competitions were mixed as well as they could be given the harsh laws which prevented easy movement of players for practices and games. Despite its lack of funding or any official support, SARU attracted a huge following for rugby amongst black South Africans across the country and particularly in the Eastern Cape. The regime's line had always been that blacks were only interested in soccer but at one point SARU had an estimated 100,000 rugby players and despite apartheid restrictions on travel still had tens of thousands watch its cup finals each year.

SACOS had affiliated groups in all the major sporting codes and ran competitions from school level to provincial representative level. Their slogan "no normal sport in an abnormal society" summed up the inextricable links between sport and politics in apartheid South Africa.

SACOS supported the sports boycott and appealed to countries not to accept apartheid sports teams representing the South African regime. Leaders of SACOS were harassed by the regime and usually denied passports to leave the country to speak in international forums. Back in New Zealand, HART and CARE enjoyed good relations with SACOS and its external wing SANROC represented internationally by Denis Brutus and later by Sam Ramsamy.

In all respects, SACOS was the sporting wing of the liberation struggle.

Despite the Gleneagles Agreement and despite international appeals for effective action to end sports links with South Africa, the Muldoon government refused to move against the 1981 Springbok tour to New Zealand. At the same time, the anti-apartheid movement had been planning for the tour since October 1979 when HART held a national "Stop the '81 Tour" conference. HART's campaign focus was a major educational/agitational campaign to educate the New Zealand public about apartheid. At that stage, public opinion polls showed majority support for the tour, especially in provincial marginal electorate seats such as Gisborne, New Plymouth, Taupo and Invercargill. Film screenings, speaking tours, deputations, delegations, petitions, posters, fundraisers, letters, pickets, marches and protests were all part of the mix designed to erode this support. The information campaign was successful, and by the time the Springboks arrived, the country was evenly split. Forty-five per cent supported the tour, 45 % opposed and 10 % refused an opinion. The country was deeply and bitterly divided.

The government's official position was opposed to the tour but they repeatedly said the decision was the rugby union's and they would not intervene. In the lead-up to the tour, Muldoon made a nationwide broadcast on the issue but instead of asking for the tour to be called off he spoke of the strong bonds between New Zealanders and South Africans who had fought together in the Second World War and whose dead were buried together overseas. It was widely and accurately seen as a nod and a wink for the tour to proceed.

The HART strategy involved forming coalitions of groups in every centre across the country. HART would provide national co-ordination while local coalitions would organise protests and anti-tour activities in their centres. Two nationwide mass mobilisations were held in the lead-up to the tour (the biggest protests in New Zealand history to that point) and the country prepared itself for the Springboks' arrival. HART tactics were for "multiple simultaneous demonstrations" around the country whenever the Springboks played. "Non-violent direct action" and civil disobedience were supported. We condoned damage to property such as pulling down fences but we did not support violence against people.

From the outset, Māori activist groups were an integral part of the movement. Fresh from local battles (over land rights and other non-sporting issues) they brought energy and determination to this particular struggle.

The first game was held in Gisborne where the Springboks, in a ceremony full of drama and controversy, were welcomed onto Poho o Rawira Marae where the first Springbok team had been welcomed in 1921. They were told they would not be welcomed by Māori again unless they came after the end of apartheid.

The most effective action taken during the tour was to stop the second game of the tour, Springboks versus Waikato, in Hamilton. A protest was organised which broke through the park fence where the game was scheduled, and 300 people ran onto the field and refused to move till the game was called off. The TV images went around the world and shocked white South Africans but boosted the morale of black South Africans. Nelson Mandela said that when the political prisoners on Robben Island heard the game had been cancelled by protests they grabbed the bars on their cell doors and rattled them around the prison. He said it was "like the sun came out".

Added impact was given to the protest because the game was to be the very first rugby match to be televised live in South Africa from overseas. When South Africans got up in the middle of the night to watch they were shocked to only see protestors. The psychological effect on both black and white South Africans was powerful and for many white South Africans it began the process of realising things had to change.

The tour continued with a huge injection of government funding for security for the remaining games. Pitched battles were fought between police and tour supporters on one side and protestors on the other side. Over two thousand arrests were made and the cases dragged through the courts for the following two years. Several people received jail terms such as Marx Jones who flew a small plane over the last test of the tour as his passenger Grant Cole dropped flour bombs and leaflets on the pitch, hitting an All Black player in the process.

The most important consequence of the tour was that the Springboks never left South Africa again to play any significant rugby games while the country maintained its race-based apartheid laws. However in New Zealand, Prime Minister Muldoon's strategy paid off and the 1981 election saw him retain power through winning the provincial marginal electorates mentioned earlier.

The much more important effect of the tour in New Zealand however was to advance the Māori political struggle. The tour brought race relations to the centre of public debate in a way which yielded tangible change. Māori activists had challenged the anti-tour movement by asking "How can you be concerned about racism thousands of miles away in South Africa but ignore it in New Zealand?" The subsequent public debate in the post tour environment helped create space for the further development of Māori nationalism. The idea of bi-culturalism gained momentum and Māori were increasingly seen as equal partners under the Treaty of Waitangi. The Waitangi Tribunal was given additional powers in the years after the tour to investigate historical breaches of the Treaty and negotiate compensation. (Previously, the tribunal only had the power to examine current breaches as they arose.) This process still continues in New Zealand with several tribes having signed settlement agreements with the government for breaches of the Treaty with others still in negotiation.

Despite everything which had transpired (and despite numerous strong appeals from HART, from governments around the world and from international organisations) a further All Black tour to South Africa was planned for 1985 and it was only abandoned a few days before the team was due to leave when an injunction against the tour was issued in the High Court. It was a legal action brought by two rugby club members who successfully argued that the tour would damage the sport of rugby were it to proceed and this in turn would be in breach of the Rugby Union's constitution.

Furious rugby die-hards then organised, in secret, a rebel All Black tour to South Africa the following year. The team called itself the Cavaliers but to South Africa and the world, they were the All Blacks. Public opinion hardened against sporting links with South Africa and no major contacts occurred until 1992 when the ANC lifted the sports boycott.

Meanwhile, internal pressure mounted on the regime with strikes and mass civil disobedience inside the country and growing international pressure through boycotts. Apartheid was no longer tenable so South Africa's political and corporate elites launched a protection operation for its capitalist economy. The ANC was the vehicle by which this would be achieved. It had become the leading liberation movement by the mid-1980s, and contacts between the ANC and South African corporates were organised and broad plans for the end of apartheid, but the maintenance of the capitalist economy, were developed.

After Apartheid: Mission Accomplished?

One objective of the ANC was to gain hegemony over all anti-apartheid institutions and especially in sport. Sport had proven to be particularly effective in pressuring white South Africa and the ANC wanted to be able to utilise this pressure politically. The major problem they faced was SACOS which was strong, well organised, deeply embedded in South African society but not affiliated to the ANC. SACOS administrators were, of course, deeply political but the organisation itself was politically unaligned. Some administrators were associated with the ANC, others the PAC or the New Unity Movement and a host of other political groupings. Moreover, SACOS resisted alignment when the ANC made political overtures. They felt it important for any sports body to be politically independent. At this point, the ANC's strategy changed and it set up its own sports body, the National Sports Council (NSC), which sought affiliation from sports bodies associated with SACOS.

A good deal has been written about the conflict which quickly grew between SACOS and ANC's sporting arm, the NSC. However, much of it is not helpful in understanding the sporting and political transformation which followed. The ANC's version of events says SACOS was adamantly opposed to negotiation at a time when white South Africa was keen to negotiate the political and sporting structures for a new South

Africa. According to this version, SACOS' policy of non-collaboration with apartheid sports bodies was an obstacle to change and as a result, SACOS affiliates formed a new organisation, the NSC, and left SACOS to wither and die. This narrative is a crude, self-serving distortion of what took place.

The simple truth is the ANC wanted to use the power of the sports boycott as a negotiating tool and needed SACOS out of the way before it could take control of sport. SACOS was a non-racial organisation in a country dominated by race-based laws. It was not opposed to negotiation but was not prepared to trade away the principle of non-racialism. SACOS rightly saw the first priority to be the establishment of well-funded development programmes in each sport so that those oppressed under apartheid would have the chance to develop their sporting talents to the full. The priority of the white sports bodies however was to regain admittance to international sport. Development plans could come later as far as they were concerned and the ANC agreed. The ANC's strategy to destroy SACOS opened perhaps the darkest chapter of South Africa's sporting history. It was unprincipled, vicious and self-serving. SACOS affiliates who did not cooperate in the NSC and the ANC strategy were actively undermined at home and in international sports bodies.

SACOS was variously accused of being anti-ANC, being dominated by Indian and so-called "coloured" South Africans and of not doing enough to organise sport in the black townships. SACOS officials and affiliates were openly challenged in some international sporting bodies by other country's delegates who asked questions such as "How could Indian South Africans claim to represent oppressed black South Africans?" and "Why was SACOS opposed to the liberation struggle?" "Why was SACOS opposed to the ANC?" These were some of the worst examples of race-based attacks led by ANC-aligned officials. It was ironic that race played a significant role in undermining a non-racial organisation.

SACOS countered that it was politically non-aligned, as any national sports body should be; that, while it did not have administrators in proportion to South Africa's population demographics, this reflected the huge difficulties of organising non-racial sport in a society dominated by apartheid laws; and that while it had development strategies for increasing sports organisation in black communities these were actively opposed by the regime and made very difficult to deliver in practice by the country's race-based laws.

Within a short period of time, however, SACOS was sidelined and key figures in ANC-aligned sports organisations went on to draw huge salaries and were feted around the world. It was a political travesty and, as time would tell, a sporting disaster. As the ANC gained control of sport, they also gained control of the sports boycott.

The ANC National Council met and endorsed the sports boycott continuing but left the issue to a subcommittee led by Sports-Minister-in-waiting Steve Tshwete. Within a few weeks, Tshwete announced the sports boycott was to be lifted on a sport-by-sport basis as the sports bodies "integrated". In most cases, this meant a white sports body would co-opt a black administrator or two onto its board, promise a development programme and get the green light from the ANC. Thus, the sports boycott was rapidly lifted with South African sports bodies readmitted to international competition through agreement with the ANC. Much was promised in sports development but precious little was ever delivered.

For HART the situation became tricky, then difficult and finally impossible. The organisation received appeals from SACOS and affiliates for international support and solidarity. Such calls also came from ANC activists appalled at the behaviour of key figures in the ANC leadership. We were urged to maintain the pressure of the sports boycott. The first tour planned to New Zealand was a South African cricket team in 1992, two years before democratic elections were scheduled.

On the one hand, we were asked if it was not the height of arrogance for HART to suggest protesting a South African team which Mandela's ANC was supporting, while on the other hand, we were being asked by sporting and political activists to maintain the boycott because Tshwete in no way represented the viewpoint of the oppressed.

HART kept its options open and, when the cricketers arrived in New Zealand, asked the team to sign a declaration of opposition to apartheid and express full support for a democratic South Africa. However, the team had brought with them none other than Steve Tshwete himself. Tshwete was "riding shotgun" for the team to ensure no protest occurred. A HART delegation met with Tshwete and it was clear that while he gave fulsome praise to HART's solidarity he said we needed to accept things had changed. Tshwete said the players would be embarrassed at signing the declaration and told the players to ignore it.

It was now game over for HART and the organisation wound up later in the same year. HART had been formed as an anti-apartheid organisation to campaign on the boycott and now with the boycott called off a new organisation would be needed.

Needless to say, the wide-ranging promises made by organisations such as the South African Rugby Board to put in place extensive and well-resourced development programmes for black players were quietly shelved or heavily scaled back once admission to international competition was established. The same people who had run apartheid sport on behalf of the regime for many decades were again in the driving seat while so many of those who had worked so hard against apartheid in sport were marginalised and discredited.

It is now more than two decades since the first democratic elections and rugby is a good example of the failure of meaningful sports development. The Springbok team competing in the 2015 Rugby World Cup competition had several "non-white" players (most of whom were left out of the big games) and the entire team was educated in private or semi-private schools. Twenty-three years on from the lifting of the sports boycott, South Africa is debating whether it should have quotas for black players in the Springboks. No such debate would be necessary if decent development programmes had been put in place as demanded by SACOS.

Sport has enormous power to transform societies. In post-apartheid South Africa, it never got a chance. Most aspects of daily life have changed little for the vast majority of black South Africans since race-based apartheid was abolished.

In New Zealand, rugby has been a unifying force between Māori and Pakeha. In South Africa, it is still a benchmark of racism and exclusion for the black majority.

Postscript

It was not until 2010, 50 years after Māori players were last excluded from tours to South Africa, that the NZRU finally apologised for the exclusion of Māori players from teams to tour South Africa during the apartheid era. The apology came after South African Minister of Sport, Makhenkesi Stofile, said both South Africa and New Zealand were guilty of excluding players on the basis of race during the apartheid era and apologies were in order. The South African union then issued an apology by media release followed shortly after by the New Zealand union.

John Minto is a former National Organiser and National Chairperson of HART (Halt All Racist Tours)

A summary of rugby exchanges between South Africa and New Zealand from 1919 to 1992

1919	Māori excluded from New Zealand Services tour to South Africa
1921	South Africa tours NZ and for the first time plays a "coloured" team—New Zealand Māori
1928	NZ tours SA but Maori players are excluded (including All Black legend George Nepia) from team
1948	NZ tour South Africa—Māori excluded
1956	Springboks tour NZ
1960	"No Māoris – no tour!" campaign fails to stop Māori excluded from 1960 All Black tour to South Africa
1965	Springboks tour NZ
1967	All Black tour to SA called off because Māori were to be excluded
1970	All Blacks tour SA with three Māori and one Pacific player included
1973	Springbok tour to NZ cancelled by Norman Kirk's Labour government on basis of likely disruption and disorder
1976	All Blacks tour South Africa in wake of Soweto uprising—29 African and Caribbean countries walk out of Montreal Olympics in protest at NZ participation
1981	Springboks tour NZ—mass demonstrations with one game abandoned through protest—Mandela says it was "like the sun came out"
1985	All Black tour to SA cancelled after a court injunction to stop the tour succeeds
1986	Rebel All Black team (the "Cavaliers") tours South Africa
1992	Sports boycott called off by the ANC

Open Letter to the President of South Africa (25 January 2008)

Tena koe Thabo Mbeki,

I understand a nomination has been put forward for me to receive a South African honour later this year, the Companions of O R Tambo

Award, on behalf of HART and the anti-apartheid movement of New Zealand for our work campaigning to end apartheid in South Africa.

I note the particular honour is conferred by the President of South Africa and awarded to "foreign citizens who have promoted South African interests and aspirations through co-operation, solidarity and support".

We are proud of the role played by the movement here to assist the struggle against apartheid and I appreciate the sentiment behind the nomination. However after the most careful consideration I respectfully request the nomination proceed no further. Were an award to be made I would decline to accept it either personally or on behalf of the movement.

New Zealanders who campaigned against apartheid did so to bring real and meaningful change in the lives of South Africa's impoverished and disenfranchised black communities. We were appalled and angered at the callous brutality of a system based on racism and exploitation of black South Africans for the benefit of South African corporations.

However while political rights have been won and celebrated, social and economic rights have been sidelined. It is now 14 years since the first African National Congress government was elected to power but for most the situation is no better, and frequently worse, than it was under white minority rule.

The number of South Africans living on less than $1 a day more than doubled to 2.4 million in the first 10 years of ANC government. Despite strong economic growth overall poverty levels have not improved and the gap between rich and poor has increased with many black families being driven more deeply into poverty. Unemployment remains high at around 26%.

It seems the entire economic structure which underpinned apartheid is essentially unchanged. Oppression based on race has morphed seamlessly into oppression based on economic circumstance. The faces at the top have changed from white to black but the substance of change is an illusion.

None of us expected things to change overnight but we did expect the hope for change to always burn brightly as people looked ahead for their children and grandchildren. This is now a pale gleam, dimmed by the destructive power of free-market economics.

My own country New Zealand preceded the ANC in adopting free-market economic reforms. Since 1984 we have experienced a particularly virulent dose of these vicious policies which have brought wealth to the few at the expense of the many.

Hundreds of thousands of New Zealand families have been driven out of decent employment into poverty where they struggle to raise families on part-time, poorly paid work. They are worse off now than they were 20 years ago. The same policies have brought the same outcomes to South Africa. For the majority life is tougher now than at any time since the ANC came to power.

The promises made by those who drove through the reforms in New Zealand were a lie just as they are in South Africa. Wherever these policies have been put in place anywhere in the world they have resulted in a reverse Robin Hood – a transfer of wealth from the poor to the rich.

When we protested and marched into police batons and barbed wire here in the struggle against apartheid we were not fighting for a small black elite to become millionaires. We were fighting for a better South Africa for all its citizens.

I take heart from the many community groups in South Africa fighting against privatisation of community assets; supporting settlements against forced removals; opposing police harassment and brutality; struggling for decent healthcare, water supplies and education; campaigning for decent pay, reasonable working conditions and affordable houses. These people, such as the Durban Shackdwellers, are looking for respect and dignity as human beings. Many carry the ideals of the Freedom Charter, once the bedrock document for ANC policy, close to their hearts.

Apartheid was accurately described as a "crime against humanity" by the United Nations and the ANC. I could not in all conscience attend a ceremony to receive an award conferred by your office while a similar crime is in progress.

Receiving an award would inevitably associate myself and the movement here with ANC government policies. At one time this may have been a source of pride but it would now be a source of personal embarrassment which I am not prepared to endure.

Yours sincerely,

John Minto

Suggested Reading

Black, D., & Nauright, J. (1998). *Rugby and the South African nation: Sport, culture politics and power in the old and new South Africas.* Manchester: Manchester University Press.

Booth, D. (1998). *The race game: Sport and politics in South Africa.* London: Frank Cass.

Chapple, G. (1984). *1981:The Tour.* Auckland: AH and AW Reed.

Desai, A. (Ed.) (2010). *The race to transform: Sports in post apartheid South Africa.* Pretoria: HSRC Press.

Maclean, M. (1998). From old soldiers to old youth: Political leadership and Aotearoa/New Zealand's 1981 Springbok rugby tour. *Occasional Papers in Football Studies, 1*(1), 22–36.

Maclean, M. (1999). Of warriors and blokes: The problem of Maori rugby for Pakeha masculinity in New Zealand. In T. Chandler & J. Nauright (Eds.), *Making the rugby world: Race, gender, commerce* (pp. 1–26). London: Frank Cass.

Maclean, M. (2000). Football as social critique: Protest movements, rugby and history in Aotearoa/New Zealand. *International Journal for the History of Sport, 17*(2/3), 255–277. http://dx/doi.org/10.1080/09523360008714136

Maclean, M. (2001). "Almost the same, but not quite...Almost the same, but not white": Maori and Aotearoa/New Zealand's 1981 Springbok tour. *Kunapipi: Journal of Postcolonial Writing, 23*(1), 69–82.

Maclean, M. (2010). Anti-apartheid boycotts and the affective economies of struggle: The case of Aotearoa New Zealand. *Sport in Society, 13*(1), 72–91.

Maclean, M. A right old bust up: Rugby union, imperial ideology, and the Springboks in New Zealand (www.englandrugby.com/mm/.../Malcolm MacLeanpaper_Neutral.pdf).

Mulholland, M. (2009). *Beneath the Maori moon – An illustrated history of Maori rugby.* Wellington: Huia Publishers.

Nauright, J. (1993). 'Like fleas on a dog' Emerging national and international conflict over New Zealand rugby ties with South Africa 1965–74 (www.la84. org/SportsLibrary/SportingTraditions/1993/.../st1001h.pdf).

Newnham, T. (1983). *Batons and barbed wire Millersburg.* Ohio: Graphic Publications.

Pollock, J. (2004). 'We don't want your racist tour': The 1981 Springbok tour and the anxiety of settlement in Aotearoa/New Zealand. *Graduate Journal of Asia-Pacific Studies, 2*(1), 32–43.

Richards, T. (1999). *Dancing on our bones: New Zealand, South Africa, rugby and racism.* Wellington: Bridget Williams Books.

Fighting Toxic Greens: The Global Anti-Golf Movement (GAG'M) Revisited

Anita Pleumarom

Preface

6 December 2014—environmental activists in Rio de Janeiro gear up for "Ocupa Golfe" (Occupy Golf) in protest of what they see as the largest environmental devastation in their city's history. The controversy is about the golf course built for the Rio 2016 Olympic Games. The site covering an area the same as 100 soccer fields is part of the previously protected Marapendi's Municipal Natural Reserve and home to approximately 300 species of animals, many of which are endangered. As is customary with Olympic-related projects with tight timelines, the area was fast-tracked for development. On the initiative of city mayor Eduardo Paes, Supplementary Law No. 125/2013 was passed in an emergency city council session to allow the golf course construction, even though the

A. Pleumarom (✉)
Tourism Investigation & Monitoring Team (tim-team), P.O. Box 51,
Chorakhebua, Bangkok, 10230, Thailand

© The Editor(s) (if applicable) and The Author(s) 2016   **151**
J. Dart, S. Wagg (eds.), *Sport, Protest and Globalisation*,
DOI 10.1057/978-1-137-46492-7_8

Public Prosecutor's Office had already considered the text to be unconstitutional. According to Ocupa Golfe, the true reason behind this golf course project is property speculation as the concession allows for the immediate construction of 23 22-storey luxury buildings. Along with the Public Prosecutor's Office, Ocupa Golf is fighting for the suspension of the environmental license and the recovery of the environment degradation caused by the construction of the Olympic golf course.[1]

After a 112-year hiatus, golf is returning to the Olympic Games in Rio in 2016 and, promptly but not surprisingly, a significant conflict has emerged caused by the new golf course created for the Olympic tournament. Not only is the city government's dubious relationship with developers a contentious point, but local residents who are struggling with intermittent water supplies in the midst of one of the worst droughts ever in Brazil uncomprehendingly watch water sprinklers being in full use to keep the Olympic course green.

The emergence of "Occupy Golf" in Rio is a convenient entry point to reflect on the Global Anti-Golf Movement (GAG'M, pronounced gag'em) that was founded in 1993 in response to the frenzied proliferation of resort and golf course development worldwide. One of its first major actions, as we will later see in this essay, was to lobby the International Olympic Committee (IOC) to ban golf permanently from the Olympics.

The following GAG'M story is not an academic discourse but rather a personal account of an anti-golf activist and researcher over a period of more than 20 years. As such, it does not claim to be "objective" or "neutral". My interest in golf courses and golf tourism began around 1990 when I was directing an action research project on tourism, development and the environment for the Bangkok-based Ecumenical Coalition on Third World Tourism (ECTWT). While studying the environmental impacts of golf courses and golf tourism as part of my work, I came in contact with many individuals and groups, mainly from the Asia-Pacific region, who later joined forces to take action against the golf boom. That is how I became a co-founder of the movement

---

[1] Waldron, I., "Three months and counting: Occupy Golf takes on the Rio Olympics", *Latin Correspondent*, 12 March 2015, http://latincorrespondent.com/brazil/three-months-counting-occupy-golf-takes-rio-olympics/ (accessed 10 July 2015).

(representing the Asian tourism activist network ANTENNA) and a member of the original GAG'M core group, along with Gen Morita of the Japan-based Global Network for Anti-Golf Course Action (GNAGA) and Chee Yoke Ling and Maurizio (Farhan) Ferrari of the Malaysia-based Asia-Pacific Peoples' Environment Network (APPEN). Besides researching and writing on golf-related themes, I helped to produce the "GAG'M Updates" that were published twice a year for information and the raising of public awareness. Notably, anti-golf activism emerged at a time when the Internet was still nascent and social media did not exist. The GAG'M Updates consisted of hundreds of pages of newspaper clippings, articles and documents that were sent by "snail mail" or faxed to us from all over the world. Today, it would be hardly possible to give a historic account of the movement in the first years of its existence without the GAG'M Updates prints because most of the information never made it to the Internet. However, there has been considerable confusion among Internet users over the appearance of a website[2] and, more recently, a Facebook page,[3] which use the name GAG'M. In fact, the original GAG'M has never had any links with these newer initiatives.

This essay will focus on the work of GAG'M from its foundation in 1993 until 1997 when the Asian financial crisis caused a sharp decline in Southeast Asia's golf industry. Even though the movement was thereafter no longer visible as before, the GAG'M core group has continued to be active as the forces of globalization have kept pushing golf towards new expropriations and local communities are still seeking support in their struggle against unwanted and damaging golf projects.

How It All Began

The idea to form an international people's alliance against golf courses and golf tourism was born in December 1992 in Phuket, southern Thailand, at an international civil society event called "Peoples Forum

---

[2] http://www.antigolf.org/english.html (accessed 10 July 2015).
[3] https://www.facebook.com/groups/8063947826/ (accessed 10 July 2015).

on Tourism" organized by the ECTWT in cooperation with Thai civic groups. The meeting was part of the Asia-Pacific "People's Plan for the 21st Century (PP21)", a comprehensive civil society programme that aimed at building a trans-border movement of hope among ordinary people, like fishermen, farmers, women, workers and other marginalized social groups.

In Phuket, tourism activists and representatives from local communities affected by mass tourism identified and discussed the widespread construction of golf resorts which was a significant common issue not only affecting Asian tourist destinations but communities worldwide.[4] Governments tended to give full support to golf course and resort development in the belief that they would raise their country's global image, attract investors and bring in tourist dollars. But often, such projects simply created skewed land use and deprived local residents of land and resources they depended on.

Thailand was experiencing a major "golf rush" at that time with more than 100 new golf course projects in the planning and construction stages. The Thai Golf Association and the Tourism Authority in Thailand, which were jointly gearing up for a "Visit Thailand Golf Year 1993", had announced that hundreds of new courses were needed to meet the demand. The government's aim was to turn Thailand into a "playground" for high-spending golfing tourists, especially from golf-addicted Japan.[5] Worldwide, there were around 25,000 courses. The USA alone had 13,600 courses for around 25 million golfers,[6] whereas Japan had approximately 2000 courses for 17 million players with an additional 1600 courses under construction or in the planning stage by the end of 1993.[7]

Three activist networks that were represented at the tourism forum in Phuket—GNAGA, APPEN and ANTENNA—jointly organized

---

[4] Anon., People's Plan for the 21st Century: Statement of the International People's Forum on Third World Tourism, 29 Nov.–4 Dec. 1992, Phuket Teachers College, Phuket, Thailand, *Contours*, Vol. 5, No. 8, Dec. 1992, pp. 18–19.

[5] Pleumarom (1994).

[6] Pearce, F., "How green is your golf?" *New Scientist*, 25 Sept. 1993, pp. 30–35.

[7] *GAG'M Update* No. 2, 1994, "Background Notes of the 2nd International Conference on Resort and Golf Course Development, Kamogawa, Japan, 21st–24th March 1994: Summary of the Country Report Japan", p. 11.

the first Global Conference on Golf Course and Resort Development in Penang, Malaysia, from 26 to 28 April 1993. At this event, delegates from Hawaii, Hong Kong, India, Indonesia, Japan, Malaysia, the Philippines and Thailand resolved that golf could no longer be considered as simply a sport, as it primarily served the interests of the elites and promoted an arrogant and wasteful lifestyle. Case studies from various countries showed that the major losers were ordinary citizens whose governments ignored the high social and environmental costs of golf courses and favoured developers even if they breached the law and committed human rights violations. The conference concluded with the launch of the GAG'M and the declaration of 29 April as "World No Golf Day" in order to convey a clear message: "In this age of increasing environmental awareness, there is no more room on Earth to destroy nature for the sake of a mere game".[8]

Concerns Raised by GAG'M

One grave concern, clearly, was the environmental impact of golf courses. In the Penang Conference statement, the courses were described as "another form of monoculture, where exotic soil and grass, chemical fertilizers, pesticides, fungicides and weedicides, as well as machinery are all imported to substitute natural ecosystems. These landscaped foreign systems create stress on local water supplies and soil, at the same time being highly vulnerable to disease and pest attacks".[9] Loss of forest and farm lands, destruction of wetlands, depletion of water resources, soil contamination from run-off of silt and toxic chemicals as well as air pollution from spraying pesticides in the courses were continuously reported from many countries.

Various issues of social justice were identified: unethical or illegal land acquisition practices for golf course development, disruption and displacement of local communities; exploding prices for land; escalating

---

[8] *GAG'M Update* No. 1, 1993, introduction, pp. 2–4.
[9] *GAG'M Update* No. 1, 1993, "Statement of the Global Anti-Golf Movement, 29 April 1993", p. 5.

rents; increasing gaps between rich and poor; human rights violations, corruption and crime. Sexism and the precarious working conditions of women caddies, particularly in Southeast Asia's golf courses, were another area of concern.[10] Caddies were so poorly paid that they often relied on golfers' tips. Thailand already had a reputation as a sex tourism destination and it was conspicuous that the predominantly male golfers required the services of young women to carry around their very heavy bags, their umbrellas and chairs, their drinks and mobile phones. Unfortunately, Thailand became the "model" with the marketing and exploitation of female caddies used to bait golf tourists spreading to other Asian countries.

While the opportunity costs were high, it was evident that local communities derived few economic benefits from golf course development and golf tourism. Since golf course construction and maintenance were extremely expensive with developers heavily depending on loans, investments and imports from abroad, most of the revenues did not trickle down to the local economy but were kept by large companies. As was pointed out in the GAG'M Statement, "At the heart of the golf industry is a multi-billion dollar industry involving transnational corporations, including agribusiness, construction firms, consultancies, golf equipment manufacturers, airlines, hotel chains, real estate companies, advertising and public relations firms as well as financial institutions".[11]

GAG'M activists also highlighted the drive by Japanese companies for golf course development in foreign countries and, related to this, the transformation of golf memberships into a valuable commodity, which resulted in widespread speculation and dubious practices. In the words of Japanese anti-golf campaigners: "the game being played most earnestly here is the money game".[12] The speculative nature of memberships and associated property transactions turned golf into a financially unsound, high-risk business. Thus, it was not surprising that many golf course and resort projects went bankrupt when the economic bubble burst in Japan at the beginning of the 1990s.

---

[10] Pleumarom (1992).

[11] *GAG'M Update*, No. 1, 1993, op. cit. p. 5.

[12] The Global Network for Anti-Golf Action (GNAGA), *GNAGA Statement of Purpose*, Tokyo 1992, p. 2.

In view of the multi-dimensional problems associated with golf, the GAG'M founding conference demanded an immediate moratorium on all golf course developments as well as an end to development aid, advertising and the promotion of golf courses and golf tourism. In addition, it welcomed the decision of the IOC to reject the inclusion of golf as an Olympic "sport" in the 1996 Atlanta Games and urged the IOC to introduce a permanent ban on golf as an Olympic "sport", reasoning that "this would amount to the legitimization and international recognition of a 'sport' which destroys the environment, creates social disruptions and which is financially unsound".[13]

On 29 April 1993, the first World No Golf Day, four GAG'M representatives presented the outcome of the Penang Conference at a press conference in Kuala Lumpur, Malaysia. Simultaneously, anti-golf activists and support groups marked the "golf-free" day with various activities in Thailand, Hawaii, Indonesia, the Philippines, Japan, India, Nepal, Taiwan, Australia, Switzerland and the UK. The activities varied from public awareness raising campaigns, discussions and debates, and the planting of trees to peaceful walks and demonstrations by farmers, students and citizens groups in India, Indonesia and Japan.[14]

The Asian "Golf War"

Media interest in golf-related problems was remarkable, with their coverage generally supportive of the issues raised by anti-golf activists. National and international newspapers and magazines—including *Newsweek*,[15] *The Guardian*,[16] *New Scientist*[17] and *The Economist*[18] —all highlighted the issues GAG'M addressed.

The Australian carried an article noting: "A glut of golf courses is spreading across Asia from Indonesia to Singapore to satisfy rich businessmen

---

[13] *GAG'M Update*, No. 1, 1993, op. cit. p. 7.
[14] *GAG'M Update*, No. 1, 1993, see various reports in section "World No Golf Day", pp. 8–15.
[15] Emerson, T., "The Anti-Golf Guerillas", *Newsweek*, 20 Sept. 1993.
[16] Sarah, S., Chatterjee, P., "Fairway to heaven?" *The Guardian*, 17 Sept. 1993.
[17] Pearce, op. cit.
[18] Anon. (1993), "Greens against greens", *The Economist*, 15 May.

and the social set. Tourism is booming but the effects on the poor and the environment are devastating".[19]

Stating that the world was increasingly "colonized by golf", the *New Scientist* opined that "golf courses are emerging as one of the most rapacious and socially divisive forms of tourist and property development".[20]

A cover story in *Asia Magazine* that featured GAG'M as demanding "rough justice" reported that: "The backlash against golf courses is really the result of a problem reaching critical mass. For when a motley crew of conservation-minded groups gathered in Penang..., the impact of Asia's golfing boom had grown severe enough to coalesce them into an anti-golf movement".[21]

Meanwhile, forceful evictions of local residents and clashes between golf course opponents and developers backed by state authorities frequently made the headlines. Of all the countries in Asia, the Indonesian experience was the one most overtly linked to the suppression of local people by its government. The campaign against the Cimacan golf resort project in West Java, which had started in 1989 and involved hundreds of families as well as supporting students and environmental groups, became famous for the peasants' fierce determination to defend their land rights. Eventually, the project grew so unpopular that it was shelved.[22] Cimacan was not an isolated case in land-scarce Java, where most Indonesian golf course sites were located. Course developers, who sought to acquire large tracts of land for their projects, inevitably clashed with peasants who had been tilling the land for generations. A serious incident happened on 14 April 1993—just a few days before the first "World No Golf Day"— when 800 people from four villages in the district of Ciawi, West Java, peacefully demonstrated against plans to transform their farm plots into a luxury golf resort, arguing that the 257 hectare site was ancestral land and should continue to be used for agriculture. Military and police forces broke up the protest and arrested 14 villagers. On 29 April, the Indonesian "Movement against Golf Course Development" (KAAPLG)

---

[19] Barnes, B. (1993), "Fairway to hell", *The Australian*, 29 March.

[20] Pearce 1993, op. cit.

[21] Allison, T. et al., "Clubbed to Death—How much longer before Asia is turned into one big golf course?" *Asia Magazine*, 15–17 April 1994, p. 12.

[22] Anon., "Golfmania", *Tapol Bulletin*, No. 117, June 1993, pp. 17–20.

was established in the capital Jakarta, and a rally was staged outside parliament. The anti-golf activists called for a nationwide moratorium on golf course development, complaining that thousands of hectares of productive agricultural lands would be lost within five years if all golf course projects would be implemented as planned.[23]

The "golf war" in Indonesia continued unabated throughout 1993, particularly in Bogor, West Java. In Cijayanti, 329 families were forced from their land to pave the way for a golf course development. In Rancamaya, some 1000 families lost their land and livelihoods for another golf project. After learning about a violent land conflict between a golf course developing company PT Light Instrumendo and farmers in Citeureup district, Bogor, West Java, GAG'M stepped in and wrote a letter to petition Indonesian President Soeharto. Expressing their dismay at police siding with the company and ignoring the unfair treatment of poor farmers who were criminalized and called communists because they had resisted eviction, GAG'M appealed to the President to use his position to help bring about a peaceful and just solution in the Citeureup land dispute.[24] The activists and farmers who had been arrested during the protests were later released. But the farmers lost their land and the golf course was eventually built.

In India, there was strong local resistance to plans to build eight golf courses in Goa, and a major concern with the large-scale Japanese Holiday Village project that included a course. Protests were spearheaded by the "Vigilant Goan Army" (JGF)—an action group fighting socially and environmentally damaging tourism– and included local women rights organizations.[25] Many women were against golf course projects with water issues being a main concern to them. On World No Golf 1993, women protesters carried empty pots as symbols to express their concern that golf courses would worsen the water shortage crisis in many villages of Goa. The experience of Thailand, where research had found that an

---

[23] Ibid., p. 19.

[24] *GAG'M Update*, No. 1, 1993, "GAG'M Letter to President Soeharto of the Republic of Indonesia, 17 November 1993", pp. 27–29.

[25] Goswami, R., "How green will these golf courses in Goa be?", *The Sunday Observer*, 9–15 May 1993.

average golf course consumed 6500 cubic meters of water per day showed this was equivalent to the daily demand of 60,000 villagers.[26]

Most significant in the Goan anti-golf struggle was the implementation of a "people's ban of developers", which worked as follows: the land earmarked for golf course projects was either owned privately or by the community, and course developers usually tried to get the Goan Government to "acquire" the land for them ostensibly for "public purposes" through the Land Acquisition Act. But these attempts were now effectively blocked by village assemblies that passed resolutions against such government measures. This way, all eight golf course projects in Goa could be stopped.[27]

Fortunately, the Asian "golf war" remained in most cases a war of words. Golf protagonists often reacted harshly to criticism, and labels for anti-golf course campaigners ranged from "anti-green greenies", "killjoys", "environmental extremists", "linksophobes", "morons" to "cold war losers".[28] However, there was also support from unexpected places. For instance, President Kim Young Sam of South Korea vowed in May 1993 to lay down his golf clubs until his term would expire in 1998 and expressed the view that golf "does more harm than good in today's South Korea, which needs to hunker down and work".[29] A survey conducted in South Korea had revealed that 80 % of the public opposed the boom in golf course construction in the country. Three months later, in China, set to become the next golf frontier at that time, the governing state council ordered a halt to golf course development as part of a campaign against wasteful spending.[30]

A particularly powerful "weapon" used against the often seemingly omnipotent golf lobby was a documentary produced in Thailand, which confronted viewers head-on with the inconvenient truth about golf and its social and environmental impacts. Entitled "Green Menace: The Untold Story of Golf" produced by Thai journalist Ing K and American

---

[26] Pleumarom (1994), op. cit., p. 54.

[27] GAG'M Update, No. 2, 1994, "Background Notes of the 2nd International Conference on Resort and Golf Course Development, Kamogawa, Japan, 21st–24th March 1994: Summary of the Country Report Goa—India", p. 15.

[28] GAG'M Update, No. 2, 1994, p. 143.

[29] Cit. in Emerson 1993, op. cit.

[30] Ibid.

environmental investigator Brian Bennett, the film was released in 1993 and won in the same year the First Prize for Environmental Documentary at the Suffolk/New York Film Festival. It was shown at other international film festivals and at a public symposium in Tokyo organized by GAG'M in connection with its 2nd International Conference on Resort and Golf Course Development in March 1994.

Some of the issues covered in the film—particularly the involvement of high-ranked military officers and crime syndicates in golf course construction, widespread land encroachment and theft of water from public water reservoirs for the maintenance of courses in Thailand—were highly sensitive; this is probably the reason why Thai authorities have never permitted it to be shown in the country's mass media.

"Green Menace" included a revealing interview with a Thai water resource engineer who quit working for golf courses after witnessing first-hand all the problems they caused. Among other things, he explained: "I saw outrageous exploitation. [One] golf course usurped a water source that was used by three to four villages of over 1,000 people. 'Go find another one, this is mine now!' Just like that. The villagers suffer, but can do nothing. The golf course owners are influential people with everything in their power, including high government officials".[31]

The excessive use of chemicals in golf courses was also addressed in the film. Golf course workers and caddies complained of often falling sick with symptoms of acute pesticide poisoning. A medical doctor confirmed that some of the chemicals used in Thailand's courses, such as the weedkiller paraquat, could cause severe illness and miscarriage. GAG'M campaigner Gen Morita who used to be a golf player himself warned viewers that golfers are usually not aware that they are "playing in poison".[32]

Golf superstar and course architect Jack Nicklaus was also interviewed in "Green Menace". In fact, Nicklaus had been involved in Thailand's best-known golf course scandal. When a Thai newspaper published a photo showing Nicklaus hitching a ride on a Thai Air Force helicopter to survey the site of the Golden Valley golf resort project some 250 km northeast of Bangkok, which he himself had designed, there were allegations of

---

[31] Ing and Bennett (1993).
[32] Ibid., p. 24.

misusing public funds. The situation escalated when Golden Valley's construction crews dynamited a state-owned mountain, cleared a forest and encroached upon Khao Yai National Park to build the golf course. When Nicklaus was asked by the filmmaker about golf-related problems in Thailand, he remained in denial and instead enthused: "You've got great land. You've got plenty of water, with the amount of rain that you have. Certainly, you don't have problems from our standpoint".[33]

Highlights of "World No Golf Year 1994"

When the news broke in December 1993 that another application had been submitted to the IOC to include golf in the 2000 Sydney Games after it had been dropped from the 1996 Atlanta Games, GAG'M submitted a petition to then IOC President Juan Antonio Samaranch, reiterating concerns about golf being a sport and calling for a permanent ban of golf from the Olympics. It said: "We realize, and resist, the commercialization of sports, which is driven by powerful profit-seeking corporate interests. Golf promotion, in particular, has been frenzied. In addition to environmental destruction, golf is deepening the gap between the rich and poor in developing countries, juxtaposing elitist and arrogant lifestyles against traditional cultures and basic needs of local people".[34]

An editorial in the Bangkok newspaper *The Nation* echoed the GAG'M call: "Keep golf out of the Sydney 2000 Games". It noted: "For an international institution as highly regarded as the Olympics to accept golf as an Olympic sport would give the business a big boost, while making a farce of its own ideals: fitness, health, fairness and the glorification of amateur athletes—playing sports for its own sake, not for money or status... From its origins as a simple, pastoral sport, golf has become corrupted. Its existence as a sport can no longer be separated from its role in destroying the environment and local communities".[35]

---

[33] Ibid., p. 22.

[34] *GAG'M Update,* No. 2, 1994, "NO to golf in the Olympics"—GAG'M Letter to Mr Juan Antonio Samaranch, President of the IOC, 30 December 1993, pp. 19–20.

[35] Anon., "Keep golf out of the Sydney 2000 Games", *The Nation,* 24 Jan. 1994.

At the 2nd International Conference on Resort and Golf Course Development, co-organized by GAG'M and local groups in Japan in March 1994, participants from 12 countries raised strong criticisms regarding the 1998 Winter Olympics in Nagano. In a symbolic gesture, the delegates bought tree saplings to be planted in Nagano City in support of the resistance shown by Japanese people to the massive developments related to the Games. The aim was to draw international attention to the way in which forests and farmland, which local communities had nurtured and safeguarded for generations, were being destroyed for this mega-sports event. The related GAG'M Press Release stated: "For just '16 days of glory' large amounts of public money are being invested in building extensive infrastructural developments, which include golf and ski resorts, the Hokuriku Shinkansen (Bullet Train) railway project, hotels, etc. that only benefits vested interests, chiefly the Kokudo Group owned by Mr. Yoshiaki Tsutumi".[36]

The strategy of GAG'M was to make its presence felt in Japan, as it was the centre of the global golf mania at that time. Campaigners from around the world came together there to raise awareness among the Japanese public and to meet with government and industry representatives in order to question and petition them concerning their involvement in resort and golf course development in foreign countries. The fact that GAG'M hardly had the means to organize a major event in expensive Japan could not deter them. American journalist James Fahn, who accompanied GAG'M delegates in Japan, wrote: "[GAG'M is comprised of] seasoned activists. But they have come together as a result of pressure from the grassroots. A striking element of any anti-golf gathering is the testimony of ordinary people, often farmers, who have bravely fought the construction of golf courses in their communities despite the powerful forces behind development. Also of note is the movement's poverty. The conference was held at a youth training centre where participants slept in dorm rooms and ate in a cafeteria along with the members of a 100-piece orchestra from a local girls' school".[37]

---

[36] *GAG'M Update*, No. 2, 1994, GAG'M Press Release on occasion of the 2nd International Conference on Resort & Golf Course Development, Tokyo 28 March 1994, p. 9.

[37] Fahn, J., "Going against the green", *The Nation*, 29 April 1994.

Before the anti-golf conference started in the small fishing village of Kamogawa in Chiba (the province that was known as "Golf Ginza"), delegates held talks in Tokyo with officials of the Japanese International Trade and Industry Ministry and the Japan International Cooperation Agency (JICA) as well as the Taisei Corporation that was building and operating a number of golf courses overseas. At that time, Japan was already experiencing a sharp decline of the golf industry, including the golf membership market, due to the country's economic crisis. This, together with a strong national citizens movement, led to the cancellation of some 700 golf course projects nationwide by the end of 1993, 924 courses (in the planning stage or under construction) by the end of 1994 and over 1000 by the end of 1995.[38] However, Japanese developers were still aggressively pushing development outside the country—for example, China, Indochina, and India in Asia, in Australia and even in Europe. A number of Japanese overseas projects were stopped by joint actions of the Japan-based GNAGA, GAG'M and local groups. Especially in Hawaii, Japanese golf companies such as the Royal Hawaiian Country Club, Sokan, Obayashi Gumi, Kato, Nansei and Shalon—many of which were linked to the Yakuza (Japanese crime syndicates)—were forced to stop and withdraw from developments.[39]

As a step forward from the first "World No Golf Day" in 1993, the GAG'M conference in Japan launched a "World No Golf Year 1994".[40] In the following months, GAG'M continued to campaign on a wide range of issues. Monitoring the expansion of tourism-related golf businesses to Indochina, Burma and China, GAG'M wrote letters to heads of state—for example, Prime Minister Khamtay Siphandone of Laos[41] and Prime Minister Vo Van Kiet of Vietnam[42]—to inform them about the dangers of golf courses and petition them to help stop these controversial

---

[38] Tsutomi Kuji (1995).

[39] Ibid., p. 13.

[40] *GAG'M Update*, No. 2, 1994, "GAG'M Statement: Help launch the World No Golf Year on 29 April 1994!", pp. 17–18.

[41] *GAG'M Update*, No. 3, 1994, "GAG'M letter to Prime Minister Khamtay Siphandone, People's Democratic Republic of Laos, 21 November 1994", pp. 70–71.

[42] *GAG'M Update*, No. 1, 1995, "GAG'M letter to Prime Minister Vo Van Kiet of Vietnam', 27 January 1995", pp. 129–133.

projects. GAG'M also joined the international boycott campaign against a "Visit Myanmar Year 1996" in military-ruled Burma and published articles to condemn land grabs and human rights violations in relation to tourist resort and golf course construction in that country.[43] Moreover, cases were publicized of land-hungry golf course builders affecting Indigenous Peoples' rights in the Philippines, Aotearoa (New Zealand) and Hawaii. Last, although not least, the spotlight was once again turned on Indonesia, where Balinese farmers, students, environmentalists and religious groups were fighting a fierce battle against the building of the Bali Nirwana Golf Resort next to Hindu sacred sites and the Tanah Lot temple.[44]

GAG'M Expansion to the USA and Europe

In the USA, GAG'M joined hands with the Environmental Coalition of Ventura County in California to take legal action in 1994 against a Japanese golf course project in Ojai Valley.[45] Links were established with citizens' groups to fight the Pebble Beach Company (PBC) that was then owned by the Lone Cypress Company, formed by the Japanese Sumitomo Bank and Taiheiyo Club. The PBC planned to develop a golf course on property they owned on the Monterey Peninsula in California, which would have resulted in the destruction of an old pine tree forest in Pescadero Canyon. Local protesters called for the protection of the pristine and unique Jeffers Forest area as a nature reserve, as they feared that the development of the forest would mean the irreplaceable loss of an ancient ecosystem and its genetic archive.[46] In 1999, Hollywood star Clint Eastwood became the major investor in PBC along with golf celebrities like Arnold Palmer and Tiger Woods. Nevertheless, the battle against the golf course expansion and the building of a 160-room luxury hotel for golfers continued until 2007, when the California Coastal

---

[43] Pleumarom, A., "A holiday in hell", *The Nation*, 29 March 1993.

[44] Warren (1998).

[45] *GAG'M Update*, No. 3, 1994, see various articles by Alasdair Coyne published in "Voice", pp. 140–143.

[46] *GAG'M Update*, No. 1, 1995, "The Pebble Beach golf course story", pp. 177–189.

Commission eventually turned down PBC's application on environmental grounds. Dave Dilworth, director of the local Helping Our Peninsula's Environment (HOPE) group and US representative of GAG'M, who had led the Pebble Beach protest from the beginning, rejoiced: "The magnificent dark green cloak covering our Monterey Peninsula is saved from Clint's and Pebble Beach Company's Chainsaw Massacre".[47]

On the US East Coast, GAG'M allied with citizens groups in New York and the Coalition for Forests who were fighting a huge golf course scheme on Long Island. The plan included the privatization of existing public golf courses and the building of several new courses on Suffolk parklands. Local residents called for the project to be terminated, arguing it would decimate scarce and significant remnants of Long Island's once-abundant forests and result in air and water contamination from the use of toxic substances in the golf courses.[48]

In Europe, the UK had been experiencing a major golf boom since the late 1980s. It possessed 2157 courses in 1990; that was an increase of more than 50 % on Europe's existing courses. In 1991, 1800 planning applications for new courses had been submitted to local authorities in the UK alone. Had all these projects been realized, an area the size of the Isle of Wight would have been swallowed up by golf courses, according to the Council for the Protection of Rural England. However, due to worsening economic conditions and increased public protests, less than 350 golf courses were actually completed between 1988 and 1993.[49]

A major victory for the British anti-golf course movement was the withdrawal of the Japanese Kosaido company which had plans to build a second golf course in an Area of Outstanding Natural Beauty near Liphook, Hampshire. Local people with access rights to this registered common land fought the scheme for five years.[50] With additional pressure

---

[47] Dilworth, D., *Monterey Pine Forests Win!* Helping Our Peninsula's Environment (HOPE), Carmel, USA, 18 June 2007.

[48] *GAG'M Update*, No. 1, 1995, see various reports, pp. 199–204.

[49] *GAG'M Update*, No. 2, 1994, "Background Notes of the 2nd International Conference on Resort and Golf Course Development, Kamogawa, Japan, 21st–24th March 1994: Summary of the Country Report United Kingdom", p. 16.

[50] *GAG'M Update*, No. 1, 1993, Press Release by Desmond Fernandes, GAG'M representative U.K., and The Bramshott Commoners Society: Commoners vs golf course development: Old Thorns, Liphook, Hants., pp. 83–84.

from national environmental groups and GAG'M, Kosaido was eventually forced to cancel its project in December 1993.

Another area in Europe specially targeted by golf course developers was the Alps region, spanning seven countries: Austria, France, Germany, Italy, Liechtenstein, Slovenia and Switzerland. There were plans to increase the number of courses from 160 in 1993 to 300 in 1995, prompting resistance from community groups and environmentalists who expressed concerns about building courses in biodiversity-rich and ecologically fragile areas, particularly in mountainous terrains featuring rare alpine flora and fauna and forests and wetlands.[51] In Switzerland, GAG'M partnered with the Working Group Tourism & Development (AKTE) and the Swiss Foundation for Landscape Conservation to highlight the "seamy" side of the golf boom, including the privatization of public forests and wilderness areas, disputes over land distribution, conversion of agricultural land and conservation conflicts.[52]

GAG'M also responded to calls from golf course critics in Slovenia. In February 1995, GAG'M wrote letters to petition the Slovenian government to heed the concerns of local citizens, environmental groups and the scientific research community regarding two proposed golf course projects in ecologically sensitive wetland areas along the Drava River near Ptuj town.[53] Both courses had been pushed forward in great haste without adequately informing the public and undertaking an Environmental Impact Assessment. Due to increasing public pressure, these projects were soon cancelled.

In April 1995, GAG'M and a number of European groups—Tourism Concern in the UK and AKTE in Switzerland among others—launched a letter writing campaign to the Prime Minister of Greece, Andreas Papandreou, to protest against the construction of a golf course and a hotel complex at the Butterfly Gorge in the southeastern part of Crete. They pointed out that the last remaining pine forest stretching from the coast to the mountains, a precious butterfly habitat and place of natural

---

[51] *GAG'M Update*, No. 3, 1994, "Alps designed as golf development region", p. 145.

[52] Ibid., " The seamy side of the golf boom", p. 149.

[53] *GAG'M Update*, No. 1, 1995, see two GAG'M letters to the Government of Slovenia, 23 February 1995 and 15 June 1995, pp. 228–231.

beauty that visitors came to enjoy, should not be destroyed for the sake of golf tourism.[54]

Anti-Golf Solidarity from Hawaii to Mexico

On the occasion of the third World No Golf Day on 29 April 1995, GAG'M launched an international campaign against the controversial "The Villages of Hokukano" golf resort project (later renamed in Hokuli'a golf resort) in South Kona, on Hawaii Island. The project was operated by the Hawaiian joint venture company Oceanside, in which Japan Airlines (JAL) held 75 % of the interest. The massive tourism scheme included over 1000 luxury villas, a private members' lodge of up to 100 units and a 27-hole golf course that was to be designed, once again, by Jack Nicklaus.

Local residents opposed the JAL-led project from the start out of fear that the nearby pristine National Marine Life Refuge at Kealakekua would be damaged and the local agricultural lifestyle degraded. Native Hawaiians (Kanaka Maoli), in particular, were up in arms because the project would have involved the loss of their ancestral land and ocean natural resources, and violated their traditional access and gathering rights to these resources for their livelihood. Furthermore, many Kanaka Maoli's archaeological features, including burial sites, would have been desecrated and, perhaps, destroyed by the construction of golfing facilities and luxury homes, around and on top of the sites.

Apart from many protest actions in Hawaii—including rallies at the Waikiki JAL Office in Honolulu—as well as citizens filing a lawsuit at the Circuit Court against the Hawaii County government that had approved the first phase of construction, GAG'M joined the local opposition movement to call for a boycott of JAL, to increase pressure on the company to withdraw from the project.[55] With JAL already hampered by financial problems and accusations of mismanagement, the Japanese Transport Minister publicly stated in September 1994 that JAL's tourism venture

---

[54] Ibid., GAG'M letter to Prime Minister Andreas Papandreou of Greece 19 April 1995, pp. 217–218.

[55] GAG'M, *Boycott JAL—Save South Kona!*, Fact Sheet, Penang 1995.

in Hawaii was a failure. Yet, Oceanside did not give up and between 1998 and 2003 spent more than US$190 million on the project's development. As late as 2003, after almost a decade of people's struggles, the court finally ruled the embattled project illegal. According to the judge, Oceanside did not have the government approvals it needed to establish a private, luxury resort residential subdivision on its property.[56]

In September 1995, GAG'M received an emergency call concerning one of the most violent "golf wars" it had ever heard of. In Tepoztlan, Mexico, a joint venture of Mexican and American developers had been given the green light to build a US$311-million resort including an 18-hole golf course, also designed by the ubiquitous Jack Nicklaus, inside of El Tepozteco National Park, in the state of Morelos, south of Mexico City.[57] Apart from the fact that the El Tepozteco Golf Club development was slated to be built in a protected area, it was estimated that the golf course would have consumed more than 4000 cubic meters of water a day, about five times the amount consumed by the entire town of Tepoztlan. In addition, concerns were raised about the potential impact of the course's heavy use of agrochemicals.

By the end of August 1995, locals opposing the project seized the town hall of Tepoztlan and took several officials hostage, hanging them in effigy. After 12 days of protest, the mayor, who had approved the controversial El Tepozteco scheme, resigned. Meanwhile, international support groups— among them a coalition of American environmentalists including Ralph Nader and executives of the US Friends of the Earth, Greenpeace USA and the Sierra Club—called on the developing companies, GTE Data Services and Jack Nicklaus' Golden Bear Course Management, to withdraw their participation from the disputed project.[58]

On 10 April 1996, squads of heavily armed police ambushed busloads of anti-golf course campaigners from Tepoztlan, who were travelling to express their grievances to the then Mexican President Ernesto Zedillo. There were shootings that left one protester dead and 20 more injured.

[56] Dayton, K., "Ruling suspends luxury project on Big Island', *Honolulu Advertiser*, 10 Sept. 2003, http://the.honoluluadvertiser.com/article/2003/Sep/10/ln/ln01a.html (accessed 12 July 2015).

[57] *GAG'M Update*, No. 1, 1996, country file Mexico, pp. 188–207.

[58] Ibid., pp. 195–197.

The incident that was video-taped and broadcast on television resulted in the arrest of six police officers allegedly responsible for the violence. There was a storm of protest with GAG'M and many other local and international action groups sending appeals to President Zedillo to carry out a thorough investigation of this incident and to bring the killers to justice. It was only in the wake of this violence that the developers cancelled the El Tepozteco project. On 29 April 1996, the fourth World No Golf Day, the protesters gathered at the project site and held a ceremony to honour Marcus Olmedo, the 62-year-old man who was killed in the shootings.[59]

Confronting Jack Nicklaus

On the occasion of the fourth World No Golf Day, GAG'M addressed an open letter to Jack Nicklaus saying that as President of the Golden Bear International Company, he had to take some of the responsibility for the El Tepozteco debacle because he had chosen to ignore all earlier appeals to withdraw from the misguided project.[60] The GAG'M letter also denounced Nicklaus' involvement in other harmful golf course developments, for example, the JAL-led golf resort project at South Kona, Hawaii; the Golden Valley project in Thailand; the Manila Southwoods Golf & Country Club in Cavite, Philippines, which had involved the forced evictions of hundreds of farmers in 1991; the Andaman Club, a golf-plus-casino resort in military-run Burma, a pariah state that was subjected to economic sanctions by the international community; and the Jack Nicklaus II golf course project near Bangalore in India, which had been scrapped in 1994 due to vigorous protests by neighbouring communities. With the letter, GAG'M was sending a clear message that it was high time for Nicklaus to acknowledge the problems his golf projects had created and needed to consider their consequences. It said: "Your career as a golf player has already made you famous and rich; you have even been given the title 'Golfer of the Century'. Therefore, we fail to understand that for an honorarium of

---

[59] Ibid., special supplement: "Update on the 'El Tepozteco' case in Mexico".
[60] GAG'M, *Open Letter to Mr. Jack Nicklaus, President of Golden Bear International, Inc.*, Penang 23 April 1996, 4 pages.

US$2 million per project as a golf course designer, you are spoiling your reputation by participating in all these ecologically and socially devastating ventures. Worse, some of these business deals have implied the involvement of unscrupulous developing companies, corrupt officials, and dictatorial regimes who have no concerns for safeguarding the environment and people's wellbeing and rights".[61] Nicklaus did not respond.

Rebutting the Golf Industry's "Green-Wash"

Due to the worldwide and growing anti-golf activism and more informed public discussion on the negative consequences of resort and golf course construction, golf's image had been considerably tarnished. Influential agencies such as the United Nations Economic and Social Commission for Asia and the Pacific (UN-ESCAP, based in Bangkok) and the Asian Development Bank (ADB) acknowledged the serious problems caused by the golf boom. In the section on tourism in ESCAP's 1995 report about the "State of the Environment in Asia and the Pacific", it was stated: "Perhaps the best example of the impacts of mass commercial tourism is provided by the promotion and massive growth of golf tourism. Golf course construction has created widespread negative social, cultural and environmental impacts, particularly in the developing countries of the region. Typical impacts include forest destruction, and air, water and soil pollution caused by the excessive use of chemicals. Furthermore, in some cases, local people have been deprived of rights to land, disrupting communities and endangering their livelihoods".[62]

The golf industry responded to such critical discourses with overzealous, well-funded public relations activities. From 1994 onwards, the number of international conferences organized by golf advocates to parade golf as a "green and clean" sport increased significantly. North American and European "green golf course experts", in the service of the golf industry, made tenacious attempts in the media and on all possible occasions to pre-empt and discredit the arguments of golf critics and

---

[61] Ibid., p. 4.
[62] UN-ESCAP and ADB (1995).

to promote the idea of environmentally friendly and pesticide-free golf courses. The European Golf Associations' Ecology Unit, funded by the Royal and Ancient Golf Club of St Andrews, the PGA European Tour and the European Golf Association and led by British ecologist David Stubbs (who was later appointed as Head of Sustainability for the London 2012 Olympic Games), stated in one of its glossy booklets, for example: "Golf is accused of being harmful to the environment—by using too many chemicals, too much water, damaging wildlife habitats and the landscape, and being generally anti-social. Most of this is technically unfounded and appears to mask socio-political objections. However, the overall effect is a very poor environmental image and this inevitably influences public opinion and regulatory bodies".[63]

GAG'M took up the challenge and intensified its efforts to expose the "green myths" being created by the international golf lobby. They argued that the drive by the US Golf Association (USGA), the European Golf Association (EGA) and other golf-promoting organizations to design and promote environmentally friendly strategies for courses were insufficient and were primarily aimed at protecting the industry's commercial interests. Proposals for purportedly ecologically sound course management—for example, improvements of landscaping and methods to reduce high consumption of water and chemicals—were repudiated as incomplete, superficial and cosmetic, as they were limited to mostly technical measures while a wide range of serious problems such as land grabs, loss of pristine biodiversity-rich areas and fertile farmlands, and displacement of communities remained unaddressed.

A high-profile "Golf and Environment" conference was held at the Pebble Beach golf course in California, USA in January 1995, which aimed to get endorsement for their "green golf" strategies from environmental groups. Local golf course critics attended and distributed a GAG'M statement to participants and organizers urging them "to tell the truth" and to present a complete picture of the environmental impacts of golf.[64]

---

[63] European Golf Association Ecology Unit, *An environmental strategy for golf in Europe*, Oxford 1995, p. 6.

[64] *GAG'M Update*, No. 1, 1995, GAG'M letter to the participants of the conference on Golf and the Environment, Pebble Beach, California, 15–18 January 1995: "Golf conference urged to tell the truth", pp. 14–16.

In fact, it was seen as a great irony—and a reflection of the reality of the golf industry—that the conference was held at the same Pebble Beach that had sparked a huge environmental conflict with its course expansion project. It was also pointed out in the letter that the adoption of international standards for golf courses as developed by the USGA was utterly insufficient and the claims made for golf's potential to comply with the principles of sustainable development were untenable.[65]

Another significant event in the history of the GAG'M was the World Congress on Sport & Environment co-organized by Spanish government agencies and the IOC in Barcelona from 20 to 23 March 1996, which included a session on environmentally friendly golf courses, with David Stubbs of the EGA Ecology Unit as a panelist. GAG'M contributed a statement to the conference, entitled "Green Fees: The Environmental and Social Costs of Golf Courses", reiterating golf's negative impacts and questioning the procedures of the industry-driven EGA Ecology Unit that had constantly played down the critical environmental and social dimensions of golf course development.[66] The Statement urged Congress participants to present an accurate and complete picture of harmful sports, golf in particular, and to consider a moratorium on all destructive sports facilities, including golf courses. It also renewed its call on the IOC to remain firm in objecting golf as an Olympic sport.[67]

Greenpeace Spain, who sent delegates to the Congress and distributed the GAG'M Statement to participants, reported that a lively debate took place between "green" golf course promoters and locals who contested golf projects' environmental sustainability, particularly in a country like Spain where people had been suffering tremendous hardship from serious drought. In 1995 alone, 10 million Spaniards, or one quarter of the total population, had faced water restrictions, whereas the golf courses

---

[65] Ibid.

[66] *GAG'M Update*, No. 1, 1996, 'GAG'M letter to the participants of the World Congress on Sport and Environment', Barcelona, 20–23 March 1996: "Green fees: The environmental and social costs of golf courses", pp. 27–31.

[67] Ibid.

in and around Madrid had consumed as much water as a community of 135,000 people.[68]

Anti-Golf Action Then and Now

Asian golf courses and related property developments faced a serious downturn in fortune during the financial crisis of 1997. Due to its speculative nature, the golf industry was a major contributor to the bubble economy, and when the bubble burst, it became one of the first victims. "Ask any golfer: what goes up must come down", said an article in *The Economist* at that time. "[Golf is] a symptom of a social, political, economic and environmental malaise whose effects are only just beginning to be felt. Many theories have been put forward to explain why the economic progress of Southeast Asia has so suddenly left the fairway: the forces of globalization; misguided economic policies; exclusive and unresponsive political systems; a pursuit of growth at the expense of everything else, including the environment and the livelihoods of the poor. The phenomenon of golf unites all these hypotheses".[69] The outcome was devastating for many property developers. In Thailand, for example, three golf courses, once valued at the equivalent of US$200 million, were on the market in November 1997 for a mere US$18 million.[70]

The crisis of "Asian golf course capitalism" also resulted in a slowdown of the anti-golf movement in the region as the most active campaign groups were based here. However, GAG'M has continued—albeit to a lesser extent—to monitor, expose and challenge golf course development and golf tourism.[71]

In a number of cases, GAG'M's involvement followed requests to support, what often began as, very localized struggles against golf course

---

[68] Ibid.; GAG'M note on the Barcelona conference, p. 32.

[69] Anon., "Golfonomics: Asia in the rough", *The Economist*, 20 Sept. 1997, p. 89.

[70] Ibid., p. 91.

[71] See e.g. Chee Yoke Ling, "Golf tourism", *Third World Resurgence*, No. 207/208, 2007; as well as the Tourism Investigation & Monitoring Team's (tim-team) bi-monthly bulletins: *New Frontiers* 1997–2009 and *Southeast Asia Tourism Monitor* 2010–2015, posted at http://www.twn.my/tour.htm (accessed 13 July 2015), as well as *tim-team Clearinghouse* Email-service 2007–2012.

projects. In 2006, GAG'M joined hands with a broad alliance of local residents and environmentalists from across the world to protest a mega-golf and casino resort in a biodiversity-rich mangrove area in North Bimini, the Bahamas.[72] Letters were written to the Bahamian authorities and the Hilton Hotel Group that was involved in developing and managing the controversial Bimini Bay Resort, to call for a stop of the project and to express full support for the local people's plan to have a Bimini Marine Protected Area established in the area.[73] Due to strong protests, the controversial golf course project was not implemented, but people have been fighting an even bigger and more destructive tourism scheme, the Resorts World Bimini, that is being developed by the Malaysian Genting company. It includes a large cruise ship pier, marinas, a casino and a luxury resort, and it is rumoured that there is also a plan to build a golf course.[74]

In 2008, GAG'M in cooperation with an alliance of local activists and civic groups in Kerala appealed to the Indian Union Minister of Tourism and Culture to heed local people's call for a moratorium on all golf course developments in the state of Kerala and for the conversion of a controversial golf course in Trivandrum into a public biodiversity park.[75] At the time of writing, the dispute in Kerala was unresolved. The Sports Authority of India (SAI) has taken over the Trivandrum Golf Course to develop it into a national golf academy. Critics say the project only benefits an influential group of bureaucrats and business people and want to open the property to the general public.

GAG'M also worked with the environmental organization SAVE in Quintana Roo, Mexico, calling on then President Felipe Calderón and relevant government agencies to implement a moratorium on all new golf course developments along the Mexican Riviera Maya and to take effective measures for the protection of the area's rich biodiversity and fragile

---

[72] See various press reports at: http://www.savebimini.org/news.html (accessed 15 July 2015);

[73] See e.g. GAG'M letter to The Rt. Hon. Hubert Alexander Ingraham, Prime Minister and Minister of Finance of The Bahamas, and Bahamas government agencies: *YES to Bimini marine protected area—NO to golf course*, Bankok, 21 November 2008.

[74] http://www.tribune242.com/news/2014/sep/22/insight-anyone-golf-not-us-say-biminites/ (accessed 15 July 2015).

[75] GAG'M letter to Smt. Ambika Soni, Indian Union Minister of Tourism and Culture: *NO to golf courses—YES to biodiversity park in Kerala*, Bangkok, 17 June 2008.

ecosystems.[76] Unfortunately, it was not possible to stop the most controversial project, the Gran Bahia Principe Resort and Golf Club developed
by a Spanish hotel group. It is a luxurious gated community including
a Robert Trent Jones II 27-holes golf course that is ranked #3 in all of
Mexico.

In 1996, the Washington-based World Watch Institute revealed in a
study on "Shrinking Fields" that the rapid proliferation of golf courses
around the world was taking food out of the poor people's mouths.[77] In
the following years, reports surfaced about worsening food shortage crises
that eventually led to food riots in some poor countries. Meanwhile, golf
courses—often mega-projects embracing hotels, holiday villas, marinas,
and shopping malls—continued to mushroom, particularly in the developing world. GAG'M welcomed government action that prioritized food
and water security for common people over golf resorts for wealthy locals
and foreigners. In China, for example, in December 2006 the central
government imposed a ban on the building of new golf courses, residential villas and race tracks on undeveloped land in order to protect China's
rapidly diminishing farmland. Similarly, some governments of Southeast
Asian nations moved to restrict the conversion of rice lands into luxury
housing, resorts and golf courses.[78] In order to draw attention to the
issue of golf courses and food security, GAG'M member tim-team and
Biothai—a network of Thai civic groups concerned with biodiversity
protection and sustainable agriculture—conducted research in Thailand,
to calculate how much rice could be produced on the land used for
golf courses in the country. It was found that existing golf courses had
significantly affected the natural resource base on which Thailand's agriculture depends, with considerable consequences for food production.[79]

In recent years, GAG'M has observed a dramatic downswing in golf in
the wake of the global financial crisis in 2007–2008. The American golf

---

[76] SAVE letter to the President of Mexico, Felipe Calderón, with support from GAG'M: *NO to golf courses—YES to biodiversity*, Akumal, Riviera Maya, Mexico, 2 August 2008.

[77] Gardner, G., *Shrinking fields: Cropland loss in a world of eight billion*, World Watch Paper No. 131, Washington 1996.

[78] Webb, S., "Golf courses, developers nibble at Asia's rice paddies", Reuters, 1 May 2008;

[79] Pleumarom, A., *Golf courses and food production* in *Thailand*, TWN-Features No. 3361, Dec. 2008, p. 5.

industry is still suffering substantially from an exodus of players and the mass closure of courses. According to the National Golf Foundation, five million Americans left the golf sport over the past decade. Roughly, every two days, a golf course shuts down in the USA, while just 11 courses were opened in 2014.[80] Also, in Japan, hundreds of golf courses lie abandoned owing to the severe over-development in the recent past. Given their large size and lack of shade, renewable energy companies have recently begun to use them as solar farms.[81]

Yet, golf protagonists remain optimistic, as in some parts of the developing world, golf course development is still on the rise. Moreover, with golf eventually becoming an Olympic sport in 2016, industry insiders like Jack Nicklaus expect a new boom time for golf.[82] In China, for instance, the decision to add golf to the Olympic programme has meant more money for golf from the state-run sport system and the prospect of building hundreds, if not thousands, new courses in the country.[83]

Golf is "a dangerous game" and stands as a symbol for environment and social justice abuses, excess, greed and self-delusion of the rich and powerful.[84] That is why GAG'M and many other groups have been actively opposing it for so many years. Golf's comeback as an Olympic sport is likely to give fresh impetus for the movement. The powerful actions of "Ocupa Golf" in Rio de Janeiro, which has drawn the attention of the international media and received support from environmental and human rights groups in and outside Brazil, are promising signs. GAG'M

---

[80] Greenfeld, K.T., "The death of golf", Men's Journal, August 2015, http://www.mensjournal.com/magazine/print-view/the-death-of-golf-20150625 (accessed 12 July 2015).

[81] Bolton, D., "Japan is turning abandoned golf courses into solar farms to solve its energy problems", The Independent, 9 July 2015, http://www.independent.co.uk/environment/japan-is-turning-abandoned-golf-courses-into-solar-farms-to-solve-its-energy-problems-10379109.html (accessed 14 July 2015).

[82] See Bloomberg interview with Jack Nicklaus, posted at YouTube 12 December 2014, https://www.youtube.com/watch?v=7ghy_A2FBY8 (accessed 10 July 2015).

[83] Richard, G., "Inclusion in Olympic programme feeds a boom time for golf in China", The Guardian, 6 March 2010, http://www.theguardian.com/sport/2010/mar/06/china-golf-courses (accessed 10 July 2015).

[84] "A Dangerous Game" is the title of a 2014 documentary from filmmaker and investigative journalist Anthony Baxter (also producer of "You've Been Trumped"), which examines the devastating impact of luxury golf resorts around the world, http://www.adangerousgamemovie.com (accessed 10 July 2015).

has lots of knowledge and experience to share with the people who are fighting against unjust and unsustainable golf-related projects. Since it came into being, it has successfully presented a golf narrative from an environmental and social justice perspective and initiated educational campaigns and actions such as "World No Golf Day", all of which have helped to increase the public's awareness on the many critical aspects of golf. Even though it was clear from the beginning that GAG'M could not achieve its professed goals to bring about a worldwide moratorium on golf course construction and a ban on golf tourism promotion, it was nevertheless able to put the powerful golf lobby on the defensive and to step up pressure on governments and developers. Through direct and legal action, many controversial projects were stopped. As the director of the documentary "A Dangerous Game" Anthony Baxter commented, it is "incredibly tough" to fight the Goliaths of golf but "when the Davids get together and they form a very impressive group of people who will fight together, then they have more of a chance".[85] Indeed, what can be learned from the GAG'M story is: when thoughtful and committed citizens and groups unite and jointly take action for a common cause, they can make a significant difference.

References

Ing K., & Bennett B. (1993). *Green menace: The untold story of Golf* (p. 8). Bangkok: Post-Production Script.

Kuji, T. (1995). *Golf and economy in Japan* (trans: Morita, G., p. 12). Tokyo: GNAGA.

Pleumarom, A. (1992). Course and effect: Golf tourism in Thailand. *The Ecologist, 22*(3), 104–110.

Pleumarom, A. (1994). Sport and environment: Thailand's golf boom reviewed. *TEI Quarterly Environmental Journal, 2*(4), 47–61.

United Nations Economic and Social Commission for Asia and the Pacific (UN-ESCAP) and Asian Development Bank (ADB). (1995). Golf Tourism

---

[85] Hogan, B., "Interview with 'A Dangerous Game' documentary filmmaker Anthony Baxter", *Ecorazzi*, 22 Juni 2015, http://www.ecorazzi.com/2015/06/22/interview-with-a-dangerous-game-documentary-filmmaker-anthony-baxter/ (accessed 12 July 2015).

in the Asian and Pacific Region. *State of the environment in Asia and the Pacific 1995* (p. 335). New York: United Nations.

Warren, C. (1998). Tanah Lot: The cultural and environmental politics of resort development in Bali. In P. Hirsch & C. Warren (Eds.), *The politics of environment in Southeast Asia: resources and resistance* (pp. 229–261). London/New York: Routledge.

'Human Rights or Cheap Code Words for Antisemitism?' The Debate over Israel, Palestine and Sport Sanctions

Jon Dart

Introduction

Since 1948 and Israel's 'War of Independence', an event described by Palestinians as 'al Nakba' (the catastrophe), different paradigms have been used to explain the state of affairs in Israel/Palestine and its populations. For the past 65 years the military superiority of the Israel Defence Force (IDF) has been coupled to a lack of meaningful support for Palestinians from neighbouring states (often informed by different interpretations of Islam), and an ongoing sense of frustration with the international 'peace processes'. These actions have left the Palestinian population subject to very high levels of poverty and unemployment, poor health services and limited educational opportunities.

The State of Israel has been variously described as a settler-colonial state, an apartheid state, an ethnocracy, a liberal democracy, with

J. Dart (✉)
Leeds Beckett University, 221 Cavendish Hall, Beckett Park Campus, Leeds, LS6 3QU, UK

© The Editor(s) (if applicable) and The Author(s) 2016
J. Dart, S. Wagg (eds.), *Sport, Protest and Globalisation*,
DOI 10.1057/978-1-137-46492-7_9

Zionism claimed by some as a national liberation movement for the Jewish people. The denial of statehood and failure of the 'armed struggle' in the 1960s and 1970s led Palestinians living in the Occupied Palestinian Territories (OPT) to launch two popular uprisings (known as 'Intifadas'[1]—1987–1991 and 2000–2005). Despite their duration and loss of life on both sides, these uprisings did not generate any meaningful advancement of or resolution to the Palestinian question.

This chapter focuses on events since 2005 and on the growing international concern over the plight of the Palestinians, be they living in Israel, the Occupied Territories, as refugees in neighbouring countries or those displaced to further afield. The chapter begins by outlining the boycott, disinvestment and sanctions (BDS) movement that emerged in 2005 and the subsequent responses of the Israeli state and its supporters. The chapter discusses how UEFAs (Men's) U21 Tournament, held in Israel in 2013, offered a platform for both supporters and critics of the Israeli state to advance their agendas and expose the relationship between sport and politics. An assessment is made of the claim that those who opposed to the hosting of the football tournament in Israeli state were guilty of antisemitism. The chapter concludes by discussing the importance of distinguishing between antisemitism and anti-Zionism.

Boycott, Disinvestment and Sanctions

In 2005, individuals and organisations in Palestinian civil society launched a co-ordinated campaign calling for BDS against the Israeli state. BDS were to be employed until the Israeli state complied with international law and agreed to recognise the fundamental rights of Palestinian people under the universal principles of human rights (Bakan and Abu-Laban 2009; Barghouti 2011; Lim 2012; Wiles 2013). The call for BDS is aimed at Israeli institutions, not individuals, and recalls the tactics used against apartheid South Africa between the late 1950s and the early 1990s. The BDS call was endorsed by over 170 Palestinian

---

[1] An Arabic word meaning 'tremor' or 'shuddering'.

individuals and organisations, including those side-lined from both state-to-state negotiations and from the armed struggle (e.g. political parties, NGOs, trade unions and community groups) and was seen as representing Palestinians in the OPT, Palestinian refugees and Palestinian citizens of Israel. The boycott was to be maintained until Israel recognised the Palestinian people's inalienable right to self-determination and fully complies with the precepts of international law by:

- Ending its occupation and colonisation of all Arab lands and dismantling the Wall (the so-called Separation and Security Wall which the Israeli government began to build on the West Bank in 2002);
- Recognising the fundamental rights of the Arab-Palestinian citizens of Israel to full equality; and
- Respecting, protecting and promoting the rights of Palestinian refugees to return to their homes and properties as stipulated in United Nations (UN) Resolution 194 (BDS 2005).

This call gained momentum following the Gaza conflict (2008–2009), the humanitarian aid flotilla sent to Gaza in 2010, the construction of the 'Peace Wall' and the continued building of illegal settlements in the OPTs.

Whilst the BDS movement draws widely from the Palestinian community, it is not supported by all Palestinians and is not been seen as the sole authority on boycott and sanctions against the Israeli state. The fractured nature of the Palestinian polity, the role of the Fatah[2]-led Palestinian Authority (PA), seen by some as a 'client state under occupation' and the different interpretations of Islam all inform the struggle for Palestinian rights (recognising Hamas[3] and the Muslim Brotherhood's prominence in Gaza), with the Fatah/Hamas distinction based less on interpretation of religious faith and more on the politics of national liberation and

---

[2] A leading secular Palestinian political party and the largest faction within the confederated multi-party Palestine Liberation Organization (PLO).

[3] A Palestinian Islamic organisation with branches across the Middle East, founded originally as an offshoot of the Egyptian Muslim Brotherhood. It won a decisive majority in the Palestinian parliament in 2006. Israel, the USA, the European Union and other states brand it as a terrorist organisation. Other states, including Russia, China and Iran do not.

'political Islam' (Tamimi 2007). What is significant is how contemporary geo-political relations and the local politics of post-colonial state-building are built upon the consequences of the partition of the region in the twentieth century.

The BDS movement does not have an agreed policy on whether there should be a one- or two-state solution to the conflict, but rather adopts a 'human rights' approach (Barghouti 2011). This absence of a position on a one- or two-state solution can be seen as a weakness in their campaigning. For all groups (pro- and anti-Zionist) the suggestion of a two-state 'solution' acts as fundamental problem, shibboleth and distraction to the conditions experienced daily by many Palestinians. The BDS campaign has repeatedly said it has not taken a position because it is focused on the three central demands (listed above), not the precise form of state or polity given the pace of change in the region. A second reason given for not adopting a clear position is that the BDS campaign needs to appeal to different audiences; adopting a position either way will likely alienate as many (or more) people who otherwise might offer their support. Some of those calling for action against the Israeli state are opposed to the 'two-state solution' because the call for 'the rights of Palestinian refugees to return to their homes' is incompatible with a two-state solution. Such debates go beyond the ending of Israeli apartheid and would require a transition into a post-Israel/Zionist political state; this is clearly beyond the capability of this chapter and, it seems, the international community.

Much of the literature calling for boycott, sanctions and/or disinvestment of the Israeli state makes comparisons between contemporary Israel and apartheid South Africa. This comparison has been extensively discussed (Greenstein 2009, 2011; Hass 2013; Lim 2012; MacLean 2014; Tilley 2012; White 2012, 2013), with much of the literature noting Naomi Klein's comment that (quoted in Loach et al. 2011: 200):

> the question is not 'Is Israel the same as South Africa?' – it is 'Do Israel's actions meet the international definition of what apartheid is?' And if you look at those conditions which include the transfer of people, which include multiple tiers of law, official state segregation, then you see that, yes, it does meet that definition – which is different than saying it is South Africa.

It is important, therefore, not to misread the comparisons made to South African apartheid. The BDS leadership have been careful to recognise that the two situations are not the same and that the South African anti-apartheid campaign should be seen as an inspiration but one that is neither directly comparable nor transferable (Barghouti 2011). That said there are echoes of the role played by trade unions in the early days of the boycott of apartheid South Africa, such as Britain's biggest trade union (Unite) who voted in 2014 to support the BDS movement.

The year before the launch of the BDS movement, Palestinian academics and intellectuals instigated the Palestinian Campaign for the Academic and Cultural Boycott of Israel (Rose and Rose 2008). In 2009, the British Universities and Colleges [trade] Union (UCU) declared its support for an academic boycott of Israeli educational institutions, but not of individuals. Over the following years, academic associations and trade unions across Europe and North America continued to pass motions in support of the Palestinians (for arguments for and against academic involvement, see Fish, Butler, Mearsheimer and Chomsky in Bilgrami and Cole 2015). There has been increased use of the legal system to sanction those who criticise Israel (which some have dubbed from 'warfare to lawfare'), based on claims that individuals are motivated by racism against Jews. Two such legal cases have been seen within the higher education sector, with supporters of BDS winning both cases: the first victory was against UCU with the judge describing the case brought by pro-Israeli supporters as a misuse of the legal process; the second against an Australian academic was dismissed for lack of standing (British Committee for Universities of Palestine 2013; Goldberg 2014). Those who suggest that boycotts/sanctions prevent 'bridge-building' is a somewhat moot claim given that when the Palestine Polytechnic University was closed by Israeli authorities in 2003 and Palestinian academics requested support from Israeli academics, none was forthcoming (Parr 2014).

There have been attempts to bring Israelis and Palestinians together through various cultural, educational and sport projects (including Daniel Barenboim and Edward Said's West-Eastern Divan Orchestra, the 'friendly' match between Israeli and Arab children organised by the Peres Centre for Peace in Jaffa, Israel and Brighton University's 'Football 4 Peace' initiative). However, BDS supporters have called for a boycott of any organisation which receives funding from the Israeli state, including the Batsheva dance company whose performances are repeatedly disrupted by 'Don't Dance with Israeli Apartheid' protestors. Film-makers and musicians have refused to appear in Israel claiming that their appearance might be interpreted as supporting government policy (*The Guardian* 2015a). However, other authors, musicians and film-makers who have continued to perform in Israel claimed in subsequent letter to The Guardian newspaper that cultural boycotts were not acceptable and that,

> Cultural boycotts singling out Israel are divisive and discriminatory, and will not further peace. Open dialogue and interaction promote greater understanding and mutual acceptance, and it is through such understanding and acceptance that movement can be made towards a resolution of the conflict. (The Guardian 2015b)

This is essentially a re-statement of the now-familiar 'bridge-building' argument: a similar position was adopted by those who performed in apartheid South Africa.

The Israeli State's Response to BDS

In response to the growing threat posed by the international boycott of Israeli organisations which were operating in the OPT, the Israeli Foreign Ministry (IFM) began, in 2005, a 'Brand Israel' public relations campaign. Known as 'hasbara' there was a recognition that 'images on television have a much greater and immediate impact on what the public abroad feels about Israel, than the arguments Israel presents' (IFM 2005; see also Schulman 2012). The IFM thought that international audiences were not willing to invest the time or energy necessary to fully understand the complex history of the region and that their collective memory was short. The IFM claimed furthermore that there was 'automatic emotional support for the Third World, anti-globalism, hatred of the United States and other such agendas ... [that] ... favour[ed] the Palestinians over Israel'. In an effort to counter this, there was a need for greater public diplomacy and use of 'high-quality printed materials and multimedia materials [...and ...] a state-of-the-art computerized system.' Israeli diplomatic staff were subsequently required to meet not only with 'government officials, the media, and Jewish communities abroad' but also 'with students and professors, ethnic and religious leaders, as well as key people in the business world and in the arts and sciences' (IFM 2005).

The Israeli state has also turned to existing projects that engaged both Israeli citizens, young Jewish people, philosemites (those with an interest in, or respect for, the Jewish people) and those more broadly supportive of Zionism in order to advance their hasbara agenda (Molad 2012). The Jewish Diaspora thus became seen as a significant ambassadorial resource for the Zionist state, exploiting the well-established social networks linked to Jewish cultural exchange programmes many of which operated on US and European university campuses (Blumenthal 2013). Several well-funded pro-Israel/Zionist groups (e.g. the American Israel Public Affairs Committee, the Anti-Deformation League, The Israel Project, the Jewish Agency, the Jewish National Fund, Shurat HaDin, StandWithUs, United Israel Appeal, We Believe in Israel, World Zionist Organisation, and the Zionist Organisation of America) were all mobilised to counter emerging pro-Palestinian narratives (Fishman 2012). Such organisations condemned any call for sanctions, arguing that culture and sport should be enjoyed without political interference and be used to bridge divides. During the campaign against apartheid South Africa, this 'bridge-building' approach had been a popular tactic amongst cultural workers and athletes although history suggests this was misguided.

The Israeli Foreign Ministry (2005) has stated there was no precise equivalent of the word 'hasbara' in English or in any other language. However, the concept of hasbara has been variously defined as propaganda, whitewashing, explaining, information providing, public diplomacy, re-branding and overseas image-building (IFM 2005; Ravid 2012; Schulman 2011, 2012). The primary cause of Israel's poor international image is seen as a failure of hasbara (i.e. 'to explain') rather than the actions of the state (Molad 2012; Gilboa 2006, 2008). Initially launched as a campaign co-ordinated by the IFM, hasbara is currently organised by the Ministry of Public Diplomacy and Diaspora Affairs (Blumenthal 2013). Other official advocacy bodies which promote Israel overseas include the

IFM, Prime Minister's Office, the IDF Spokesman's Office, the Ministry of Tourism and Culture and the Jewish Agency—each of which contains a dedicated hasbara unit (Blumenthal 2013) and which operates its own Facebook, Twitter, YouTube and Flickr accounts (Molad 2012). As part of their pro-Israel campaign, pro-Zionist supporters have sought to compare BDS activity with the Nazi-organised boycott of Jewish businesses in 1930s Germany. However, others see the hasbara activity as little more than pro-Israeli propaganda and an attempt to whitewash the issue of human rights in Israel.

UEFA's Men's' U-21 Tournament, 2013

In 1974, the IFA was expelled from the Asian Football Confederation, following lobbying by Arab and Muslim FIFA members, who had refused to play Israel. Twenty years later, having played international football only fitfully during the interim, Israel was admitted to full membership of the UEFA, the governing body of European football. The UEFA Men's Under-21 Championship is a biannual tournament that has operated in its current format since 1978. In January 2011, UEFA awarded this event to Israel. Although international sporting events have been held in Israel, including the Paralympic Games in 1968, Europe's basketball 'Final Four' event in 2004, and the quadrennial Maccabiah Games which brings together thousands of Jewish athletes from across the world ('the Jewish Olympics'—see Galily 2009), the UEFA tournament would be a high-profile sporting event and offer a significant opportunity for Israel to promote itself on the international stage.

In the lead-up to the tournament both the German Football Association (FA) and English FA worked with the Israeli Football Association (IFA) to deliver various 'football for all' and anti-racist initiatives (Willenzik 2013; Ynetnews 2013). Staging the tournament was described as an opportunity for Israel to attract media/public interest 'for the right reasons' (Dann 2013). According to one of the German team (centre-forward Pierre Michel Lasogga), it would allow Israel to 'show a different side of itself from the politics and what you see on the television' (quoted in Masters 2013). Various pro-Zionist organisations (such as StandWithUs, the Zionist Federation and the Fair Play Campaign Group) suggested that English supporters should write to their U-21 players to wish them luck and an enjoyable experience in Israel, but not to refer to anything political, including the attempted boycott of this tournament (Lipan 2013).

The tournament was promoted by the IFA, UEFA and the Israeli state as an opportunity to promote mutual respect and tolerance on and off the field and to portray the Israeli state as a democracy (Willenzik 2013). The Israeli U-21 squad, it should be noted, did contain six Israeli-Arab players (Taylor 2013) reflecting the 17–20 % of Israeli-Arabs (the exact number depends on whether or not Palestinians living in East Jerusalem and Druze in the Golan Heights are included). However, it should also be noted that this part of the population (i.e. Palestinians living within pre-1967 Israel borders), occupy a very different position to Palestinians who live as refugees in neighbouring countries, the Palestinian diaspora and those living in the West Bank and Gaza (Shor and Yonay 2010; Sorek 2005, 2007).

In the period immediately prior to UEFA's choice of the venue, fighting between the IDF and Hamas in the Gaza Strip resulted in 139 deaths (6 Israelis and 133 Palestinians) and over 1000 people wounded (240 Israelis/840 Palestinians—see Beaumont 2015). The main sports stadium in Gaza was hit by Israeli missiles who claimed it was a Hamas rocket launch site; four youngsters who were playing football in the stadium at the time were all killed (Ogden 2012).

The announcement of Israel as the host led to a number of individuals and organisations to call upon UEFA to revoke their decision (Russell 2012; Tutu 2013; Warshaw 2013). French/Malian footballer Freddie Kanoute started a petition which was initially signed by a number of high-profile players, although some of these later distanced themselves from the campaign (Ogden 2012). The Palestinian FA, the Palestine Solidarity Campaign and the Muslim Public Affairs Committee (UK) all called for the tournament to be moved. Twenty-four British MPs put forward an 'Early Day Motion 640' titled 'Racism in Football and European Football Tournament in Israel' and detailed how,

That this House congratulates the Football Association for its Kick It Out campaign against racism in football; registers with profound disapproval, however, that the FA is prepared to participate in the European Under-21 football tournament to be played in Israel in June 2013, even though Israel is geographically not in Europe and is a country which has policies of racial apartheid against Palestinians; and therefore calls on the Government to support the Red Card Israeli Apartheid campaign which calls for this European football tournament to be played in Europe (Russell 2012).

The British Guardian (centre-left) newspaper published a letter from Archbishop Desmond Tutu (2013), politicians and others, under the heading 'UEFA insensitivity to Palestinians' plight', which stated their shock at UEFA's decision to award the tournament to Israel and outlined their concern that Israel would use the event to 'whitewash its racist denial of Palestinian rights and its illegal occupation of Palestinian land'. Immediately before the tournament opened, Pro-Palestinian supporters demonstrated outside the UEFA Congress in London and later disrupted a private dinner held for the UEFA delegates and invited guests (Warshaw 2013), with a further demonstration taking place in Amsterdam prior to the UEFA Champions League Final.

When Michel Platini (the head of UEFA) met with pro-Palestine campaigners, at FIFA headquarters, he indicated that he would 'think about' moving the tournament (Abunimah 2013). However, after thinking about it, he stated that UEFA and the IFA were responsible only for football matters and could not be held responsible for the politics of governments. Two years earlier when Platini had met Palestinian football officials, he had implied that he would suspend Israel's membership of UEFA because of the restrictions they were imposing on Palestinian football players, but had changed his mind. Platini had originally indicated that,

> We accepted them [Israel] in Europe and furnished them the conditions for membership and they must respect the letter of the laws and international regulations otherwise there is no justification for them to remain in Europe. Israel must choose between allowing Palestinian sport to continue and prosper or be forced to face the consequences for their behaviour. (Palestinian Information Centre 2010)

The IFA described hosting the UEFA tournament as an 'amazing opportunity' to promote the country and its football, not just to

Europe, but to a global audience (Sanderson and Hart 2011). The event would allow the Israeli national team to participate at the highest level and improve the country's football infrastructure (Sanderson and Hart 2011). Prior to the opening game, UEFA announced that 100,000 tickets had been sold which was celebrated as a marker of the tournaments 'success', given that sales for the 2011 tournament (held in Denmark) had been in the region of 50,000 (Daskal 2013). Average attendance during the 2013 matches was around 11,500 (this was approximately 70 % take-up of the total tournament capacity). A cumulative global audience of over 120 million followed the tournament via television, with 1.9 m visits and 7.6 m page views made to UEFA.com during the tournament; in addition some 100,000 fans linked to UEFA's U-21 Facebook page with 9000 following @UEFAUnder21 on twitter (UEFA 2013).

At the end of the tournament, Platini congratulated the IFA for organising a successful tournament which had seen no political violence or reports of demonstrations linked to the event; for Platini, this vindicated UEFA's decision not to move the tournament. Platini reiterated that Israel had the same right as the other 53 UEFA member nations to bid and host its tournaments and that UEFA's decision had been 'to do what is good for football and not for politics' (Warshaw 2013). The IFA subsequently made a bid for Jerusalem to be one of the 13 venues for the 2020 European Championship (Euro 2020). Their bid generated similar responses to the 2013 U-21 tournament with Palestinian soccer clubs and NGOs writing to Platini and the newspapers to express their concerns (*The Independent* 2014). In their bid document, the IFA accepted that the political situation in Israel was complex. In the end, Jerusalem was not selected because UEFA stated they received insufficient information on the budget and documentation relating to renovation work. However, BDS supporters claimed the bid was unsuccessful because of the recent Gaza conflict ('Operation Protective Edge'—a bombardment and invasion of the Gaza Strip by the IDF). This seven-week conflict in southern Israel/Gaza in 2004 saw UEFA prevent Israeli teams from playing any home matches in the Champions League and Europa League with the matches moved to Cyprus. During this period of open conflict, the International Tennis Federation, also citing safety concerns, relocated Israel's Davis Cup tie against Argentina to the USA (Reynolds 2014).

An American Football tournament and an international swimming championship were also moved out of the country.

'Sports Sanctions Against Israel Are Hypocritical and Antisemitic': Israel's Defenders Come Out Fighting

The standard reaction to the call for action against the Israeli state is to claim it is antisemitic: Why is Israel subject to different standards to those applied to the rest of the world? Why is Israel the target rather than its regional neighbours? Why does the BDS movement fail to highlight the (lack of) rights of Palestinian refugees in neighbouring Arab countries?

Supporters of the Israeli state might accept that Israel is guilty of some human rights violations in its treatment of Palestinians, but, they argue, when compared to many other states, such violations are negligible. Muslim countries in particular are identified as failing to offer basic human rights, especially to women and homosexuals. Across much of Africa and the Middle and Far East, child labour is widespread, whilst female genital mutilation (FGM) is practised in North Africa; Pakistan has a high incidence of female infanticide, whilst homosexuality is punishable by death in Iran. In Saudi Arabia, women are restricted in their ability to drive a car, prevented from travelling without authorisation from a male relative, banned from working in a number of professions and have very limited sporting opportunities—yet there is no comparable groups calling for BDS against these countries. In response to the charge 'Why Israel?', BDS activists might question why Israel appears to be given exceptional treatment in allowing to continually flout international law (including numerous UN resolutions), the exceptional levels of funding it receives from successive US governments, its repeated and disproportionate use of violence against a civilian population and the long-standing nature of the struggle for Palestinian rights.

While recognising that sporting activities can improve relations between communities, it can also be argued that by admitting Israel into European sporting competitions, UEFA was directly condoning the institutionalised racism of the Israeli state. Critics maintain that the Israeli government does not conform to European values and that its apartheid policies should disqualify the IFA from membership of UEFA. Moreover, by excluding Israel from European sporting activities Israel will 'have' to address its apartheid behaviour and racist structures. Given that, Israeli teams currently play in European club and country competitions one might look to UEFA to take a lead. However, UEFA stated that neither it, nor the IFA, is responsible for the politics of the region or the decisions of the Israeli state. Michel Platini's claim that 'I don't do politics, I do football' (Nieuwhof, 2013) is another echo of the arguments used to support the maintenance of sporting links with apartheid South Africa.

When FIFA recognised the state of Israel (in 1929, initially as the Eretz Israel Football Association) and Palestine (in 1998) and allowed them to join FIFA they, perhaps unwittingly, became involved in much wider geo-political game. The IFA now find itself caught between the Israeli state and UEFA/FIFA and within an environment in which everything is viewed, from the Israeli government's standpoint, through the lens of national security. Although the IFA has tried to work with its Palestinian counterpart and with those Arabs who live within Israel's pre-1967 borders, they admit they have no power in matters of national security and the safety of the Israeli population (AP, 2013). The Israeli state has been careful not to be seen to interfere with football, as demonstrated by their lack of direct involvement in the U-21 tournament. Whilst UEFA/FIFA has the ability to exclude the IFA from its competitions, they have not (yet) done so, not least because to do so would lead to accusations of double standards and antisemitism.

Due to the IFA's (and Israel's) desire to integrate itself into European sporting competitions, BDS campaigners have maintained that the standards of human rights sought (if not always achieved) in Europe, should apply to all members. BDS supporters insist that, rather than judging Israel by different standards to those applied to another country, on the contrary, Israel is failing to achieve a minimum level of commitment to the UN's Declaration of Human Rights. Ongoing human rights abuses of Palestinians (be it those living in Israel, the OPT or as refugees) have been well documented by organisations including the International Red Cross, the UN, Amnesty International and Human Rights Watch, as well as Israeli/Palestinian human rights organisations (including Adalah, B'tselem and Breaking the Silence).

Antisemitism and Anti-Zionism: An Important Distinction

There is a frequent tendency to confuse Judaism with Zionism. The former British Conservative MP, Louise Mensch, exemplified this conflation when she sought to express her concern over rising antisemitism. She tweeted that she 'would mute the word "Zionist" from her feed because it was used to attack Jews'. Responding to the question on whether she would mute Theodore Herzl, Mensch replied: 'Who? If he uses Zionist, then yes. Cheap code word for Jew. Anti-Semitism. Not having it' (Media Mole 2014). Herzl, a Hungarian Jewish intellectual, was one of the founders of the modern Zionist movement and author of key texts used by the Zionist movement.

The Israeli state is actively seeking archaeological evidence to bolster its claim of a 2000-year presence in the land (El Haj 2001; Sand 2009, 2012). Archaeological evidence is being used to generate debate on what constitutes 'Jewish'—a religion/race/tribe/people/nation and/or an ethnicity. What is more easily identifiable are the nature and consequences of the Zionist political project which emerged in central Europe in the nineteenth century in response to increasing violent antisemitism. Its aim was the creation of a national homeland for the Jewish diaspora (Avineri 2014; Beinart 2013; Chomsky 1999; Gilbert 2008; Rose 2004; Sand 2009, 2012; Shalaim 2014). Drawing upon socialist ideas, early Zionist groups sought to increase Jewish migration to the region, an activity which peaked in the 1930s and 1940s. The subsequent establishment and maintenance of the State of Israel has led some post-Zionists to question, now that it has fulfilled its original remit, whether a single state, one that is secular and liberal, neither Jewish nor Arab in character, might now be a realistic option (Chan et al. 2002; Morris 2007; Pappe 1999, 2007, 2011, 2014; Segev 2000; Shalaim 2014).

The influence of post-Zionism has been marginal when compared with the Neo-Zionist narrative which sees the land of Israel as the natural home of the Jewish people and one 'promised to them by god' (i.e. a biblical mandate). This right-wing and nationalistic ideology is underpinned by religious faith and an assertion that Jews and Arabs cannot live in peaceful coexistence, thus leaving its supporters to call for the expulsion of all Arabs from the Land of Israel. The two events which fuelled the growth of Neo-Zionism were the 'Six-Day War' (1967) which left Israel in control of the Sinai Peninsula, Gaza Strip, Golan Heights and West Bank, and the election in 1977 of the first government to be formed by Likud, an alliance of right-wing factions, forged four years earlier. This government allowed religious activists to take advantage of the 'spoils of war'—particularly across Judea and Samaria (Zertal and Eldar 2007). The religious settlers have moved so far to the political right that they now describe any Jew (Israeli or Diaspora) who opposes settlement of these lands as an antisemite, and a non-Jew who opposes settlement as a Nazi (see Misgav 2015).

The German journalist, Wilhelm Marr, has been cited as the first to use the term 'anti-Semitism' in 1879 in founding the anti-Jewish organ-isation the League of Anti-Semites (Antisemiten-Liga) (Rattansi 2007). Meer and Noorani (2008) detail how Marr wanted to identify the Jews as a distinct race and identified the term as sounding like a new, scientific concept rather than simple (old-fashioned) religious bigotry. There is a long history of anti-Jewish prejudice for specific historic reasons with Christian Europe grounding much of their prejudice on the claim that Jews were responsible for the death of Christ. Other manifestations of antisemitism were due to specific historic conditions which informed the creation of tropes based on economic, racial, social, cultural, religious and/or ideological stereotypes (Beller 2007; Browning 2005; Dee 2012; Finkelstein 2008; Herf 2014; Julius 2012).

Although used throughout the 1990s, the term 'new' antisemitism dates back to 1921 with the 'roots of the idea deep[er] and ... older than the State of Israel' (Klug 2013: 469). It is suggested that because Israel has increasingly proclaimed itself as a Jewish state, and some Jews have elected to identify with Israel, long-standing antisemitism is being embodied and displaced onto the Jewish state. Hostility towards Zionism (antisemitism 'by proxy,' see Klug 2003) allows criticism of Israel to be interpreted as both anti-Zionist and antisemetic. The Israeli government's 'hasbara' activity has sought to discredit the BDS campaign by adopting this position. It claims that in place of traditional (old) forms of anti-semitism which targeted individual Jews and their communities, (new) attacks on 'the Jews' are being made via the state of Israel. This approach is partially premised on the claim that Europeans have—and continue to hold—a long-standing (for some, pathological) hatred of the Jews (Klug 2013; Tait 2013; Weinthal 2014). The new/disguised form of antisemitism is seen in the actions of those who seek to undermine Israel's legitimacy, deny its right to exist and demonise the Israeli/Jewish state (the '3D' mnemonic of delegitimisation, double standards and demonisation, see Sharansky 2004).

The claim that the individual Jew has been replaced by the state of Israel becomes increasingly relevant when reflecting on the upturns in antisemitism during conflicts in Israel/OPT (Berg 2015). Reports of antisemitism increased in 2014 which some explained as a reaction to the incursion by the IDF into the Gaza Strip ('Operation Protective Edge'). The UK parliament commissioned research on the antisemitic discourse associated with these incidents. Gidley (2014) highlights the importance of the context when assessing if an act is antisemitic or not, taking as an example the displaying a Palestinian flag outside the Israel embassy. This, he suggests, should not be seen as antisemitic, but, if displayed outside a 'kosher deli' or synagogue, it would become an act of antisemitism. Gidley (2014: 5) also suggests that if the slogan 'child murderer' were directed at Israel it would be 'potentially legitimate criticism' but could also provoke sensitivities because of the link to the long-standing Jewish blood libel trope. There are many 'grey areas' with genuine disagreements and misunderstandings arising; however, Gidley (2014: 13) concluded that most of the demonstrations against Israel were not antisemitic, but whilst words and deeds may have no antisemitic intent, they could be seen as 'objectively' antisemitic in their impact. The conclusion of the parliamentary study highlighted the important role the media (both mainstream and Jewish) should play in reporting responsibly and avoiding any hyperbole which might increase tensions. It also stressed the need for 'mainstream Britain to understand and take seriously the insecurity felt by some in the [Jewish] community'.

Klug (2013: 470) states that 'If Zionism is seen as the only alternative to anti-Semitism, then it follows that hostility to Zionism (or to the state of Israel as the expression or fulfilment of Zionism) must be anti-Semitic'. There are many reasons why people might feel hostility towards Zionism or Israel and as Peace (2009: 117) notes, 'the fact that anti-Zionism and antisemitism are not interchangeable does not mean that anti-Zionism or anti-Zionist cannot be anti-Semitic'. As a political expression, opposing Zionism should not necessarily or automatically be seen as racist or antisemitic. The Israeli state and its supporters have deliberately sought to exploit antisemitism by conflating Zionism with Judaism and thus suggesting that any criticism of Zionism is antisemitic. This has proved successful in closing down almost all criticism of the State of Israel. What is evident is how the conflation of anti-Zionism with antisemitism has drawn attention away from other forms of racism (such as growing Islamophobia and anti-Romany/traveller racism in Europe), with researchers beginning to focus on the relationship between the rise in Islamophobia and antisemitism (Meer 2013; Meer and Noorani 2008; Romeyn 2014; Werbner 2013).

There are many countries with poor human rights and to campaign against the presence of Israel in international sport is not to claim that they necessarily lead the table of human rights abusers. It is improbable that any political activist would actually compile a list of human rights abusers, rank them and start campaigns against those at the top; individuals who chose to 'get involved' do so for many different reasons (Barr and Drury 2009; Barrows-Friedman 2014; Verhulst and Walgrave 2009; Woods et al. 2012). Ryall (2006) has discussed the limits of sport's responsibilities to human rights and whilst no country has a 'perfect' human rights record and if 'perfect' criteria were adopted, no country would be allowed to host a sports event. As Dein and Calder (2007) point out, how one might draw a line in a non-hypocritical fashion is a highly complex, philosophical task. However, the inability to draw such a line does not prevent individuals from taking action, just because they cannot solve all the world's injustices at the same time. As Marqusee (2014a, b) has noted, the logic of not 'singling out' a country and campaigning against injustice would leave every struggle isolated from international sport.

Whilst there are multiple reasons why individuals choose to oppose the actions of the Israeli state, it is recognised there are similarly multiple motives for acts of genuine antisemitism; therefore, one needs to look at the motive—is it anti-Zionist or antisemitic? Klug (2013: 478) suggests looking at the 'company they keep' and their history/pattern of (anti-) racist activity; what literature do they produce, who constitutes its membership and what are the wider political connections of their organisation? Poulton and Durell's (2014) discussion of the meaning and contested uses of the term 'Yid' in English football fan culture shows this particular term to have multiple meanings: Tottenham Hotspur supporters, for example, give the word a positive meaning, affirming themselves as 'Tottenham Yids', rather as African American rappers have endorsed the word 'nigger'. Used as both an ethnic epithet and term of endearment they conclude that the cultural context and intent behind the term is crucial.

Supporters of the Israeli state have sought to delegitimise those campaigning for 'better' (i.e. equal) human rights for Palestinians by resorting to ad-hominem attacks. This type of 'moral blackmail'—that is, calling anyone who opposes the current treatment of Palestinians antisemitic or a 'self-hating Jew'—does little to address the fundamental question of addressing Palestinian rights. Ill-considered use of the term antisemitism devalues it and should not be abused by those unwilling to address Israel's treatment of the Palestinian people. In order to maintain credibility for their campaign and refute the conflation of the issues, supporters of BDS will need to uphold a clear and absolute distinction between anti-Zionism and antisemitism, and distinguish between individual Jews and the action of the State of Israel.

Whether ones sees Israel as an ethnocratic, an apartheid or a colonial-settler state, supporters of Palestinians need to consider whether sport is an Israeli 'weak spot'. Some have claimed that sanctions/boycott of Israeli science, technology, education and (non-sporting) culture would have a greater impact as these activities are more highly valued (Banks 2013). Despite the trope that 'Jews don't like sport', a survey by Israel's Central Bureau of Statistics showed the most common leisure activity was sport with Israeli men more engaged than women (55 % compared to 24 %) and watched more sports events (Lior 2012). Excluding Israel from sports governing bodies and tournaments would have a greater symbolic than economic impact on the country. Whilst, as a forum for debate, sport should not be seen as a substitute for formal political institutions, it is nevertheless well placed to contribute to a programme of political, moral and cultural awareness-raising in the best tradition of public diplomacy and soft power. Sport can be used to introduce international audiences to the politics of the region and perhaps lead them to reflect on where they stand on the issue of Israel/Palestine. It has been the cumulative impact of many different activities on international public opinion which has led the Israeli state to invest significantly in its hasbara apparatus. The next question would be whether individuals are willing to engage in debate to better understand the issues and ultimately what, if any, action they chose to take.

Conclusion

Securing the UEFA U-21 tournament was a major accomplishment for the IFA and Israeli state. Unlike those nations which have sought to use sports events to advance its economic growth, the priority for Israel was to present a positive international image and seek greater acceptance, especially within European sporting circles. Hosting the tournament allowed the Israeli state to present an alternative image to the ongoing conflicts in southern Lebanon and the OPT of the West Bank and Gaza Strip. The success of the U-21 tournament claimed by UEFA and the IFA will give the latter greater leverage within the former. Although the size of Israel will limit its ability to host a truly mega-sport event, securing smaller international events will allow it to present itself as a 'normal' country and to gain greater acceptance in the international (sporting) community.

It is not yet clear whether the BDS movement will be able to gain a greater sympathy around the world for the cause of the Palestinian. Zionist ideology has created a racist and genocidal environment for Palestinians. Palestinians living in Israel, the OPT and refugee camps have for the past 60 years experienced discrimination, mass unemployment, subsistence wages, poor living conditions, inadequate health services, sub-standard transport, housing shortages and inferior education. Pro-Palestinian supporters are increasingly accused of antisemitism for protesting against systematic ethnic cleansing, house/village/community demolitions, collective punishments, (illegal) Jewish-only settlements, travel restrictions, an illegal 'Peace Wall', state torture, detention without trial and assassinations (Corrigan 2009).

It remains an open question as to whether boycotts and sanctions are a legitimate means of protest against occupation, racism and colonisation. Those opposed to BDS argue that building bridges and maintaining links are more constructive activities. Whilst the merits and faults of BDS will continue to be debated, the Israeli state is busy providing funds and granting building permits for illegal settlements in the OPT which has led to the disappearance of a two-state solution. As 'peace talks' between two very unequal partners come and go, 'facts' are being established on the ground. It is recognised that in writing about a region experiencing constant political turmoil one can be overtaken by events. Meanwhile, taking no action or ignoring the problem is a political choice and to suggest that 'politics has no role to play' is taking a political stance, with implications, irrespective of whether or not they are intended or desired.

References

Abunimah, A. (2013). Platini "thinking about" whether Under 21 tourno will still go ahead in Israel after protest at UEFA HQ. *Electronic Intifada*. Available at: http://electronicintifada.net/blogs/ali-abunimah/platini-thinking-about-whether-under-21-tourno-will-still-go-ahead-israel-after. Accessed 21 Mar 2015.

Avineri, S. (2014). *Herzl: Theodor Herzl and the foundation of the Jewish State*. London: Weidenfeld & Nicolson.

Bakan, A., & Abu-Laban, Y. (2009). Palestinian resistance and international solidarity: The BDS campaign. *Race & Class, 51*, 29–54.

Banks, I. (2013). Why I'm supporting a cultural boycott of Israel. *The Guardian*. Retrieved from http://www.guardian.co.uk/books/2013/apr/05/iain-banks-cultural-boycott-israel

Barghouti, O. (2011). *BDS: The global struggle for palestinian rights*. Chicago: Haymarket Books.

Barr, D., & Drury, J. (2009). Activist identity as a motivational resource: Dynamics of (Dis)empowerment at the G8 direct actions, Gleneagles, 2005. *Social Movement Studies, 8*(3), 243–260.

Barrows-Friedman, N. (2014). *In our power: U.S. students organize for justice in palestine*. Charlottesville: JustWorld.

BDS. (2005). Palestinian civil society call for BDS. Available at: www.bdsmovement.net/call. Accessed 21 Mar 2015.

Beaumont, P. (2015). UN accuses Israel and Hamas of possible war crimes during 2014 Gaza conflict. *The Guardian*. Retrieved from http://www.theguardian.com/world/2015/jun/22/un-accuses-israel-and-hamas-of-possible-war-crimes-during-2014-gaza-war

Beinart, P. (2013). *The crisis of Zionism*. New York: Picador.

Beller, S. (2007). *Antisemitism: A very short introduction*. Oxford: Open University Press.

Berg, S. (2015). UK anti-Semitism hit record level in 2014, report says. *BBC News*. Available at: http://www.bbc.co.uk/news/uk-31140919. Accessed 21 Mar 2015.

Bilgrami, A., & Cole, J. (Eds.) (2015). *Who's afraid of academic freedom?* Columbia: Columbia University Press.

Blumenthal, M. (2013). Israel cranks up the PR machine. *The Nation*. Retrieved from www.thenation.com/article/israel-cranks-pr-machine/

British Committee for Universities of Palestine. (2013). Newsletter 63, April. Available at: http://www.bricup.org.uk/documents/archive/BRICUP Newsletter63.pdf. Accessed 21 Mar 2015.

Browning, C. (2005). *The origins of the final solution: The evolution of Nazi Jewish Policy September 1939–March 1942*. London: Arrow.

Chan, S., Shapira, A., & Derek, J. (2002). *Israeli historical revisionism: From left to right*. London: Routledge.

Chomsky, N. (1999). *Fateful triangle: The United States, Israel and the Palestinians*. London: Pluto Press.

Corrigan, E. (2009). Is anti-Semitism anti-Semitic? Jewish critics speak. *Middle East Policy, XVI*(4), 146–159.

Dann, U. (2013). Transcending soccer, European Under-21 championship kicks off in Israel. *Haaretz*. Retrieved from http://www.haaretz.com/life/sports/.premium-1.527856

Daskal, O. (2013). Israeli Football – The real story. *Soccer Issue*. Available at: http://www.soccerissue.com/2013/06/06/israeli-football-the-real-story/. Accessed 21 Mar 2015.

Dee, D. (2012). *Jews and British sport*. Manchester: Manchester University Press.

Dein, E., & Calder, G. (2007). Not cricket? Ethics, rhetoric and sporting boycotts. *Journal of Applied Philosophy, 24*(1), 95–109.

El Haj, N. A. (2001). *Facts on the ground: Archaeological practice and territorial self-fashioning in Israeli society*. Chicago: University of Chicago Press.

Finkelstein, N. (2008). *Beyond chutzpah: On the misuse of anti-Semitism and the abuse of history*. London: Verso.

Fishman, J. (2012). The BDS message of anti-Zionism, anti-Semitism, and incitement to discrimination. *Israel Affairs., 18*(3), 412–425.

Galily, Y. (2009). The contribution of the Maccabiah Games to the development of sport in the State of Israel. *Sport in Society: Cultures, Commerce, Media, Politics, 12*(8), 1028–1037.

Gidley, B. (2014). 50 days in the summer: Gaza, political protest and antisemitism in the UK. *Sub-report commissioned to assist the All-Party Parliamentary Inquiry into Antisemitism.* Available at: http://www.antisemitism.org.uk/wp-content/uploads/BenGidley50days inthesummerAPPGAAsubreport-1.pdf. Accessed 21 Mar 2015.

Gilbert, M. (2008). *Israel: A history*. London: Black Swan.

Gilboa, E. (2006) 'Public Diplomacy: The Missing Component in Israel's Foreign Policy.' Israel Affairs, 12 (4): 715–747.

Gilboa, E. (2008) Searching for a Theory of Public Diplomacy. The Annals of the American Academy of Political and Social Science, 616, pp. 55–77.

Goldberg D. (2014). Australian court drops racism case against professor who backs BDS. *Haaretz.* Available at: http://www.haaretz.com/jewish-world/. premium-1.605672. Accessed 21 Mar 2015.

Greenstein, R. (2009). Class, nation, and political organization: The anti-Zionist left in Israel/Palestine. *International Labor and Working-Class History, 75,* 85–108.

Greenstein, R. (2011). Israel/Palestine and the apartheid analogy: Critics, apologists and strategic lessons. In A. Lim (Ed.), *The case for sanctions against Israel* (pp. 149–158). London: Verso.

Hass, A. (2013). What does 'Israeli Apartheid' mean, anyway? *Haaretz.* Available at: http://www.haaretz.com/news/features/.premium-1.562477. Accessed 21 Mar 2015.

Herf, J. (Ed.) (2014). *Anti-Semitism and anti-Zionism in historical perspective: Convergence and divergence*. London: Routledge.

Israeli Foreign Ministry. (2005). What "Hasbara" is really all about. Retrieved from http://mfa.gov.il/MFA/ForeignPolicy/Issues/Pages/What%20Hasbara %20Is%20Really%20All%20About%20-%20May%202005.aspx

Julius, A. (2012). *Trials of the diaspora: A history of anti-Semitism in England*. Oxford: Oxford University Press.

Klug, B. (2003). The collective Jew: Israel and the new antisemitism. *Patterns of Prejudice, 37*(2), 117–138.

Klug, B. (2013). Interrogating 'new anti-Semitism'. *Ethnic and Racial Studies, 36*(3), 468–482.

Lim, A. (Ed.) (2012). *The case for sanctions against Israel.* London: Verso.

Lior, G. (2012). What do Israelis do in their free time? *YNetNews.* Retrieved from http://www.ynetnews.com/articles/0,7340,L-4318880,00.html

Lipan, J. (2013). London protest planned against Israel hosting Uefa Under-21s. *Jewish Chronicle.* Retrieved from http://www.thejc.com/news/uk-news/107852/london-protest-planned-against-israel-hosting-uefa-under-21s

Loach, K., O'Brien, R., & Laverty, P. (2011). Looking for Eric, Melbourne Festival and the cultural boycott. In A. Lim (Ed.), *The case for sanctions against Israel* (pp. 199–202). London: Verso.

MacLean, M. (2014). Revisiting (and revising?) sports boycotts: From rugby against South Africa to soccer in Israel. *International Journal of the History of Sport, 31*(15), 1832–1851.

Marqusee, M. (2014a). Further on BDS and 'singling out' Israel. *Links: International Journal of Socialist Renewal.* Available at: http://links.org.au/node/3672. Accessed 21 Mar 2015.

Marqusee, M. (2014b). "If not now, when?" On BDS and 'singling out' Israel. *MikeMarqusee.com.* Available at: http://www.mikemarqusee.com/?p=1489. Accessed 21 Mar 2015.

Masters, J. (2013). How German football is embracing Israel. *CNN.* Retrieved from http://edition.cnn.com/2013/06/13/sport/football/football-germany-holocaust-israel

Media Mole. (2014). Louise Mensch accidentally calls the father of modern Zionism an anti-Semite. *New Statesman.* Available at: http://www.newstatesman.com/media-mole/2014/08/louise-mensch-accidentally-calls-father-modern-zionism-anti-semite. Accessed 21 Mar 2015.

Meer, N. (2013) 'Racialization and religion: Race, culture and difference in the study of anti-Semitism and Islamophobia', *Ethnic and Racial Studies,* 36 (3): 385–398.

Meer, N., & Noorani, T. (2008). A sociological comparison of anti-Semitism and anti-Muslim sentiment in Britain. *Sociological Review, 56*(2), 195–219.

Misgav U. (2015). Settler group's new video: Money-grabbing leftists playing into European Nazis' hands. *Haaretz.* Available at: http://www.haaretz.com/news/national/.premium-1.642526. Accessed 21 Mar 2015.

Molad. (2012). Israeli Hasbara: Myths and facts. A report on the Israeli hasbara apparatus 2012. Available at: http://www.molad.org/images/upload/researches/79983052033642.pdf. Accessed 21 Mar 2015.

Morris, B. (2007). *Making Israel*. Ann Arbor: University of Michigan Press.

Nieuwhof, A. (2013). Move football tournament out of Israel, says growing campaign. *The Electronic Intifada*. Available at: http://electronicintifada.net/blogs/adri-nieuwhof/move-football-tournament-out-israel-says-growing-campaign. Accessed 21 Mar 2015.

Ogden, M. (2012). Premier League players call on Uefa to remove Israel as European U-21 hosts. *Daily Telegraph*. Available at: http://www.telegraph.co.uk/sport/football/news/9715577/Premier-League-players-call-on-Uefa-to-remove-Israel-as-European-U-21-hosts.html. Accessed 21 Mar 2015.

Palestinian Information Centre. (2010). Platini threatens to annul Israel's membership of UEFA. Available at: www.middleeastmonitor.com/news/middle-east/1551-platini-threatens-to-annul-israeli-membership-of-uefa#sthash.E7BR1ojc.PYogXbbs.dpuf. Accessed 21 Mar 2015.

Pappe, I. (1999). *The Israel/Palestine question: A reader*. London: Routledge.

Pappe, I. (2007). *The ethnic cleansing of Palestine*. Oxford: Oneworld Publications.

Pappe, I. (2011). *The forgotten Palestinians: A history of the Palestinians in Israel*. New Haven: Yale University Press.

Pappe, I. (2014). *The idea of Israel: A history of power and knowledge*. London: Verso.

Parr C. (2014, February 6). Opponents of Israel boycott marshal forces across the US. *Times Higher Education*, pp. 18–19.

Peace, T. (2009). UN antisemitism nouveau? The debate about a 'new antisemitism' in France. *Patterns of Prejudice, 43*(2), 103–121.

Poulton, E., & Durell, O. (2014). Uses and meanings of 'Yid' in English football fandom: A case study of Tottenham Hotspur Football Club. *International Review for the Sociology of Sport*. Epub ahead of print 16 October. doi:10.1177/1012690214554844.

Rattansi, A. (2007). *Racism: A very short introduction*. Oxford: Oxford University Press.

Ravid, B. (2012). Think tank: Israel's poor international image not the fault of failed hasbara. *Haaretz*. Available at: http://www.haaretz.com/news/diplomacy-defense/think-tank-israel-s-poor-international-image-not-the-fault-of-failed-hasbara.premium-1.490718. Accessed 21 Mar 2015.

Reynolds, T. (2014). Israel's tennis team forced to play 'home-court' Davis Cup qualifier in Florida. *Haaretz*. Available at: http://www.haaretz.com/life/sports/1.615311. Accessed 21 Mar 2015.

Romeyn, E. (2014). Anti-Semitism and Islamophobia: Spectropolitics and immigration. *Theory, Culture & Society, 31*(6), 77–101.

Rose, J. (2004). *The myths of Zionism*. London: Pluto Press.

Rose, H., & Rose, S. (2008). Israel, Europe and the academic boycott. *Race & Class, 50*, 1–20.

Russell, B. (2012). Racism in football and European football tournament in Israel. Early day motion 640. Available at: http://www.parliament.uk/edm/2012-13/640. Accessed 21 Mar 2015.

Ryall, E. (2006). Cricket, politics, and moral responsibility: Where do the boundaries lie? In L. Howe, H. Sheridan, & K. Thompson (Eds.), *Sport, culture and society: Philosophical reflections* (pp. 8–45). Boston: Meyer & Meyer.

Sand, S. (2009). *The invention of the Jewish people*. London: Verso.

Sand, S. (2012). *The invention of the land of Israel: From Holy Land to homeland*. London: Verso.

Sanderson, P., & Hart, P. (2011). Israel already planning for 'amazing opportunity.' *UEFA.com*. Available at: http://www.uefa.com/memberassociations/association=isr/news/ newsid=1646923.html. Accessed 21 Mar 2015.

Schulman, S. (2011). Israel and 'Pinkwashing.' *The New York Times*. Available at: http://www.nytimes.com/2011/11/23/opinion/pinkwashing-and-israels-use-of-gays-as-a-messaging-tool.html?_r=2and. Accessed 21 Mar 2015.

Schulman, S. (2012). *Israel/Palestine and the Queer International*. Durham: Duke University Press.

Segev, T. (2000). *The seventh million: The Israelis and the Holocaust* (trans: Watzman, H.). New York: Owl Books.

Shalaim, A. (2014). *The iron wall: Israel and the Arab world*. London: Penguin.

Sharansky, N. (2004). 3D test of anti-Semitism: Demonization, double standards, delegitimization. *Jewish Political Studies Review, 16*, 3–4.

Shor, E., & Yonay, Y. (2010). 'Play and shut up': The silencing of Palestinian athletes in Israeli media. *Ethnic and Racial Studies, 34*(2), 229–247.

Sorek, T. (2005). Between football and martyrdom: The bi-focal localism of an Arab-Palestinian town in Israel. *British Journal of Sociology, 56*(4), 635–661.

Sorek, T. (2007). *Arab soccer in a Jewish state: The integrative enclave*. New York: Cambridge University Press.

Tait, R. (2013). British guile of disguised anti-Semitism, says Israeli minister. *Telegraph*. Retrieved from http://www.telegraph.co.uk/news/worldnews/middleeast/israel/10074775/British-guilty-of-disguised-anti-Semitism-says-Israeli-minister.html

Tamimi, A. (2007). A political conflict. *The Guardian*. Available at http://www.theguardian.com/commentisfree/2007/jul/20/apoliticalconflict.   Retrieved 10 Apr 2015.

Taylor, L (2013). England enter a politically loaded European under-21 championship. *The Guardian*. Available at: http://www.guardian.co.uk/football/2013/jun/03/england-under-21-championship. Accessed 21 Mar 2015.

The Guardian. (2015a). Letter: Over 100 artists announce a cultural boycott of Israel. *The Guardian*. Available at: http://www.theguardian.com/world/2015/feb/13/cultural-boycott-israel-starts-tomorrow. Accessed 21 Mar 2015.

The Guardian. (2015b). Israel needs cultural bridges, not boycotts – Letter from JK Rowling, Simon Schama and others. *The Guardian*. Available at http://www.theguardian.com/world/2015/oct/22/israel-needs-cultural-bridges-not-boycotts-letter-from-jk-rowling-simon-schama-and-others. Accessed 1 Dec 2015.

The Independent. (2014). Letters: Leave Jerusalem off Euro 2020 host city list. Available at: http://www.independent.co.uk/voices/letters/letters-leave-jerusalem-off-euro-2020-host-city-list-9301342.html. Accessed 21 Mar 2015.

Tilley, V. (Ed.) (2012). *Beyond occupation: Apartheid, colonialism and international law in the occupied Palestinian territories*. London: Pluto Press.

Tutu, D. (2013). Letters: UEFA insensitivity to Palestinians' plight. *The Guardian*. Available at: http://www.theguardian.com/football/2013/may/27/uefa-insensitivity-to-palestinians-plight. Accessed 21 Mar 2015.

UEFA. (2013). Tournament review. Under21 Championship. Israel 2013. *UEFA*. Available at: http://www.uefa.com/under21/news/newsid=2006357.html. Accessed 21 Mar 2015.

Verhulst, J., & Walgrave, S. (2009). The first time is the hardest? A cross-national and cross-issue comparison of first-time protest participants. *Political Behavior, 31*(3), 455–484.

Warshaw, A. (2013). Platini praises Israel and stands firm on their right to host. *Inside World Football*. Available at: http://www.insideworldfootball.com/world-football/europe/12/46-platini-praises-israel-and-stands-firm-on their-right-to-host. Accessed 21 Mar 2015.

Weinthal, B. (2014). Why Europe blames Israel for the Holocaust: Post-1945 anti-Semitism. *The Jerusalem Post*. Available at: http://www.jpost.com/Jewish-World/Jewish-Features/Why-Europe-blames-Israel-for-the-Holocaust-Post-1945-anti-Semitism-339571. Accessed 21 Mar 2015.

Werbner, P. (2013). Folk devils and racist imaginaries in a global prism: Islamophobia and anti-Semitism in the twenty-first century. *Ethnic and Racial Studies, 36*(3), 450–467 (Special Issue: Racialization and Religion: Race, culture and difference in the study of Antisemitism and Islamophobia).

White, B. (2012). *Palestinians in Israel: Segregation, discrimination and democracy.* London: Pluto Press.

White, B. (2013). *Israeli Apartheid: A beginners guide* (2nd ed.). London: Pluto Press.

Wiles, R. (Ed.) (2013). *Generation Palestine: Voices from the boycott, divestment and sanctions movement.* London: Pluto Press.

Willenzik, J. (2013). Soccer Championship gives Israel chance to show off diversity and multiculturalism. Retrieved from http://www.aijac.org.au/news/article/soccer-championship-gives-israel-chance-to-show-

Woods, M., Anderson, J., Guilbert, S., et al. (2012). 'The country(side) is angry': Emotion and explanation in protest mobilization. *Social & Cultural Geography, 13*(6), 567–585.

Ynetnews. (2013). Israel, England teams unite against racism. Retrieved from http://www.ynetnews.com/articles/0,7340,L-4390214,00.html

Zertal, I., & Eldar, A. (2007). *Lords of the land: The war over Israel's settlements in the occupied territories, 1967–2007.* New York: Nation Books.

Chicago 2016 Versus Rio 2016: Olympic 'Winners' and 'Losers'

Kostas Zervas

Critical Olympic Studies

The study of the political and economic aspects of the Olympic Games has increased in the recent years (see Girginov 2010; Lenskyj and Wagg 2012; Giulianotti et al. 2015; Boykoff 2014). These new and markedly critical Olympic studies have contributed significantly to our understanding of the Games (and mega sporting events generally) and their impact on society—particularly on the host city and nation. The study of anti-Olympic campaigns holds a key role in this wider academic research. One of the main reasons that academics/researchers study the anti-Olympic groups and movements around the world is indirectly to investigate the IOC and the Olympic Games themselves. The bulk of the 'anti-Olympic' research projects that have been carried out in this area (notably by the Canadian academics and activists Helen Lenskyj and Christopher Shaw—see

K. Zervas (✉)
Sport, Health and Nutrition, Leeds Trinity University, Brownberrie Lane, Horsforth, Leeds LS18 5HD, UK

© The Editor(s) (if applicable) and The Author(s) 2016   **209**
J. Dart, S. Wagg (eds.), *Sport, Protest and Globalisation*,
DOI 10.1057/978-1-137-46492-7_10

Lenskyj 2000, 2002, 2008; Shaw 2008) have opened up the discussion about previously neglected aspects of the Games and have shed light on their impact on a huge swathe of the host society. Yet, these findings have rarely been appreciated, or even acknowledged, by the IOC and the local Olympic Games organising committees (LOCOGs). The growing opposition towards the Olympics and, lately, the diminished interest shown by cities in hosting the Games (both summer and winter) demonstrate that the same problems and controversies reappear, and expand to include wider areas of economic and social life. It is evident, after several decades of research into the Games, that the constitution of the IOC, the organisational framework of the Olympics and the strategic planning of the hosting cities/nations, routine promises and assessments notwithstanding, do not result in beneficial social, environmental or economic impacts. On the contrary, as we may conclude from previous research, they may contribute to the widening of economic inequality, facilitate corruption and, thus and bearing in mind the scandals engulfing world football's governing body Fédération Internationale de Football Association (FIFA) in 2015 (Jennings 2015), bring elite sport further into disrepute.

The negative impacts of hosting the Games—gentrification, democratic lack, public money spending—have been well documented in recent years (see Lenskyj 2012; Shin and Li 2013; Giulianotti et al. 2015). In many of these cases, the data that prove these controversial aspects of Olympics have been compiled by anti-Olympic movements or local monitoring groups themselves, which are usually formed to oppose bids and monitor Olympic preparation. Whether the researcher takes part, as an activist, or observes and records the story, this aspect of Olympic research has been proven to be quite significant in our understanding of the Olympic Games and the hosting of their quadrennial events. Through their campaigns these activists, who constitute part of the anti-Olympic groups/movement, have brought to light key questions on fundamental aspects of the life in our societies in relation to the choices that we make. Behind the main arguments posed in every campaign, we can identify key issues in respect of deep-seated economic, political, environmental and social choices. These issues, highlighted by individuals and groups in bidding and hosting cities, usually remain unnoticed, and are often overshadowed by the huge publicity campaigns that attend the Olympic 'super-spectacle'. It is only until recently that, in some instances, these issues have gained a wider audience and became part of the wider public discussion.

The last three Olympic Games up until the time of writing (Spring 2016)—Athens 2004, Beijing 2008 and London 2012—are important to consider for several reasons. In all three cases the countries involved (their governments and mainstream media especially) embraced the cause and actively engaged in promoting, preparing and organising the Games. Each host city had its own agenda. In the first case, Greece was (at that time) an economically growing country, part of the EU and the Eurozone, aiming to get the rewards of its fiscal adaptation to the Euro, gain leverage and perhaps some self-esteem from hosting the Games in what many Greeks still regarded as their rightful home. Greece was the site of early Olympics, and thus the repository of enduring Olympic mythology (see, for example, Golden, 2012) with Athens having staged the first modern Olympics in 1896. In 2008 China was an emerging world super-power aiming to demonstrate its prowess by hosting the Games in Beijing, and sought to renounce, once and for all, its communist origins and demonstrate its right to be a leading player on the world economic stage. Great Britain, a traditional power both in political and sporting terms, was aiming to re-assert its political and sporting position (Grix and Houlihan 2014), and inspire national pride, by bringing the Games to London in 2012 for a third time. It is history that will decide upon the long-term success of their macro-political ventures, but we can already notice serious concerns over the direct, or short-term, effect which the Olympics had on the lives of local citizens. Notably, most of the data on the consequences of these three Olympiads come from 'anti' groups and individuals, as there was never a meaningful political debate, or a wider discussion about the role of Olympics in these countries. In summary, monitor and activist groups in Athens, Beijing and London all reported three (among several other) key issues, which had negative effects, directly or indirectly, on the lives and future of local habitants: *gentrification, democratic lack* and *the questionable dispersal of public funds* (see Zervas 2012; Shin and Li 2013; Giulianotti et al. 2015).

The issue of 'Olympic regeneration', or simply gentrification (broadly speaking, the reconstruction of urban neighbourhoods, which results in increased property values and the displacing of lower-income families and small businesses), with all its negative consequences (evictions, rise in rents, the privatisation of public space) has been evident in Olympic cities since the 1980s and, more specifically, the Los Angeles Games of 1984 and were highlighted in Helen Lenskyj's (2002) book on the Sydney Olympics of 2000. There is a common practice on the part of local Olympic organisers and their associates, referred to by Giulianotti et al. (2015) as 'festival capitalism', whereby sectors of cities are privatised, commercialised and gentrified, through hosting mega events and, thus, as part of major investment and regeneration projects. This is part of what Brenner and Theodore (2002: 349) have identified as 'the strategic role of cities in the contemporary remaking of political-economic space'.

The precise results of this re-making are usually ambiguous, as we rarely find out what happens to the pre-existing inhabitants and businesses that are affected by the decision to host the Games. In the case of Athens 2004, this issue was addressed, by the Greek anti-Olympic pressure group Anti-2004, with the further expansion of the already densely populated urban area which was a long-standing issue, with long-term effects on the lives of all Athenians. The economic and environmental consequences of urbanisation threatened specifically the medium- and low-income inhabitants of the areas of the city being developed, who did not have proper healthcare, housing and access to green spaces. In particular, the previously industrial Maroussi district, where the Athens Olympic Sports Complex was situated, was developed, via the Olympics, into a space characterised by 'public and private sector offices, international and national hi-tech firm headquarters, banking, private hospitals and insurance firm headquarters, a "festival market place" (cinemas, amusement parks, cafes, restaurants, large scale shopping malls), up-market housing and gated communities' (Maloutas et al. 2009: 19).

In Beijing, the so-called environmental improvement projects targeted the *chengzhongcun*, or villages-in-the-city, inhabited by migrant tenants who far outnumbered local Beijing permanent residents (Shin and Li 2013). According to the same study, approximately 350,000 residents were evicted as a result of the Olympic regeneration projects. In London, subsequently, the consequences of winning the right to host the 2012 Olympics would directly affect those living in the East End, the area in which the most disadvantaged citizens were located. The building preparation resulted in evictions, relocations and the destruction of public green space. The future of the local residents and of the numerous small businesses in East London was drastically disturbed, with the dramatic increase in the cost of leaving—informal estimates placed annual household churn rate of pre-Olympic London at around 30 % (Giulianotti et al. 2015). As with Athens, this practice of displacement was part of a process already well in train. Writing well ahead of the Games, writer Anna Minton argued that the expropriation and forced migration of the people living in London's East End Dockland area had been taking place since the 1980s, bringing sky scrapers, office blocks, high-end flats and other such spaces, guarded by 24-hour private security (see Minton 2009: 9–14). A review of Minton's book carried the evocative headline 'They stole our streets and nobody noticed' (Behr 2009). Minton's book was re-issued in 2012 with an added chapter, in which she noted that the Olympic Park of 2012 was now a 'private new town, outside of local authority control (Minton 2012: xi–xxviii). The above examples lead to the same conclusion made by most previous researches concerning Olympic cities (see Roche 2002; Lenskyj 2000, 2008; Horne and Manzenreiter 2006).

The second issue raised by activists in the last three Olympics is the lack of transparency and openness in the decision-making processes of organisation and planning. The undemocratic, and sometimes authoritarian, behaviour by or on behalf of the Organising Committees of the Olympic Games (OCOGs) is something that has been witnessed and recorded throughout the Olympic history (Hoberman 1986; Simson and Jennings 1992; Lenskyj 2000). The stories of authoritarianism vary in each case, but they all have a common denominator: *the IOC requirements*. The justification in each instance is that the Olympic organisers and bidding committees are 'forced' to operate in strict privacy, and sometimes above the law, in order to comply with the IOC requirements and tight schedules. But is it permissible for a sporting event that is supposed to bring joy to people and promote unity and peace across international community to require those people to forgo some of their civil rights? And how can the Olympics 'legitimise' a committee to override established rules and legislation? The answer to both questions is the same: the modern Olympics are constituted on an authoritarian basis by the IOC. Its rules and requirements, set by the Olympic Charter, require and receive the suspension of state laws once the Olympic contract is signed. So, in that context, the OCOGs are 'legitimised'—and bound—to operate outside democratic boundaries. The local organising committees in Athens and Beijing especially, and in London, acted accordingly. In Athens, the state's Supreme Court for civil liberties—'Symvoulio tis Epikratias'—silently ceased dealing with Olympic-related appeals against Olympic decisions. In Beijing several civil codes—and penalties—were introduced to implement 'civilised' behaviour (Shin and Li 2013). These codes aimed to prohibit 'uncultured' behaviour, such as the use of rickshaws, in the city centre and close to Olympic venues. After the Games these restrictions were relaxed. In London, under the terms of the London Olympic Games and Paralympic Games Act of 2006 effectively banned on advertising and street trading (trading on a highway or other public place), and other activities, such as anti-Olympic protest, in the vicinity of Olympic events (James and Osborn 2011).

The third main argument, posed by anti-Olympic campaigners over the recent Olympiads, is the issue of overspending and especially, the allocation of public funding towards 'bread and circuses' rather in social welfare. As figures show, the cost of hosting the Olympic Games has risen dramatically over the last decades. It also increases exponentially between the time the bid was framed and estimates were presented and the time when bills came to be paid (London 2012 is a good example: see Weaver 2006). It has become a common practice—in the same context of 'festival capitalism' (Giulianotti et al. 2015)—that governments facilitate and subsidise public space and amenities in order to attract private investments into the hosting cities. In the case of Athens 2004, the campaigners claimed, and in some cases provided evidence, that some of the public investments for the 2004 Games were not made for Olympics per se, but in order to increase the cash flow to the private sector, resulting in so-called white elephants (for example, taekwondo arena) and purposeless pharaonic complexes (the Olympic village) (See Smith 2012; Bloor 2014). Similarly, in Beijing, the Olympics played a big part in the wider project of *chengzhenhua* (townification). Through this process and via staging the Olympics, Chinese authorities facilitated and subsidised the urbanisation of wider Beijing, which subsequently led to the so-called Ghost cities of China (see Shepard 2015). In London, the Lea Valley, an area of waterways, gardens and untended open land in London's East End—the most deprived part of the city—was offered to construction companies for large malls and high-spec apartments. In all these cases the Olympics acted as the catalyst to attract private investments and subsequently growth/redevelopment, by transferring public money and space. The hypothesis was that the fruits of these investments would outmatch the initial sacrifice. Sadly for the inhabitants, this hypothesis seems inaccurate in the case of these three cities discussed here. The role of the Olympics in creating a channel for diverting public money into the private sector has been acknowledged in several studies of Olympics. Shaw (2008) has explained the links between the 'Vancouver Organising Committee for the 2010 Olympic and Paralympic Winter Games' (VANOC) and the private sector and noted 'the obvious similarities between the corporate behaviours of wholly discredited multinational corporations, such as Enron and WorldCom', and its 'corporate governance' (2008: 93).

Whether the regeneration/gentrification and privatisation projects are successful, or not, the impact on the working-class citizens is always negative. Contrarily, the profits for the high-end economic elite are guaranteed and increase with each Olympiad. While the Olympic sponsors, partners and affiliated contractors make record profits, the hosting countries and especially the citizens have to deal with the increasing cost of hosting, subsidising and privatising that comes with the Games. An interesting aspect of Olympics which probably is underestimated in the existing research is the fact that they contribute significantly in widening the gap between the rich and the poor in hosting cities. Beyond the costs and impacts of hosting, the Olympics systematically fuel economic inequality, probably the biggest social problem worldwide. A recent study from Brookings Institution on economic inequality in the USA, named Atlanta (the last US city to host the Olympics) as the city with the biggest gap between the rich and the poor (Brookings 2013). Furthermore, according to the Organisation for Economic Co-operation and Development's inequality index, the top 10 countries with the least economic inequality—Scandinavia (3), Benelux (3), Slovenia, Slovakia, Iceland and Czech Republic—have neither hosted nor bid for the Olympic Games for more than half a century. On the other hand, among the countries with the highest economic inequality are the USA, Greece, Spain, Japan, Turkey, South Korea and the UK (for the UK, see Proud 2015 and *Observer* 2015), the countries that have hosted and/or shown the greatest interest in hosting the Olympics. Of course, I do not suggest that a sporting event that is hosted every four years is responsible for the economic inequality in these countries. The point is that, usually, the choice of bidding for the Olympics, as a means of economic development and 'regeneration', is clearly associated with political and economic models that lead to further inequality.

This chapter will now focus on 'No Games Chicago (NGC)', a group of individuals that stood politically against the will (and the millions of dollars) of the City of Chicago, US president Barack Obama and some of the world's most iconic athletes, including Michael Jordan and Ian Thorpe. In their 'journey' 'NGC' managed to successfully challenge the arguments posed by Chicago Olympic Bidding Committee and spark a debate in the city of Chicago over the future of its citizens. This discussion is a part of an ethnographic research project conducted in 2009. Since Chicago did not get the Games, this particular part of the project came to an end. Now, on the eve of Rio 2016 (instead of Chicago 2016), the significance of the 'NGC' campaign is even more relevant. Examining their role I believe contributes to our further understanding both of the Olympics and of the importance of agency in relation to the causal effects of our fundamental choices. There has been massive opposition in Brazil to the hosting of Rio 2016, which follows the protests against the amount of money expended on the FIFA World Cup tournament in 2014 (see Christopher Gaffney's chapter in this book); it is yet unknown how history will decide on the 'winners' and the 'losers' of these Games.

No Games Chicago

On 2 October 2009, Copenhagen's 'Bella Centre' conference hall hosted the IOC's 121st session which would decide the host city for the 2016 Summer Olympic Games. The final four candidate cities were Rio de

Janeiro, Chicago, Madrid and Tokyo. The decision to award the Games to Rio de Janeiro did not come as a surprise, given it was favourite amongst the bookies. A big surprise, though, was Chicago's elimination at the first round, as the city's bid was also strongly favoured.

Chicago's plan to host the 2016 Olympiad was premised on the idea of using most of the existing city's sporting facilities, venues and parks, along with the creation of new ones, all in close proximity to each other, thus creating a highly concentrated cluster of event locations close to the city centre represented by the slogan 'A Games in the heart of the city' and which was used in the Chicago 2016 official promotional brochures. By that, the organisers were promising easily accessible venues for the spectators and the athletes. The 2016 Chicago bid was launched on the initiative of the city's mayor, Richard Daley, scion of the Chicago's leading political family and the son of a previous mayor of the city, who believed that, apart from the obvious benefits, the Olympics could bring major investments in Chicago. This increased investment would contribute to improving the city's poor economic situation and help to counterbalance its budget's deficit. According to the official 'bid book', the Chicago Games would create a profit of around $500 million, and they would also give a long-term boost to the city's economy, through investments and tourist development. More specifically, according to Chicago 2016s bid book, most of the Olympic facilities would be located close to the city's waterfront, using existing buildings and parks and through a major regeneration project that would improve some of the city's most stagnated parts. In addition to the use of existing facilities, a temporary Olympic stadium would be built in Washington Park, on city's Southside, and a cycling track built in Douglas Park, in the western part of the city. The questionable choice of utilising city's parks for building the required Olympic venues was justified by Olympic campaigners as a cost-effective and less bureaucratic way to obtain the required permits, as parks were under the authority of the city's Public Park District.

Apart from these facilities, the organisers proclaimed that the Olympics in Chicago would utilise 'the Olympic Movement's power to unite all humanity' (a now routine Olympic claim—see Giardina et al. 2012) and would 'help America reach out to build and renew bridges of friendship with the world'. Additionally, the Chicago 2016 Olympics would create a legacy, which would 'inspire young people to reach for a better life' (Chicago2016). All these proclamations were, of course, little different from those used by the other three bidding candidates: Rio 2016 was proposing that the Games would be held within the city's boundaries, would create short- and long-term financial profits and would inspire Brazilian youth (Rio2016 2009). The Madrid 2016 bid was also a plan for an inner-city Olympics that would result in financial benefits and would inspire the Spanish nation's youth (Madrid2016 2009) and, unsurprisingly, Tokyo 2016 was promising 'the most compact and efficient Olympics ever', 'in the heart of the city' (Tokyo2016 2009), which would create profit and economic benefits and would inspire the youth of Japan. All these puffed-up, pompous declarations were the same, or very similar, to those used by London 2012, Beijing 2008, and Athens 2004.

So, why was Chicago's bid recognised as a strong favourite if all other bids were promising very similar things? Beyond the standard bid books, documents and Olympic plans, Chicago's bid was supported by some of the most eminent American personalities with global profile. Amongst the most notable were Oprah Winfrey, the African American media personality and businesswoman, voted many times as one of the most influential persons of the planet by the Time magazine; basketball player Michael Jordan, one of the most recognised sports personalities in the world; swimmer Michael Phelps, already, at 24, the holder of 18 Olympic gold medals; and lastly and most importantly, the US president Barack Obama. In the last case, Obama's support for his hometown's bid was not just the typical backing offered by a US president. Obama's global impact, as a personality and a political figure, was probably Chicago's 2016 'joker card', by which the organisers were hoping to cover the bid's weaknesses, either in infrastructure (which favoured Madrid), or in culture (which, given the attractiveness of tourist spots such as Copacabana Beach, favoured Rio).

Back in Copenhagen, five days before the IOC's decision, the delegations started arriving at the Danish capital. At that point, all four candidate cities were competing in a head-to-head race, with Chicago and Rio having a slight advantage according to experts and booking agencies (according to the bookmakers Chicago was a clear favourite with 8/11, followed by Rio 7/4, Madrid and Tokyo on 12/1). Outside of

Chicago itself, the press (global, US and European) was generally taking an uncommitted stance. No one at that time was willing to risk a prediction. BBC's, Matt Slater (2009), was noting:

> ...The IOC's heart calling for Copacabana but its head is worrying about crime and passing up the riches on offer in Chicago, a confusion that might just let in Madrid or Tokyo. Could that decision be made a little easier by the presence in Copenhagen of the world's most powerful man? Can Barack, Chicago's top trump, risk so much political capital on anything other than a slam dunk? (Slater 2009)

And then, four days before the IOC's decision, on 28 September, 'the world's most powerful man' decided to go to Copenhagen to speak in front of the IOC on behalf of the Chicago 2016 bid. The news of Obama's visit to Copenhagen changed the ambience drastically in favour of Chicago. Even the city's appearance changed: the shops were selling shirts labelled 'Copenhagen loves Obama' and American flags; people were talking about his arrival, where he would go, what he would say. Most of the global press covering the announcement to be made in Copenhagen estimated that this last-minute call from the US president was a decisive turn towards a Chicago win. *The Guardian*'s correspondent Owen Gibson (2009) admitted a day before the final decision: '...Obama's arrival appears to have given momentum back to Chicago'. And on the day of the decision Gibson noted: 'Obama's late, perfectly timed decision to attend the vote has robbed Rio's attempt to make Olympic history by bringing the Games to South America for the first time of crucial momentum' (Gibson 2009a, 2009b).

On the afternoon of 2 October 2009, in the central square of Copenhagen, the scene was set for the 'Olympic countdown'. The presenter announced the first two cities that had failed to receive sufficient votes to go forward in the competition to host the 2016 Olympics: Tokyo and Chicago. In fact, Chicago was the first city to be knocked out, having got the least votes in the first round (18 votes for Chicago, 22 Tokyo, 26 Rio and 28 Madrid; source: IOC). The Games were, eventually, awarded to Rio de Janeiro. But what happened to the favourite? How did Chicago failed even to pass the first round? Why were the predictions so wrong?

While the members of Chicago 2016 bid committee were preparing their presentation to the IOC and waiting for President Obama, four Chicagoans were roaming the streets of Copenhagen in order to deliver a different message. Tom, Martin, Rhonda and Jason were the representatives of a group called 'NGC'. This group was undertaking their final activities in a campaigning coalition that was formed by several citizens of Chicago who had come together to oppose the city's 2016 Olympic bid. 'NGC's' campaign was launched in January 2009 at a public forum about the Chicago 2016 bid, and since then, they had actively opposed the idea of bringing the Olympics to Chicago. They organised a series of events (including protests and public meetings) across the city of Chicago and actively attempted to start a dialogue with Chicago 2016 and the IOC.

The reasons why NGC were opposing their city's bid were explained in their website and leaflets, which they were trying, at any given opportunity, to pass to the IOC and anyone else interested (NGC 2009a, 2009b). Their message was fully in line with contemporary anti-Olympic critiques. Tom Tresser, a leading light in NGC, argued that the Games would bankrupt the city, destroy public parks,; displace poor and working-class families from neighbourhoods next to the venue sites, and provide dollars to Mayor Daley's political machine, such as to bring the 'destruction of independent politics for a generation'. The state, county and city were all financially broke and having to cut back on essential services; the Games would make money, as they always did, but only for the IOC, construction companies, consultants and corporate sponsors. NGC called instead for funds to be spent on health clinics, schools, public transit and roads and on improving and expanding public parks and services (Tresser 2009).

In the end, NGC accused Mayor Daley of being utterly corrupt and acting with authoritarian and undemocratic behaviour. Most of all, the NGC's campaign was aimed at starting a debate, within Chicago, on the utility of the Olympic Games, and ultimately to challenge Mayor Daley and his 'Machiavellian' practices. NGC suggested that they were representing around half of the Chicago population, which had not approved the city's bid and which had never been asked about it.

NGC campaign focused its strategy on two main goals: directly communicate and try to convince the individual IOC members not to vote for Chicago, and to limit the level of Chicagoans public support towards the 2016 bid. To achieve these two goals NGC members worked both within the Chicago area and abroad. On three separate occasions members of NGC attempted to meet with IOC delegates and present them with their arguments as to why Chicago should not be awarded the Olympic Games. The first attempt was in April 2009, when the IOC's evaluation team came in Chicago to inspect the bid. NGC activists attempted to greet the IOC delegation on several occasions and demonstrated opposite their hotel. That was the first major breakthrough for the campaign, as although they did not get the opportunity to address the IOC members directly, they did manage to gain the necessary local media attention and publicise their main concerns to a wider audience. In July of the same year, members of NGC travelled to Lausanne along with Chicago 2016, as a counter-delegation to visit the IOC headquarters. At that visit they delivered their 'book of evidence, a box full of documents supporting their claims on 'Why Chicago should not get the Games', to the IOC. They successfully managed to meet high-ranking officials and requested to meet the IOC president himself, to explain him their arguments. According to the NGC members, they received an assurance that their 'book of evidence' would be distributed to all the IOC members and would be considered in parallel with the official Chicago 2016 bid book. The same three members, along with a fourth member who worked simultaneously for Chicago 2016, were the ones who visited Copenhagen on the eve of IOC decision not to award the Games to Chicago. For the final 70 days prior to the IOC decision day, on 2 November 2009, NGC activists were sending daily newsletters to the IOC members updating their evidence and facts from Chicago.

In Chicago the NGC members focused their efforts on raising awareness about the negative aspects of Chicago 2016 bid and sought to limit its public support. Apart from the traditional ways of campaigning—holding demonstrations, forums, open events—the NGC group was probably the first anti-Olympic movement that fully utilised what was (at that time an emerging) social media. Apart from regularly updating its Facebook page, the group utilised Twitter, by posting daily news, streaming their events live through Myspace and waged a 'virality' war against the official Chicago 2016 campaign over whose news would come up first in digital search engine listings. According to polls conducted by the IOC in Chicago, public support towards the bid dropped from 67 % in spring 2009, to 47 % in September 2009 (a month before the decision) with 84 % of Chicagoans disapproving the use of public funds for the Olympics (*Chicago Tribune* 2009). One likely influence on this drop in public support is the NGC campaign, given that they were the leading source of opposition to the bid.

'NGC' attempted to open the discussion on an issue that concerns every society; its right to take part in the decision-making process. They encountered a well-organised team of politicians, businessmen and PR experts who constituted the official 'Chicago 2016' bid and despite the problems, the prohibitions and the closed doors they encountered throughout their campaign, they succeeded in getting their voice heard. Following in the wake of similar notable social movements that opposed the Olympic Games in their city, umbrella groups such as 'Bread Not Circuses' in Toronto (Lenskyj 2000) and 'No Games 2010' in Vancouver (Shaw 2008), NGC provided invaluable information about the potential mega-event to local organisers and the community, wherever megaevents are planned, or hosted. Beyond its academic significance, the case of Chicago 2016 bid reveals how grassroots activity, if uniting around a single cause, despite the lack of financial and personal resources and limited communication channels, can take on, and beat, a very powerful organisation (IOC), its supporters and the world's most powerful man (US president Barack Obama).

Conclusion

The development of a critical research approach to hosting major sporting events and more specifically on the biggest event of them all, the Olympic Games, has significantly informed our understanding and explanation of the essence of the Olympics and Olympism. The contribution of critical Olympic studies in the field of critical sport studies has been noteworthy: arguably starting in the early 1980s with Tomlinson and Whannel's *Five Ring Circus* (1984), in the 1990s (Simson and Jennings 1992; Jennings 1996), through to the 2000s with the notable works of Helen Lenskyj (2000, 2002, 2008), to the plethora of papers, chapters and books over recent years. Today, the displacement of people in order to host Beijing 2008 (Broudehoux 2012) and the link between hosting the 2004 Olympics and the depth of Greece's subsequently economic collapse (Karamichas 2012; Zervas 2012) do not seem paradoxical and can be analysed through the lens of critical Olympic studies. We owe much to those academics who write against the mainstream, and more importantly, to those individual activists in the hosting and bidding cities, who stand against the Olympic machine, its authoritarianism and censorship. Those who form activist groups and coalitions and raise their voices against the Olympic Industry embody the Marxian notion of 'praxis'—acts which shape and change the world.

References

Behr, R. (2009). They sold our street and nobody noticed [review of Anna Minton's *Ground Control The Observer* 5th July. Available at: http://www.theguardian.com/books/2009/jul/05/ground-control-anna-minton-review. Accessed 29 Feb 2016.

Bloor, S. (2014). Abandoned Athens Olympic 2004 venues, 10 years on – In pictures. http://www.theguardian.com/sport/gallery/2014/aug/13/abandoned-athens-olympic-2004-venues-10-years-on-in-pictures. Accessed 29 Feb 2016.

Boykoff, J. (2014). *Activism and the Olympics: Dissent at the games in Vancouver and London*. New Brunswick: Rutgers University Press.

Brenner, N., & Theodore, N. (2002, July). Cities and the geographies of "actually existing neoliberalism". *Antipode, 34*(3), 349–379.

Broudehoux, A. M. (2012). The social and spatial impacts of Olympic image construction: The case of Beijing 2008. *In The Palgrave handbook of Olympic studies* (pp. 195–209). Basingstoke: Palgrave Macmillan.

Chicago Tribune. (2009). Olympic opposites: 'Poll finds Chicagoans split on whether they back the bid. http://articles.chicagotribune.com/2009-09-03/news/0909030167_1_olympics-poll-results-poll-respondents. Posted 3rd September 2009. Accessed 3 Mar 2016.

Giardina, M. D., Metz, J. L., & Bunds, K. S. (2012). Celebrate humanity: Cultural citizenship and the global branding of 'multiculturalism'. In H. J. Lenskyj & S. Wagg (Eds.), *The Palgrave handbook of Olympic studies* (pp. 337–357). Basingstoke: Palgrave Macmillan.

Gibson, O. (2009a, October 1). 2016 'Olympics: Celebrities fly in to make Chicago's case'. *The Guardian* Accessed 13 Oct 2009. Available from: http://www.guardian.co.uk/sport/2009/oct/01/2016-olympics-chicago-copenhagen

Gibson, O. (2009b, October 1). Barack Obama stardust lifts Chicago's chances but vote will go the wire. *The Guardian*, p. 15. Available at: http://www.the-guardian.com/sport/2009/oct/02/olympic-games-2016-host-city-barack-obama. Accessed 29 Feb 2016.

Girginov, V. (Ed.) (2010). *The Olympics: A critical reader*. London: Routledge.

Giulianotti, R., Armstrong, G., Hales, G., & Hobbs, D. (2015). Global sport mega-events and the politics of mobility: The case of the London 2012 Olympics. *The British Journal of Sociology, 66*(1), 118–140.

Grix, J., & Houlihan, B. (2014). Sports mega-events as part of a nation's soft power strategy: The cases of Germany (2006) and the UK (2012). *The British Journal of Politics and International Relations, 16*(4), 572–596.

Hoberman, J. (1986). *The Olympic crisis: Sport, politics and the moral order*. New York: Aristide D. Caratzas.

Horne, J., & Manzenreiter, W. (2006). Sports mega-events: Social scientific analyses of a global phenomenon. *Sociological Review, 54*(Suppl. 2), 1–187.

James, M., & Osborn, G. (2011). London 2012 and the impact of the UK's Olympic and Paralympic legislation: Protecting commerce or preserving culture? *Modern Law Review, 74*(3), 410–429.

Jennings, A. (1996). *The new lords of the rings: Olympic corruption and how to buy gold medals*. London: Pocket Books.

Jennings, A. (2015). *The dirty game: Uncovering the scandal at FIFA*. London: Cornerstone.

Karamichas, J. (2012). A Source of Crisis?: Assessing Athens 2004. In *The Palgrave Handbook of Olympic Studies* (pp. 163–177). Basingstoke: Palgrave Macmillan.

Lenskyj, H. J. (2002). *Best games ever? The social impacts of Sydney 2000*. Albany: SUNY Press.

Lenskyj, H. J. (2012). *Best Olympics Ever? The Social Impacts of Sydney 2000*. Albany: SUNY Press.

Lenskyj, H. J. (2000). *Inside the Olympic industry: Power, politics, and activism*. Albany: SUNY Press.

Lenskyj, H. J. (2008). *Olympic industry resistance: Challenging Olympic power and propaganda*. Albany: SUNY Press.

Lenskyj, H., & Wagg, S. (Eds.) (2012). *The Palgrave handbook of Olympic studies*. Basingstoke: Palgrave Macmillan.

Madrid2016. (2009). *Madrid 2016 candidature official web site [internet]* Accessed 12 Oct 2009. Available from: http://www.madrid2016.es/

Maloutas, T., Sayas, J., & Souliotis, N. (2009, August 23–25). *Intended and unintended consequences of the 2004 Olympic Games on the sociospatial structure of Athens.* Paper Prepared for the 2009 ISA-RC21 Sao Paulo Conference: "Inequality, Inclusion and the Sense of Belonging", Available at: www.fflch.usp. br/centrodametropole/ISA2009/assets/papers/11.7.pdf. Accessed 29 Feb 2016.

Minton, A. (2009). *Ground control.* London: Penguin.

Minton, A. (2012). *Ground control* [reissue with added chapter on 'the true Olympic legacy']. London: Penguin.

NGC. (2009a). No Games Chicago [internet]. Available from: http://nogames. wordpress.com/. Accessed 13 Oct 2009.

NGC. (2009b, October 1). Press release from No Games Chicago.

Observer (2015). The vast gap between rich and poor in our capital is a crisis for us all. http://www.theguardian.com/commentisfree/2015/mar/08/observer-view-on-london. Posted 8th March 2015. Accessed 2 Mar 2016.

Proud, A. (2015, May 4). Inequality is ruining Britain – So why aren't we talking about it more? *The Telegraph.* http://www.telegraph.co.uk/men/thinking-man/11578214/Inequality-is-ruining-Britain-so-why-arent-we-talking-about-it-more.html. Accessed 2 Mar 2016.

Rio2016 (2009). *Rio de Janeiro 2016 candidature official web site* Accessed 12 Oct 2009.Available from: http://www.rio2016.org.br/en/

Roche, M. (2002). *Mega events and modernity: Olympics and expos and the growth of global culture.* London: Routledge.

Shaw, C. (2008). *Five ring circus, myths and realities of the Olympic games.* Gabriola Island: New Society Publishers.

Shepard, W. (2015). *Ghost cities of China.* London: Zed Books.

Shin, H. B., & Li, B. (2013). Whose games? The costs of being "Olympic citizens" in Beijing. *Environment and Urbanization,* 0956247813501139.

Simson, V., & Jennings, A. (1992). *The lords of the rings: Power, money and drugs in the modern Olympics.* London: Simon and Schuster.

Slater, M. (2009). *Chicago calling or roll on Rio?* BBC Sport [internet]. Available from: http://www.bbc.co.uk/blogs/mattslater/2009/08/chicago_calling_or_roll_on_rio.html. Accessed 12 Oct 2009.

Smith, H. (2012). Athens 2004 Olympics: What happened after the athletes went home? http://www.theguardian.com/sport/2012/may/09/athens-2004-olympics-athletes-home. Accessed 29 Feb 2016.

Tokyo2016. (2009). *Tokyo 2016 candidature official web site [internet]* Accessed 12 Oct 2009. Available from: http://www.tokyo2016.or.jp/en/

Tresser, T. (2009). Setting the Record Straight on 2016. Available at: https://nogames.wordpress.com/campaign/documents/about_nogames/.   Posted 13th September. Accessed 1 Mar 2016.

Weaver, M. (2006, November 21). Olympic costs could cross £3bn line. *The Guardian*. Available at: http://www.theguardian.com/society/2006/nov/21/communities.sport. Accessed 3 Mar 2016.

Zervas, K. (2012). Anti-Olympic campaigns. In H. J. Lenskyj, & S. Wagg (Eds.), *The Palgrave handbook of Olympic studies* (pp. 533–548). Basingstoke: Palgrave Macmillan.

"The Olympics Do Not Understand Canada": Canada and the Rise of Olympic Protests

Christine M. O'Bonsawin

The selection of Montreal as host of the 1976 Olympic Summer Games appeared to be a straightforward decision for the International Olympic Committee (IOC). After all, Montreal had previously hosted a highly successful 1967 International and Universal Exposition—Expo '67—and Canada had showed a remarkable aptitude for staging international and domestic multi-sport events. Moreover, the city of Montreal has long been heralded as "the cradle of Canadian sport."[1] Nonetheless, in the lead-up to the 1976 Olympic Games, amidst gloomy headlines about soaring deficits, extreme security measures, construction delays, and deaths, and in the midst of a national unity crisis, many Canadians seemingly lost interest, or quite simply opposed the arrival of the Olympic Games on national soil. An East German editor went so far as to question "[w]hy did Canada want the Games? Why did your federal government

C.M. O'Bonsawin (✉)
Department of History, University of Victoria, P.O. Box 1700 STN CSC, Victoria, BC, V8W 3P4, Canada

© The Editor(s) (if applicable) and The Author(s) 2016
J. Dart, S. Wagg (eds.), *Sport, Protest and Globalisation*,
DOI 10.1057/978-1-137-46492-7_11

agree to them if it didn't want to help pay?"[2] In responding to the East German's query, Canadian sportswriter Doug Gilbert suggested that "[t] he Olympics do not understand Canada and Canada does not understand the Olympics. As a result, 22 million of us have been going around for five years with the vague feeling something is wrong with the 1976 Games."[3] Something was certainly wrong with Montreal, and perhaps something was seriously wrong with Olympic hosting in Canada, more generally.

The primary focus of this chapter, then, is on Olympic protest associated with Canadian-hosted Olympic Games, including the 1976 Montreal Olympic Summer Games, the 1988 Calgary Olympic Winter Games, and the 2010 Vancouver Olympic Winter Games. This study further highlights Canadian Olympic bids that were unsuccessful mainly because there existed considerable opposition within Canada to the respective bid groups. Accordingly, the chapter is organized around three trends of Olympic protest, as manifested in the evolution of Olympic protest behaviour over the last half-century, and then applied to the Canadian context. First, the array of international political clashes that occurred in the lead-up to the 1976 Montreal Olympics is examined in a discussion of state-centric Olympic protest, as these Games represent one of the most significant episodes of political protest in the history of the Olympic movement. Second, protest events in the lead-up to the 1976 Montreal and 1988 Calgary Olympic Games as well as the 1996 Toronto Olympic bid are analysed in a discussion on domestic-orientated protest. Lastly, the chapter examines Canada in the era of transnational Olympic protest. In this latter discussion, Olympic protest related to the 1988 Calgary and 2010 Vancouver Games, as well as to the 2008 Toronto bid are contextualized within the broader scope of transnational Olympic protest. Over the last half-century, Canada and its domestic actors have played an exceedingly important role in the evolution of Olympic protest behaviour. As such, widespread occurrences of Olympic-related protest activities throughout Canada have significantly persuaded state-centric, domestic-orientated, and transnational Olympic protest trends.

State-Centric Olympic Protest in Canada

Throughout the Cold War era, the Olympic Games provided a viable opportunity to propagate state-centric political contention and nationalistic agendas.[4] Olympic protest during this time primarily existed at the state level. According to Donald Macintosh and Michael Hawes, the bans and boycotts of the Cold War era

> emerged in the pragmatic and nationalistic period that followed World War II. This perspective characterizes the international system as an anarchic environment in which independent sovereign states are in constant competition for power and influence. In this formulation, the central goal of every state is the pursuit of its national interest and the maximization of its power "relative" to the other states in the system.[5]

In positioning sovereign states in constant competition, this theory of political realism assumes that states are dominant actors (state-centric); states will use force to improve their position in the international system; a hierarchy exists in world politics whereby military interests dominate economic, social, and technical concerns (albeit they are all important to the international system).[6] Consequently, throughout the Cold War era, political contention with regard to the Olympic Games was primarily expressed through the enactment of political bans and boycotts.

On August 16, 1976, IOC President Lord Killanin circulated a letter to all members of the IOC stating, "[e]ver since 1896 there have been politics in the Olympic Games, but never on the scale of Montreal."[7] The IOC's clash with the Canadian government in the lead-up to the 1976 Montreal Olympic Games was, arguably, the result of Canada's desire to use the Games to enhance its political and economic standing in the international system. The IOC had awarded the Games to Montreal on the premise that all IOC accredited National Olympic Committees (NOCs) and International Federations (IFs) be granted entry into Canada to compete in the Olympic Games.[8] At this time, the IOC recognized Taiwan, also known as the Republic of China (ROC), as the sole NOC representative of China. There was significant disagreement within the IOC concerning which China to recognize, and there appeared to be support for the (re)admission of the People's Republic of China (PRC) into the Olympic family.[9] Matters were further complicated in October 1970, merely five months after the Games had been awarded to Montreal, when the Canadian government adopted its one-China policy, thereby recognizing the PRC in its new foreign policy design. According to Dongguang Pei, in an official parliamentary debate, the Government of Canada acknowledged that the question of which China to admit to the 1976 Olympics would be left in the hands of the IOC, as "these decisions have to be made by the International Olympic Committee. We are the host for the games but we do not decide who participates."[10]

Tensions quickly escalated in May 1976 when the Government of Canada revealed that "we recognize the PRC and we are not under the guise proposing to import into our foreign policy a two-China policy."[11] In the following days, the US Olympic Committee, with the support of American President Gerald Ford, threatened to boycott the Games if Taiwan was not permitted to participate, and the IOC held a secret vote to determine whether it should cancel the Games altogether. Much to the dismay of Montreal Olympic organizers, negotiations between the IOC and the government of Canada resumed even as athletes from both Taiwan and the PRC travelled to Canada to compete in the Olympic Games. Two days before the Games were scheduled to open, the Canadian government agreed that ROC athletes could participate in the Olympics as representatives of Taiwan and not China. The IOC decided to change the name of this NOC from ROC to the Olympic Committee of Taiwan for the two-week duration of the Games. With this compromise in place, the USA abandoned its boycott threat. Taiwan and the PRC, on the other hand, proved dissatisfied with the proposed resolution. The Montreal Olympic Games opened on July 17, 1976, in the complete absence of Chinese representation.[12]

Another difficulty was posed by the white-minority governments of southern Africa. With the expulsion of South Africa and Rhodesia from the Olympic movement, it appeared the IOC and thus Montreal organizers had sidestepped the possibility of an African boycott of the 1976 Olympic Games. However, the Supreme Council for Sport in Africa (SCSA) took the matter a step further and requested that the IOC impose additional sanctions on any nation that had competed against South Africa or Rhodesia, even if such competitions remained outside the realm of Olympic sport. The New Zealand All-Blacks national rugby union team had drawn significant attention to itself following a tour of South Africa in April 1976. Consequently, the SCSA called on the IOC and the Government of Canada to bar New Zealand from the Montreal Olympics. While the IOC claimed that matters regarding rugby remained outside IOC jurisdiction,[13] Canada insisted that the issue needed to be resolved by those parties directly involved, including the IOC, SCSA, and New Zealand. The Government of Canada had long held a position that it unreservedly opposed the apartheid policies of South Africa and Rhodesia; however, in the lead-up to the Montreal Olympics, the Canadian government remained impervious to anti-apartheid initiatives and SCSA appeals. For example, in January 1976, the SCSA openly criticized Canada for sending a team to the World Softball Championships in New Zealand despite the fact that the SCSA had called for a worldwide boycott of the event because of South Africa's participation.[14]

Canada's failure to take or support action against apartheid was arguably bound by understandings of political realism. As Macintosh and Hawes explain, "the central goal of every state is the pursuit of its national interest and the maximization of its power 'relative' to the other states in the system."[15] Involving itself in the anti-apartheid struggle, as it had in the two-China affair, would certainly not have fulfilled the national interests of Canada, nor would it have maximized Canada's power in the international system. Furthermore, "New Zealand was a Commonwealth member in good standing, with allies both in the Commonwealth and among other nations around the world. Barring New Zealand from the Montreal Games was beyond the reach of any political pressure that the black African nations could bring to bear on the IOC."[16] In the end, the SCSA successfully convinced 30 African NOCs to withdraw their teams from Montreal; of the African nations, only Senegal and Ivory Coast elected to stay and participate in the Olympic Games.[17]

As this discussion highlights, independent sovereign states, such as Canada, the PRC, Taiwan, the USA, and many African nations, used the Montreal Olympic platform to advance national interests, and thus maximize their position within the international system. These accounts of state-centric political boycotts and bans dominate academic and public debates surrounding the Montreal Olympics. It is important to note that the Montreal Olympic Games were also plagued by domestic-orientated Olympic protest activities, which have received significantly less attention in scholarly and popular accounts.

Domestic-Oriented Olympic Protest in Canada

By the late 1960s and early 1970s, domestic-orientated protest and political demonstrations had assumed an important position within the Olympic framework. As M. Patrick Cottrell and Travis Nelson maintained in 1994, "demonstrations are now by far the most dominant form of Olympic protest, and thus the actors using them include both transnational actors...and domestic actors within the host state using the Olympics to enter transnational space and to form new transnational connections."[18] Thus, Olympic protest activities at this time moved beyond state-centric boycotts and bans to include domestic protests and demonstrations. A shift away from a predominantly state-centric boycott model to one that included political demonstrations certainly provided domestic actors and activists with the opportunity to challenge the political priorities and repressive policies of host states, and to convey new information to national and international audiences.[19]

In regard to the 1976 Montreal Olympic Games, Bruce Kidd argues that "the impulse to stage grandiose Games and the failings it led to were magnified by the clash of nationalisms which preoccupied and polarized Canadian society at this time."[20] The federal government's decision not to finance the 1976 Montreal Games was demonstrative of national and regional tensions of the period. Throughout Quebec, there was mounting support for its separation from Canada. Prime Minister Pierre Elliot Trudeau wished to ensure Quebec's position within the federal fold of Canada and was apprehensive about creating the perception that he was offering Quebec more "handouts."[21] In October 1970, five months after Montreal was awarded the Olympic Games, political tensions in Quebec dramatically escalated, as the separatist group Front de Libération du Québec (FLQ) kidnapped several political diplomats, killing one, Quebec's deputy premier Pierre Laporte. At the urging of Quebec

Premier Robert Bourassa, Trudeau invoked the War Measures Act to deal with the "self-styled revolutionaries" of the FLQ.[22] The October events, commonly referred to as the October Crisis, deepened hostilities between Francophone and Anglophone Quebec and served to advance the provincial separatist movement. Moreover, at this time, Quebec was actively involved in conflicts over Indigenous land rights. In the early to mid-1970s, significant provincial resources were devoted to treaty negotiations with the Cree and Inuit (and eventually Naskapis and Innu) groups in northern Quebec. In 1975, following a series of legal decisions in the Quebec courts, the James Bay and Northern Quebec Agreement was implemented.[23] The enactment of this comprehensive land claim ultimately established a framework for modern "treaty" making in Canada and, in many regards, redefined the future of Indigenous–state relationships in the country.[24]

There is evidence to suggest that through its extreme security measures, Montreal Olympic organizers proactively averted any real threat of domestic protests from occurring. As Dominique Clément maintains, "[t]he 1976 Summer Olympics marked a turning point in Olympic history: it was the first highly visible security operation, which has since become the norm for Olympic games [and]…It was the largest peacetime security operation in Canadian history."[25] The Royal Canadian Mounted Police (RCMP) Security Service implemented several domestic programs designed to collect, analyse, and share information between various levels of government, and identified several "legitimate" threats to national security, including national liberation groups, Quebec separatists, Indigenous extremists, and Black nationalists. According to Stuart Russell:

> In Montreal at least 16,000 soldiers, as well as large contingents of security, provincial and city police officers, were deployed to protect an almost equal number of athletes and VIPs from possible "terrorist attacks". A year before the Games began the city police launched a massive and lengthy "clean-up" operation, arresting homeless people as well as raiding gay bars and baths; they arrested and charged hundreds of men to drive "undesirables" underground. Several months before the Games opened, police visited hundreds of homes, unions, ethnic groups and radical groups to foment fear.[26]

Shortly before the Games opened, the organizing committee dismissed approximately 20 employees who were considered risks to national security because of their affiliations with left-wing organizations. Following the Olympic Games, two of these individuals filed a joint complaint with Quebec's Human Rights Commission. The case failed to proceed because the Supreme Court of Canada ultimately determined that in this situation, and in the interest of "national security," Crown privilege was paramount.[27]

The RCMP Security Service expressed concern that Indigenous extremists within Canada would draw inspiration from the militant American Indian Movement in the USA. Despite the fact that Indigenous groups posed no substantial threat to the Games, Montreal organizers denied Indigenous groups the opportunity to be directly involved in the Olympic cultural program in the lead-up to the Games. The Indians of Quebec Association had proposed the "Indian Days" celebration be part of the Olympic cultural program, which would include performers and exhibits from the vast regions of Canada. However, a member of the organizing committee informed George Hill, the national coordinator for Canadian Indian participation, that "'a feather show'[28] was out of the question because of the possibility of demonstrations."[29] Olympic organizers worried that Indigenous people would use this event as an opportunity to make provocative statements about their own political, social, and economic plights in Canada.[30] In the absence of any concrete threats from Indigenous extremists, national liberation groups, Quebec separatists, or Black nationalists, the RCMP Security Service ultimately "concluded that there was no evidence to indicate that a terrorist organization planned to attack the Montreal Olympics. But they had been wrong before. The lack of any explicit threat did not forestall the implementation of an impressive security operation."[31]

In response to widespread problems facing the Olympic movement, particularly concerning large-scale boycotts, the Association of National Olympic Committees (ANOC) gathered in Mexico City in November 1984 to discuss the repeated cases of state "non-participation" in the Olympic Games.[32] The ANOC subsequently issued the "Mexico Declaration" proclaiming loyalty to the Olympic Charter, reaffirming the Games as the cornerstone of the movement, inviting ANOC members to participate in the success of the Seoul and Calgary Olympics, and recommending that the organizing committees adopt measures in compliance

with the Olympic Charter. The 152 NOC delegates agreed to "participate with success" in the 1988 Seoul Olympic Summer Games and the
1988 Calgary Olympic Winter Games.[33] At its 89th Session, the IOC
accepted the ANOC's "Mexico Declaration" and went so far as to express
concern, "particularly about the athletes whose lives are adversely affected
by non-participation in the Olympic Games due to political considerations."[34] Accordingly, the IOC adopted a code of sanctions in the event
that an NOC withdrew after making its final commitment.[35]

It therefore appeared the IOC had effectively undermined the state-
centric model of international political bans and boycotts through the
adoption of its 1984 resolutions. In doing so, however, it had failed to
acknowledge, and thus respond to, a considerable increase in domestic
and transnational political demonstrations within Olympic spaces. There
were certainly significant political, social, legal, and economic upheav-
als in Canada throughout the 1980s, which greatly influenced Olympic
organizing efforts for the 1988 Calgary Winter Games. Notably, the
Constitution of Canada was patriated in 1982[36] and entrenched within it
was the Charter of Rights and Freedoms. Quebec was the only province
not to endorse the Constitution. Following the failures of the 1987 Meech
Lake Accord and the Charlottetown Accord of 1992,[37] Quebec's position
within Canada became more precarious still. In regard to Indigenous
rights, sections 25 the Charter of Rights and Freedoms section 35 of
the Constitution Act acknowledged Aboriginal treaty and land rights,
as well as those that may so be acquired.[38] Constitutional entrenchment
provided the necessary foundation for acknowledging Indigenous rights
in Canada; however, due to uncertainty and ambiguity in the language,
the newly repatriated Constitution did very little to protect the inherent
rights of such peoples.

As various scholars have thoroughly discussed elsewhere, to draw
attention to their political plights, the relatively small Lubicon Nation
from northern Alberta strategically and effectively exploited the Olympic
medium.[39] To briefly summarize, in the two "Treaty 8" negotiations
between the Canadian government and Indigenous Nations in 1899
and 1900, federal government negotiators had effectively overlooked the
Lubicon people as this group resided (and continue to live) in a remote
region of northern Alberta. As early as the 1930s, the Lubicon Nation

was seeking formal entrance into Treaty 8. Following a discovery of oil in their territories in the 1950s and with the establishment of all-weather roads by the 1970s, the livelihood of the Lubicon people was drastically altered. Between the 1930 and 1980s, the federal government repeatedly disregarded Lubicon requests.[40] With seemingly few options left, in 1986, newly elected Chief Bernard Ominayak of the Lubicon Nation called for a "boycott" of the 1988 Calgary Olympic Games. Lubicon Olympic boycott efforts ultimately metamorphosed into a series of political demonstrations.

In May 1986, Calgary Olympic organizers signed a $5.5 million sponsorship agreement with Petro-Canada Inc.[41] The Lubicon Nation criticized Olympic organizers for partnering with a corporation actively involved in the destruction of Indigenous territories, including the lands of the Cree Nation, of which the Lubicon were a part. The Lubicon requested that organizations, groups, and individuals support their campaign by peacefully protesting the Petro-Canada sponsored torch relay as it travelled across Canada. In November 1987, the torch run commenced in St. John's, Newfoundland. On this occasion, members of the Native Peoples' Support Group of Newfoundland and Labrador coalition organized a relatively small demonstration in response to what one protester proclaimed to be "a dramatic case of injustice" for the Lubicon people.[42] Other noteworthy protest events along the relay route included a demonstration in Montreal by the Mohawks of Kahnawà:ke, a rally in Ottawa organized by the Grand Chief of the Assembly of First Nations, and a protest event at the Ontario–Manitoba border coordinated by the Peguis Cree First Nation. In declaring support for the Lubicon Nation at the rally in Ottawa, Grand Chief George Erasmus explained, "what we are doing is protesting against corporate supporters…who have made profits at the expense of the Lubicon people. There is a lot of hypocrisy when Petro-Canada is exploiting the flame to make profits, while destroying the Lubicon lands and livelihood."[43] Domestic avenues provided the Lubicon with an opportunity to widen national alliances and garner domestic support. Chief Ominayak, however, understood that to garner the necessary support, the Lubicon needed to move beyond domestic spaces and into global spheres.

In regard to environmentalism, John Karamichas explains that the IOC was particularly slow in responding to environmental concerns. In fact, the IOC did not seriously concern itself with environmental matters until 1994.[44] In the case of the 1988 Calgary Olympics, environmental concerns arose in local regions in 1983 when it was suddenly announced that the organizing committee had changed the Alpine ski venue from Mount Sparrowhawk to Mount Allan, a decidedly smaller hill outside Calgary. The Calgary Olympic Committee preferred Mount Allan since it was a considerably cheaper site, and because the province had already committed approximately $15 million to developing infrastructure in the region.[45] Environmentalists cited threats to wildlife habitat at Mount Allan, mainly involving bighorn sheep, as a primary concern (if the Alpine events were to be moved to Banff because of Chinook winds at Mount Allan, environmentalists further pointed out risks to the resident grizzly bear population in the Banff region).[46] Moreover, following the announcement that a section of Highway 40 south of Mount Allan would be open for the duration of the Olympic Games, the Alberta Wilderness Association (AWA) initiated court action against the government of Alberta in 1987. The AWA insisted that opening this stretch of highway for the two-week duration of the Games posed significant threats to wintering elk and moose.[47] Environmental groups from outside the local and provincial regions eventually became involved in the situation. For example, the Vancouver-based environmental group Friends of the Wolf threatened to coordinate a boycott of the Games if the province proceeded with plans to kill wolves in the Grande Cache-Willmore region north of Jasper Provincial Park.[48] The head of the organization, Paul Watson, went so far as to threaten a mid-air interception of any aircraft actively engaged in a wolf cull.[49]

As Harvey Locke explains, environmental groups had played a significant role in the defeat of three previous Calgary bids to host the Winter Olympics, including bids to host the 1964, 1968, and 1972 Games.[50] In regard to the 1972 Winter Olympics, the city of Denver withdrew its pledge to host the Olympic Games following a public referendum that called attention to environmentally destructive Olympic practices. By 1986, the IOC had declared environmentalism to be the third pillar of

Olympism.[51] As Karamichas explains, however, during this same period, "controversies in Calgary over project locations and [environmental] impacts were quickly put aside by means of the orchestration of 'a speedy end to community consultation and the democratic process.'"[52] The IOC would not actually add a paragraph on environmentalism to its Olympic Charter until 1996.

By 1986, a separate anti-Olympic protest movement was developing elsewhere in the country. At this time, community activists in Toronto began challenging the Toronto Ontario Olympic Committee's bid to host the 1996 Olympic Summer Games. Esteemed scholar and activist, Helen Jefferson Lenskyj, has written extensively on Toronto's unsuccessful bid to host the 1996 Olympic Summer Games (as well as its unsuccessful bid to host the 2008 Olympics, discussed below).[53] In regard to the 1996 Toronto bid, anti-Olympic efforts were concentrated on domestic spheres. The Bread Not Circuses (BNC) coalition which dominated these efforts comprised anti-poverty activists, trade unionists, women's groups, community groups, and individuals who opposed Toronto's bid.[54] As Lenskyj further explains, organizations such as BNC, Artists/Environment Forum, and the Abused Women's Shelters expressed concerns about the adverse impacts the Olympics might have on their communities.[55] In highlighting the importance of housing, jobs, childcare, and a safe and clean city, members of BNC made media appearances, distributed and mailed out flyers (including its weekly "Bread Alerts" flyer), and sold its Anti-Bid Book at local demonstrations. As Lenskyj asserts, "Bread Not Circuses did much to raise Toronto residents' awareness to the potential problems associated with the hosting of the Olympics."[56] Anti-Olympic entities, such as BNC, have historically played important roles in generating public debate, and ultimately defeating well-financed Olympic bids.[57] As is made clear below, this coalition would once again play a central role in defeating Toronto's bid to host the 2008 Olympic Games.

In the debate over Calgary, the Lubicon Nation expanded its Olympic boycott efforts in 1986 by declaring an international boycott of the Olympic sponsored, *The Spirit Sings* exhibition. The exhibition to be hosted by the Glenbow Museum was highly problematic on a number of fronts.[58] First, the primary objective of the exhibition was to bring together samples of Indigenous material culture from collections around the world. From the perspective of the Lubicon, it was absurd that curators would solicit 665 artefacts that had mostly been stolen from Indigenous peoples more than 300 years ago, and not establish a process of repatriation to return the artefacts to their rightful owners. Second, this $2.6 million exhibition was co-sponsored by Shell Canada Ltd. (who contributed $1.1 million), the Government of Canada, and the Olympic Arts Festival (who put in $300,000 each). The Lubicon highlighted the hypocritical nature of accepting sponsorship support from two institutions actively involved in the expropriation of Indigenous territories for an exhibit intended to "celebrate" Indigenous cultures. The Lubicon Nation launched an aggressive letter-writing campaign requesting that international museums not lend artefacts to the Glenbow Museum for *The Spirit Sings* exhibition. According to the Lubicon, 23 of the 110 international institutions refused to send artefacts. Furthermore, the Lubicon received official support from national organizations such as the Assembly of First Nations, the Indian Association of Alberta, the Métis Association of Alberta, and the Grand Council of the Crees in Quebec. Moreover, international organizations such the World Council of Indigenous Peoples, the National Congress of American Indians, the European Parliament, and the World Council of Churches openly affirmed support for this small Cree community.[59] Thus, the Lubicon Nation had strategically and cleverly used the Olympic medium to enter transnational spaces and establish transnational connections.

Canada in the Era of Transnational Olympic Protest

As we have seen, by the late 1980s, domestic groups in host nations were actively using the Olympic platform with the intention of entering international realms and ultimately establishing international alliances.[60] According to Cottrell and Nelson, at the turn of the twenty-first century, the majority of Olympic protest centred around larger issues of transnational concern,

> [a]lthough there are still significant domestic protests by, for example, Aborigines in Australia and various minority groups in China, there are also strong and significant protests by anti-globalization groups, environmental

groups, animal rights groups, and other organizations with more general and transnational concerns than had been typical of Olympic protests prior to the turn of the century. Overall then, an evolution in the types of Olympic protest most typical at the Olympic Games has been accompanied by a broadening in the scope of issues that dominate protest activity.[61]

The IOC has unremittingly insisted that the Games are apolitical, and host states have proactively supported the IOC in its effort to eliminate anti-Olympic resistance and thus limit anti-government and anti-Olympic political demonstrations.[62] As Jules Boykoff correctly asserts, however, "[t]he Olympic Games are shrouded in an apoliticism that is in fact eminently political."[63]

Between 1998 and 2003, anti-Olympic and community activists in Canada were particularly busy as Olympic boosters on opposite sides of the country organized their respective Olympic bids. The Toronto 2008 Bid Committee and the Vancouver 2010 Bid Corporation used IOC instituted strategies of protest management to quell criticisms from Olympic activists and critics. In her book entitled, *Olympic Industry Resistance: Challenging Olympic Power and Propaganda*, Lenskyj provides substantial evidence to support claims that Olympic organizers and boosters were actively involved in the suppression of anti-Olympic perspectives. Media outlets such as the Canadian Broadcasting Corporation and the *Toronto Star* (a bid sponsor) published incorrect bid information and failed to represent the interests of anti-Olympic and community activists impartially. For example, representatives from BNC were denied the opportunity to meet with the IOC Inspection Team when it arrived in Toronto in March 2001. After BNC members arrived at the meetings unannounced, the bid committee eventually allowed four (of twelve) members to speak to IOC officials (for less than five minutes each).[64] Their testimonies were perhaps quite convincing. On July 13, 2001, the IOC awarded the 2008 Olympic Summer Games to Beijing. As C. Michael Hall points out, compared to other Olympic bids, "Toronto has been fortunate to have a non-profit public interest coalition, Bread Not Circuses (BNC), actively campaigning for more information on the bid proposal and for government to address social concerns."[65]

In the case of the 2010 Vancouver Olympic Winter Games, anti-Olympic dissent commenced long before the city had won the right to host the Games. As Lenskyj points out, immediately following the announcement that Beijing had been awarded the 2008 Olympic Games, members of BNC had begun working with anti-Olympic groups in Vancouver.[66] As early as 2001, Olympic activists were citing concerns over social inequalities. First and foremost, the "No Games 2010" coalition drew attention to the fact that the Olympic Games would be hosted on non-surrendered—or stolen—Indigenous territories.[67] Secondly, Olympic critics expressed environmental concerns and explained that the Winter Games posed a substantial threat to vulnerable mountainous regions.[68] Lastly, Olympic activists pointed out that significant taxpayer money was being redirected to staging what was in effect a two-and-a-half week party, rather than to support already underfunded social programs.[69]

Five months before the IOC announced the successful candidate city, the City of Vancouver was forced to conduct an Olympic bid plebiscite vote due to emerging citizen concern about hosting the Olympic Games. Vancouver residents were asked whether they supported or opposed the city of Vancouver's participation in the hosting of the 2010 Olympic and Paralympic Games. Olympic activists were dissatisfied with the plebiscite process for various reasons. First, the plebiscite vote, organized by the city, was non-binding. Second, only residents of Vancouver and those who owned property in Vancouver were eligible to vote. Olympic opponents pointed out that residents of Whistler, where the Games would also be staged, were not afforded this opportunity, nor were British Columbians at large. The province had committed significant tax dollars to the Games, yet the vast majority of British Columbians would not receive any of the professed Olympic "benefits." Lastly, Olympic activists pointed out that a grossly disproportionate sum of resources had been committed to the two campaigns, with the "yes" side receiving substantially more financial resources than the "no" side.[70] Total voter turnout was approximately 40 percent and resulted in a vote that was 63.4 percent in favour of hosting the Olympic Games (in other words, 26 percent of eligible voters were in favour).[71] The Premier of British Columbia, Gordon Campbell, nonetheless, proclaimed the results to be "decisive" and "a powerful 'yes.'"[72] In July 2003, the IOC awarded Vancouver the right to host the 2010 Olympic and Paralympic Winter Games.

Opposition to the Vancouver Olympic Games quickly reordered into what may be described as a global justice movement organized around "an ongoing series of alliances and coalitions, whose convergences remain contingent."[73] As suggested by Boykoff, anti-Olympic organization in Vancouver was essentially "'a convergence of movements' around 'the Olympic movement.'"[74] Olympic activists united under a complex and varied array of issues of transnational concern, including the environment, poverty and homelessness, colonialism and Indigenous rights, women's rights, the criminalization of the poor, lesbian, gay, bisexual, and transgender (LGBT) rights, urban militarization, public debt, Olympic corruption, and corporate invasion.[75] Groups such as No One is Illegal, the Anti-Poverty Committee, the Impact of Community Coalition, Streams of Justice, the Power of Women Group, No 2010 Olympics on Stolen Land, Van.Act! (a pressure group campaigning for housing rights), the Native Youth Movement, and various other organizations partnered with the No Games 2010 coalition. Olympic resistance to the 2010 Vancouver Olympic Games was so extensive that it is virtually impossible to catalogue all the protest and dissonance activities for the purposes of this discussion. Nonetheless, it is possible to conceptualize three underlying themes of social justice on which Olympic activism primarily coalesced: Indigenous rights, environmentalism, and poverty and homelessness.[76]

Cottrell and Nelson correctly point out that Indigenous groups in Canada have strategically used Olympic protest tactics with the purpose of entering transnational spaces, and ultimately establishing transnational support.[77] The Lubicon Nation serves as a clear example. In the lead-up to the 2010 Vancouver Olympic Games, Indigenous groups made various claims that were domestic in nature, although the fundamental injustices inflicted on Indigenous peoples in Canada were undoubtedly global in scope. In truth, the 2010 Vancouver Olympic Games were to be hosted on non-surrendered Indigenous lands. Anti-Olympic organizations united under the "No Olympics on Stolen Native Land" campaign, which drew attention to historical and legal matters that were specific to Indigenous land rights in Canada (or lack thereof). The narrative of disempowerment and dispossession, however, remained consistent with that of colonial encounters throughout the globe, as settler states continue to deny the political, legal, territorial, and thus human rights of Indigenous peoples.[78] In essentially all facets of Olympic resistance, activists underscored the denial of Indigenous land, and thus human rights in Canada.[79]

In June 2002, representatives of the St'at'imc and Secwepemc First Nations presented the IOC with an official submission, which outlined human rights abuses against Indigenous peoples throughout Canada. In requesting that the IOC not award the Games to Vancouver, St'at'imc and Secwepemc representatives proclaimed:

> Although Canada prides itself as one of the countries with the highest living standards in the world according to the UN Human development index, when the same indicators were applied to aboriginal people by the federal department of Indian and Northern Affairs, we only ranked 47th. The same is true for Vancouver being declared the city with the best living standard in the world, our people are the poorest in town, many living on the East side under deplorable social and economic conditions. This is what happens when we as aboriginal people lose our link to the land, alcoholism and youth suicides are only indicators for underlying problems... As indigenous peoples we have to oppose the Vancouver-Whistler Olympic bid as long as regressive and destructive environmental practices [and] policies that undermine and do not recognize indigenous rights are in place.[80]

Olympic events were not hosted on the St'at'imc and Secwepemc territories; however, this did not mean that their lands were free from encroachment or development. For example, in October 2004, the province announced its Spirit of 2010: British Columbia Resort Strategy and Action Plan, thereby encouraging developers and international investors to "maximize opportunities" created by the 2010 Olympics Games.[81] The Nippon Cable Company (of Japan) certainly maximized such opportunities by immediately announcing a $285 million expansion of Sun Peaks Resort, which is located on non-surrendered Secwepemc territory. One year later, the Austrian National Ski Team reached an agreement with Sun Peaks Resort to use this venue as its training facility in the lead-up to the 2010 Olympic Games.[82] When the Austrian team arrived at the Sun Peaks Resort to commence training in 2007, Secwepemc representatives requested that they respect their land rights and boycott this venue. The Austrian continent ignored Secwepemc appeals and seemingly sidestepped the controversy altogether. Rather than abandon its training plans at Sun Peaks, the Austrian skiers simply invited a Secwepemc representative from the Little Shuswap Lake Indian Band to open formally their training season.[83]

Olympic opponents, environmentalists, and Indigenous peoples were certainly concerned about the possibility of encroachment and development on vulnerable mountainous regions.[84] In the planning stages of the Games, there was significant protest concerning the expansion of the Sea-to-Sky Highway. It is important to note that activists were not protesting improvements to this treacherous stretch of highway per se; rather, they were opposing the construction of an additional 2.4-km stretch of roadway through ecologically sensitive wetlands when environmentally sound alternatives were available.[85] Following many peaceful demonstrations, legal proceedings, numerous arrests, and even the death of Pacheedaht (Nuu-chah-nulth) elder Harriet Nahanee (Tseybayoti) following her incarceration,[86] Kiewit Corporation eventually received an injunction to remove protesters forcefully so it could resume development plans. Nahanee understood that she had a cultural obligation as well a legal right to protect unceded Indigenous lands, including the ecologically sensitive wetlands at Eagleridge Bluffs.[87] With the arrival of the Olympic Games in British Columbia, it appeared that unceded Indigenous lands and resources were reopened for business, despite the fact that a treaty process was underway in the province. As argued elsewhere, however, "alterations to the landscape have everlasting effects, and the rights of countless Indigenous groups in British Columbia have been permanently, and negatively, altered as a result of the Olympic presence on non-surrendered Indigenous territories."[88]

For Indigenous peoples, loss of land, and of access to the land, has resulted in the migration of countless individuals and families to urban centres, where many are forced to live in deplorable social and economic conditions. In the lead-up to the 2010 Olympic Games, 15,000 people lived in the Downtown Eastside (DTES) of Vancouver—roughly 30 percent were Indigenous (compared to a national Indigenous population of approximately 5 percent). For decades, the DTES has had one of the highest poverty levels in North America with nearly 75 percent of the population living below the poverty line. For this reason, the DTES is commonly referred to as Canada's "poorest postal code." Between 2003 and 2010, there were large-scale evictions and dislocations of the urban poor, which resulted in a dramatic increase in Vancouver's homeless population.[89] To make matters worse, in 2006, the city launched Project Civil City and instituted new by-laws, making it illegal to sleep outdoors or beg for money (amongst many other restrictions). Furthermore, the city passed the Assistance to Shelter Act in 2009, allowing police to force homeless people into shelters. The city had essentially assisted Olympic organizers by instituting a process for criminalizing the poor, which was essentially a form of social cleansing, according to Olympic activists.[90] In a chapter entitled, "Space Matters: The Vancouver 2010 Olympic Games," Boykoff explains that it was in the DTES where activists from various social justice groups converged to contest social injustices that were being created as a result of the Olympic Games.[91] The DTES became an important space for Olympic activism, resistance, and protest. It was a space where Olympic activists could protest the repressive policies of the host state, in this case Canada, and ultimately enter transnational spaces and make transnational connections.

Conclusion

In August 2015, Toronto Mayor John Tory announced that the city would not bid for 2024 Olympic Summer Games. Once again, Olympic activists and critics were credited with having prevented a Toronto bid from moving forward. As scholar and public educator Janice Forsyth explained in the lead-up to the Mayor's announcement, "[t]here are now 'No Olympics' campaigns in Toronto, where there is a solid history of grassroots activism…Organized and informed citizen groups are exposing the way private interests dig deep into the public purse to fund corporate development, saddling the public with massive debt while leaving important city projects on the backburner for years, and sometimes decades."[92] Olympic activism in Toronto in 2015 extended beyond social advocacy groups, such as the 1988 BNC coalition, to include financial professionals who argued that there now existed 20 years of proof that the Olympics are a bad business deal for cities.[93] Returning to the sentiments expressed by sportswriter Doug Gilbert in 1975, one year before the Montreal Olympics, it was suggested that "Canada does not understand the Olympics." Conversely, one might argue that Canada has always understood the Olympic Games. For a half-century, domestic actors in Canada have arguably fully understood the immense burdens imposed by hosting the Olympic Games. As cities such as Boston, Oslo, Krakow, Stockholm, Hamburg, and Lviv withdraw their respective bids to host the Olympic Games, the international community seems to be coming to the realization that the Olympics are a bad business deal for cities, and thus nations.

Why did Canada ever want the Olympic Games? As demonstrated in this chapter, Canada and its domestic actors have played an influential role in the evolution of Olympic protest behaviour over the last half-century. The 1976 Montreal Olympic Games serve as one of the most noteworthy examples of state-centric Olympic protest, as many sovereign states, including Canada, sought to use the Games to enhance their political and economic standing in the international system. The 1976 Montreal and 1988 Calgary Olympics, and the 1996 Toronto bid initiative all provided opportunities for domestic-orientated protest. Social advocacy and special interest groups, including the Lubicon Nation, used the Olympic platform to challenge the political priorities of the Canadian state, and ultimately convey important information to national and global audiences. By the turn of the century, the majority of Olympic protest was based on larger matters of transnational concern, such as Indigenous rights, environmentalism, poverty, and homelessness. As such, transnational issues dominated Olympic protest activities in Canada throughout the 2008 bid process and the 2010 Vancouver Olympic Games. In 1975, Gilbert had claimed that the international community seemed puzzled by Canada's "curious blend of federal, provincial, municipal and trade union politics, the lack of interest in Olympism and our unique idea that an Olympiad which doesn't pay for itself is somehow, a failure."[94] It might be said that such characteristics continue to define the political landscape of Canada, and thus its relationship to the Olympic movement. As such, Gilbert might have been inaccurate in his suggestion that Canada does not understand the Olympic Games; however, he was absolutely right in his assessment that "the Olympic Games do not understand Canada."

Notes

1. Some of these multi-sport events included the 1930 Empire Olympiad in Hamilton, the 1954 British Empire Games in Vancouver, and the 1967 Pan American Games in Winnipeg. See Bruce Kidd, "The Cultural Wars of the Montreal Olympics," *International Review for Sociology of Sport* 27, no. 2 (1992): 151–162.
2. See Ibid., 152.
3. Ibid.
4. Senn (1999).
5. Macintosh and Hawes (1992).
6. Donald Macintosh and Michael Hawes, "The IOC and the World of Interdependence."
7. 'Future Policy,' Letter (Lausanne, 1976), ID Chemise: 204899, CIO JO.1976S-PETI, SD 1: Petitions Août, International Olympic Committee Archives.
8. Pei (2006).
9. Bairner and Hwang (2010).
10. See Pei, "A Question of Names," 24.
11. Ibid., 24.
12. Ibid.
13. Rugby union was played in the summer Olympics of 1900, 1908, 1920, and 1924. A pitch invasion during the 1924 Olympics in Paris, coupled with the difficulty of attracting sufficient teams, had led to its subsequent exclusion.
14. Macintosh and Hawes (1994).
15. Macintosh and Hawes, "The IOC and the World of Interdependence," 30.
16. Macintosh and Hawes, *Sport and Canadian Diplomacy*, 71.
17. There is considerable debate concerning the number or African nations that withdrew from the Montreal Olympic Games. In the official records of the IOC, twenty-two African countries are listed as "[c]ountries which actually withdrew" and another eight African countries are listed as "[t]hose who did not arrive for any particular reasons but made entries." See "Olympic Games, Montreal—NOC Withdrawal," Letter (Lausanne, 1973), CIO JO.1976S-PETI, SD 1: Petitions Août, International Olympic Committee Archives.

18. Cottrell and Nelson, "Not Just the Games?" 739.
19. Ibid.
20. Kidd, "The Cultural Wars of the Montreal Olympics," 153.
21. Whitson (2005).
22. Pierre Trudeau, "Pierre Trudeau's War Measures Act Speech During the October Crisis," CBC Television News (October 16, 1970), http://www.cbc.ca/archives/entry/october-crisis-trudeaus-war-measures-act-speech (accessed November 17, 2015).
23. Rynard (2000).
24. The word treaty is in quotations because the comprehensive land claims process in Canada is often referred to as a "modern treaty process." However, it must be noted that the word treaty is not employed in any of the text or titles of these agreements, including the James Bay and Northern Quebec Agreement. The legal implications for the Canadian state in doing so would be extensive.
25. Clément (2015).
26. Stuart Russell, "And the Winner is…," *Alternative Law Journal* 19, no. 3 (June 1994): 119.
27. Ibid.
28. An examination of the 1975 and 1976 Olympic Charters suggests that the requirement of the organizers to portray the host's national culture in the Games' opening ceremony was underplayed at the time, although it was embedded in the Charter by the 1990s. In the 1970s, directions for the cultural program (including the opening ceremony) seems to have been quite narrow.
29. Daniel Drolet, "Indian Co-ordinator Says Festival Ignored," *The Saturday Citizen* (July 26, 1976), 4.
30. Forsyth (2002).
31. Clément, "The Transformation of Security Planning for the Olympics," 11.
32. The National Olympic Committees, *Mexico Declaration* (November 1984), http://library.la84.org/OlympicInformationCenter/Olympic Review/1984/ore206/ORE206e.pdf (accessed November 5, 2015).
33. "Olympic Delegates Sign Pledge to End Political Boycotts," *Mohave Daily Miner* (November 9, 1984), 10.

34. "Minutes of the 89th IOC Session," Minutes (Lausanne, December 1–2, 1976), International Olympic Committee Archives, 192.

35. IOC sanctions consisted of: immediate suspension of financial and technical aid, suspension of scholarships, and suspension of economic aid to attend the Games; the NOC would be unable to be nominated as an Olympic host for eight years; other NOCs would be forbidden from maintaining relations with the suspended NOC; and the possibility of additional sanctions imposed by the IOC Executive Board. See "Minutes of the 89th IOC Session," 181–182.

36. That is, under Canada's Constitution Act of 1982, which amended the British North America Act of 1867, Canada took possession of its own constitution from Britain.

37. Following Quebec's rejection of Canada's patriated Constitution in 1981, the federal government sought to obtain Quebec's support of the Constitution Act of 1982. The 1987 Meech Lake Accord proposed to strengthen provincial powers as well as declare Quebec a "distinct society." Following the failure of the Meech Lake Accord, the federal government and provinces devised the Charlottetown Accord in 1992 to obtain Quebec's consent to the Constitution Act of 1982. Canadian voters ultimately rejected the Accord in a federal referendum.

38. *Constitution Act, 1982*, Schedule B to the Canada Act (UK), 1982, c 11.

39. Goddard (1991), Ferreira (1992), Wamsley and Heine (1996), and O'Bonsawin (2013).

40. Goddard, *Last Stand of the Lubicon Cree* and Ferreira, "Oil and Lubicons Don't Mix."

41. "Sponsorship Agreement," Contract (Lausanne, Switzerland, 1986), C-J02—1988/142, SD2: Petro Relay Contract, International Olympic Committee Archives.

42. Karen Booth, "Lubicon Supporters Protest Torch Run," *Windspeaker* 5, no. 16 (1987), 3, http://www.ammsa.com/node/16359 (accessed November 21, 2015).

43. Jamie McDonell, "Signs Read 'Share the Blame,'" *Windspeaker* 5, no. 21 (1987), http://www.ammsa.com/node/16479 (accessed November 21, 2015).

44. Karamichas (2012).

45. "Row Brewing Over Calgary Olympics Downhill Site," *The Ottawa Citizen* (April 27, 1983), 50.
46. Edward Flattau, "Calgary Olympics Face an Uphill Fight," *The Milwaukee Journal* (March 4, 1984), 15.
47. "Outdoor Briefs: Elk Ranging Threatened," *The Spokesman Review* (December 13, 1987), 44.
48. "Wolf Fans Threaten Olympics Boycott," *The Spokesman Review* (November 7, 1986), 12.
49. "Wolf-control Plan Stirs Olympic Boycott Threat," *Spokane Chronicle* (November 6, 1986), B8.
50. For example, the National and Provincial Parks Association of Canada was actively involved in protecting Banff National Park from environmental destruction, and actively protested against the Calgary Olympic Association's bid to host the 1964 Winter Olympics. Calgary's bid to host the 1972 Winter Games was vigorously protested by the Canadian Society of Wildlife and Fishery Biologists and World Wildlife Fund, once gains citing concern for the protection of Banff National Park. See Harvey Locke, "Civil Society and Protected Areas: Lessons from Canada's Experience," *The George Wright Forum* 26, no. 2 (2009): 101–128 and Russell Field, "Who Invited You? Party Crashers or Unwelcomed Guests: The Legacy of Social Protest at the 2010 Winter Olympics," *Rethinking Matters Olympics: Investigations into the Socio-Cultural Study of the Modern Olympic Movement*, ed. Robert K. Barney, Janice Forsyth, and Michael K. Heine (London: International Centre for Olympic Studies, 2010): 192–202.
51. Karamichas, "The Olympics and the Environment."
52. John Karamichas, *The Olympic Games and the Environment* (London: Palgrave Macmillian), 27.
53. See Lenskyj (1992, 1994, 2000, 2002, 2008).
54. Lenskyj, "Buying and Selling the Olympic Games."
55. Lenskyj, "More than Games".
56. Ibid., 86.
57. Lenskyj, "International Olympic Resistance."
58. O'Bonsawin, "Indigenous Peoples and Canadian-Hosted Olympic Games."

59. Wamsley and Heine, "Don't Mess with the Relay."

60. Cottrell and Nelson, "Not Just the Games?"

61. Ibid., 740.

62. The policy of "protest management" was instituted at the 2002 Salt Lake City Olympic Games, during a period of heightened security threats following the events of 9/11. See Cottrell and Nelson, "Not Just the Games?"

63. Boykoff (2014).

64. Ibid.

65. Michael Hall (2001).

66. Lenskyj, *Olympic Industry Resistance*.

67. O'Bonsawin (2010).

68. Lenskyj (2012).

69. Jules Boykoff, "The Anti-Olympics," *New Left Review* (January/February 2011): 41–59.

70. It was reported that the "yes" side campaigned on approximately $1 million and received significant support from the media as well as potential corporate Olympic sponsors/partners. Conversely, the "no" side relied entirely on volunteers and campaigned on a few thousand dollars that was raised through fundraising efforts. See Lenskyj, *Olympic Industry Resistance*.

71. Boykoff, "The Anti-Olympics."

72. "Voters Support Vancouver Olympic Bid," *CBC News* (February 24, 2003), http://www.cbc.ca/news/canada/voters-support-vancouver-olympic-bid-1.393853 (accessed December 2, 2015).

73. See Boykoff, "The Anti-Olympics," 46.

74. Ibid.

75. "No Olympics on Stolen Land: Why We Resist the 2010 Winter Olympics," *No 2010 Olympics* (n.d.).

76. Field, "Who Invited You? Party Crashers or Unwelcomed Guests."

77. Cottrell and Nelson, "Not Just the Games?"

78. O'Bonsawin (2015).

79. Indigenous peoples and groups were involved in various forms of direct action during the planning stages of the Games, including: Secwepemc and St'at'imc statements to the IOC in 2002 and 2003; Indigenous protest at the Olympic countdown ceremony in February

2007; the theft of the Olympic flag from City Hall in March 2007 by members of the Native Warrior Society during the visit of the IOC evaluation team; attendance of Indigenous people from BC at the 515 Years of Indigenous Resistance gathering in Vicam, Sonora, Mexico in October 2007; the Secwepemc protest of the Austrian ski team in November 2007; the creation of the Native Anti-2010 Resistance organization in December 2010; and multiple protests along the torch relay route through 2009 and 2010. See "Anti-2010," *No 2010 Olympics* and Christine M. O'Bonsawin, "Igniting a Resistance Movement: Understanding Indigenous Opposition to the 2010 Olympic Torch Relay," *Critical Dialogues on the Olympic and Paralympic Games*, ed. Janice Forsyth and Michael Heine (London: International Centre for Olympic Studies, 2012): 99–104.

80. "Anti-2010—Information Against the Olympic Industry, No Olympics on Stolen Native Land," *No 2010 Olympics* (January 2009), 10.

81. British Columbia, *British Columbia Resort Strategy and Action Plan* (Victoria, British Columbia, 2004), 1.

82. Maya Rolbin-Ghanie, "'It's All About the Land': Native Resistance to the Olympics," *The Dominion: News From the Grassroots* (March 1, 2008), http://www.dominionpaper.ca/articles/1738 (accessed December 4, 2015).

83. Rolbin-Ghanie, "It's All About the Land."

84. Lenskjy, "The Winter Olympics: Geography is Destiny?"

85. O'Bonsawin (2014).

86. In January 2007, for refusing to obey a court injunction not to return to the wetlands at Eagleridge Bluffs, Harriet Nahanee was incarcerated in the notorious Surrey Pretrial Services Centre, which is a male-dominated correctional facility infamous for its overcrowded conditions, frigid temperatures, and 'a noted hell-hole for women in poor health.' Her already precarious health deteriorated there and she died shortly after her release. See O'Bonsawin, "Showdown at Eagleridge Bluffs," 82–83.

87. Beyond anti-Olympic activities surrounding Sun Peaks and Eagleridge Bluffs, there was also significant environmental activism concerning massive clear-cuts in the Callaghan Valley (the Whistler

Olympic Centre), clear-cuts at Cypress Mountain (north of Vancouver), and concern over the massive amounts of concrete, gravel, sand, and asphalt that would being used in transit and road development. See "Anti-2010," *No 2010 Olympics.*
88. O'Bonsawin, "Showdown at Eagleridge Bluffs," 86.
89. "Anti-2010," *No 2010 Olympics.*
90. No Olympics on Stolen Land, *No 2010 Olympics.*
91. Boykoff, *Activism at the Games.*
92. Janice Forsyth, "Toronto's 2024 Olympic Bid Process Suffers from a 'Democracy Deficit,'" *Huffington Post* (August 27, 2015), http://www.huffingtonpost.ca/janice-forsyth/toronto-2024-olympic-bid_b_8045240.html (accessed December 7, 2015).
93. Sarah-Joyce Battersby, "This Time, Bay St. Joins Toronto's Anti-Olympic Team," *Toronto Star* (August 28, 2015), http://www.thestar.com/news/gta/2015/08/28/this-time-bay-st-joins-torontos-anti-olympics-team.html (accessed December 7, 2015).
94. Kidd, "The Cultural Wars of the Montreal Olympics," 152.

References

Bairner, A., & Hwang, D.-J. (2010). Representing Taiwan: International sport, ethnicity and national identity in the Republic of China. *International Review for the Sociology of Sport, 43*(3), 231–248.

Boykoff, J. (2014). *Activism at the games: Dissent at the games in Vancouver and London* (p. 21). New Brunswick: Rutgers University Press.

Clément, D. (2015). The transformation of security Planning for the Olympics: The 1976 Montreal Games. *Terrorism and Political Violence*, 1–25.

Ferreira, D. A. (1992). Oil and Lubicons don't mix: A land claim in Northern Alberta in historical perspective. *Canadian Journal of Native Studies, 12*(1), 1–35.

Forsyth, J. (2002). Teepees and tomahawks: Aboriginal cultural representation at the 1976 Olympic Games. In K. B. Wamsley, R. K. Barney, & S. G. Martyn (Eds.), *The global nexus engaged: Past, present, future interdisciplinary Olympic studies* (pp. 71–76). London: International Centre for Olympic Studies.

Goddard, J. (1991). *Last stand of the Lubicon Cree.* Vancouver: Douglas & McIntyre.

Karamichas, J. (2012). The Olympics and the environment. In H. J. Lenskyj & S. Wagg (Eds.), *The Palgrave handbook of Olympic Studies* (pp. 381–393). Basingstoke: Palgrave Macmillan.

Lenskyj, H. J. (1992). More than games: Community involvement in Toronto's bid for the 1996 summer Olympics. In R. K. Barney & K. V. Meier (Eds.), *Proceedings: First international symposium for Olympic research* (pp. 78–87). London: International Centre for Olympic Studies.

Lenskyj, H. J. (1994). Buying and selling the Olympic Games: Citizen participating in the Sydney and Toronto bids. In R. K. Barney & K. V. Meier (Eds.), *Critical reflections on Olympic ideology* (pp. 70–77). London: International Centre for Olympic Studies.

Lenskyj, H. J. (2000). *Inside the Olympic industry: Power, politics, and activism.* Albany: State University of New York Press.

Lenskyj, H. J. (2002). International Olympic resistance: Thinking globally, acting locally. In K. B. Wamsley, R. K. Barney, & S. G. Martyn (Eds.), *The global nexus engaged* (pp. 205–208). London: International Centre for Olympic Studies.

Lenskyj, H. J. (2008). *Olympic industry resistance: Challenging Olympic power and propaganda.* Albany: State of New York Press.

Lenskyj, H. J. (2012). The Winter Olympics: Geography is destiny? In H. J. Lenskyj & S. Wagg (Eds.), *The Palgrave handbook of Olympic studies* (pp. 88–102). Basingstoke: Palgrave Macmillan.

Macintosh, D., & Hawes, M. (1992). The IOC and the world of interdependence. *Olympika: The International Journal of Olympic Studies, 1,* 30.

Macintosh, D., & Hawes, M. (1994). *Sport and Canadian diplomacy.* Montreal/Kingston: McGill and Queen's University Press.

Michael Hall, C. (2001). Imaging, tourism and sports event fever. In I. Henry & C. Gratton (Eds.), *Sport in the city: The role of sport in economic and social regeneration* (pp. 166–184). New York: Routledge.

O'Bonsawin, C. M. (2010). 'No Olympics on stolen Native land': Contesting Olympic narratives and asserting Indigenous rights within the discourse of the 2010 Vancouver Games. *Sport in Society, 13*(1), 143–156.

O'Bonsawin, C. M. (2013). Indigenous peoples and Canadian-hosted Olympic Games. In J. Forsyth & A. Giles (Eds.), *Aboriginal peoples and sport in Canada* (pp. 35–63). Vancouver: UBC Press.

O'Bonsawin, C. M. (2014). Showdown at eagleridge bluffs: The 2010 Vancouver Olympic Games, the Olympic sustainability smokescreen, and the protection of indigenous lands. In J. Forsyth, C. O'Bonsawin, & M. Heine (Eds.), *Intersections and intersectionalities in Olympic and Paralympic studies* (pp. 82–88). London: International Centre for Olympic Studies.

O'Bonsawin, C. M. (2015). From Black power to Indigenous activism: The Olympic movement and the marginalization of oppressed peoples (1968–2010). *Journal of Sport History, 42*(2), 200–219.

Pei, D. (2006). A question of names: The solution to the 'Two Chinas' issue in modern Olympic history—The final phase, 1971–1984. In N. B. Crowther, R. K. Barney, & M. K. Heine (Eds.), *Cultural imperialism in action: Critiques in the Global Olympic Trust* (pp. 19–31). London: International Centre for Olympic Studies.

Rynard, P. (2000). 'Welcome in, but check your rights at the door': The James Bay and Nisga'a agreements in Canada. *Canadian Journal of Political Science, 33*(2), 211–243.

Senn, A. E. (1999). *Power, politics, and the Olympic Games.* Human Kinetics: Champaign.

Wamsley, K. B., & Heine, M. (1996). 'Don't mess with the relay—It's bad medicine': Aboriginal culture and the 1988 Winter Olympics. In R. K. Barney, S. G. Martyn, D. A. Brown, & G. H. MacDonald (Eds.), *Olympic perspectives* (pp. 173–178). London: International Centre for Olympic Studies.

Whitson, D. (2005). Olympic hosting in Canada: Promotional ambitions, political challenges. *Olympika: The International Journal of Olympic Studies XIV*, 29–46.

'The Atos Games': Protest, the Paralympics of 2012 and the New Politics of Disablement

Stephen Wagg

This chapter is about the demonstrations that took place against the French IT company Atos in Britain during 2011 and 2012. The protests were led by disabled people and their supporters and were directed to draw attention both to the company's government contract to assess the work capabilities of disabled people and to Atos' simultaneous sponsorship ('partnership' in the Olympic lexicon) of the Paralympics, the next Paralympics being due in London in September 2012. The chapter will outline the political background to these protests first of all by tracing the fashioning of a new governmental approach to disablement back to the 1980s; it will then describe the enactment of new 'welfare-to-work' policies for the disabled and the political furore that resulted; thirdly, it will discuss the attitudes to disablement that were aired around the time of the Paralympics of 2012 and the protests, running parallel to this debate,

S. Wagg (✉)
Leeds Beckett University, 221 Cavendish Hall, Beckett Park Campus,
Leeds, LS6 3QU, UK

© The Editor(s) (if applicable) and The Author(s) 2016   **257**
J. Dart, S. Wagg (eds.), *Sport, Protest and Globalisation*,
DOI 10.1057/978-1-137-46492-7_12

by disability activists over Atos and government policy; finally, it will attempt to judge the consequences for the politics of disablement of the Paralympics—a vital consideration since both the British Prime Minister and the Chairman of the London Organising Committee of the Olympic and Paralympic Games argued that these Games had had a transforming effect on public attitudes towards the disabled.

The Budget Cutter's Bible: Disablement, Welfare and British Politics in the 1980s and 1990s

As part of the welfare consensus established in Britain after the Second World War, the government formally embraced responsibility for the disabled, chiefly under the terms of the National Assistance Act of 1948, which was based on the Beveridge Report of 1942 (Beveridge 1942) and formally abolished the Poor Laws, which had stood since the sixteenth century. Under the new dispensation, disabled people now qualified for a range of benefits and services. Disabled people were viewed and understood via what a number of writers have since called the 'medical model' of disability (see, e.g., Brisenden 1986) wherein disabled people were seen as individuals with problems that called, primarily, for medical attention, some of it, of course, involving hospitalisation. This model is generally seen to have been affirmed in a government report of 1971 (Harris et al. 1971) which distinguished between *impairment* (lacking all or part of a limb), *disablement* (the loss or reduction of functional ability) and *handicap* (disadvantage or the restriction of activity caused by disability) (Oliver and Barnes 2012: 17). In 1970, the UK parliament passed the Chronically Sick and Disabled Persons Act which gave rights to disabled people and resulted from a private member's bill introduced by Labour MP Alf Morris. It was one of the last measures passed under the Labour government that lost office later that year; on Labour's return to office in 1974, Morris became the first minister to be given specific responsibility for the disabled.

These conventions of policy and perspective were undermined in three, ultimately inter-connected ways. First, in the late 1960s, political disquiet was expressed about institutionalised care—initially in relation to the elderly. In 1967, Barbara Robb of the care organisation AEGIS (Aid to the Elderly in Government Institutions) published a damning report documenting the poor treatment of geriatric and/or mental patients on the wards of the National Health Service (NHS), then less than 20 years old, and sympathetic politicians, chiefly in the House of Lords, took up the issue (Robb 1967)[1]. Second, the general (and understandable) concern about institutionalised treatment was augmented by a growing critique of the medical model among disabled people themselves. This critique was developed principally by disabled writers and activists and a new model—the social model—emerged, in tandem with a growing political movement for the rights of disabled people. This movement took its place alongside other similar identity-based movements (feminism, anti-racism, and the campaign for gay rights, e.g.) and borrowed, politically, from them. For example, a pioneer of the social model of disability was Vic Finkelstein, a South African disabled following a pole vaulting accident at the age of 16. Finkelstein had been jailed in his native country for his opposition to apartheid and, on his release, was given a five-year banning order under South Africa's Suppression of Communism Act: 'He would later claim that this limited his activities little more than the restrictions already placed on him as a disabled person' (Sutherland 2011). Thus, with others, Finkelstein now argued that the 'disability' lay not in the misfortune of individuals, but in the organisation of society. Thirty years later he recalled taking part in discussions about a television programme about disablement: he had argued that:

> the programme ought also to look at important key issue for us—e.g. that society is disabling us and therefore it is society that has to change, not disabled people. I remember at one meeting a person who had been involved for some time in the so-called disability world, the professional world, protesting "But what you're saying is revolutionary. It'll never happen. People will never regard disability as something that is created by society. Disability is something you're born with or when you have an accident. It's part of you and people need to intervene to help you. You need

---

[1] AEGIS operated between 1946 and 1976. Its papers are held by the British Library of Political and Economic Science. Details at: http://archiveshub.ac.uk/data/gb97-aegis

professional services." So, in wanting a television programme that inter-preted the nature of disability in social terms, that it's not disabled people who need to change but actually the non-disabled world that needs to change, this was called revolutionary! This experience impressed upon me just how challenging many non-disabled people regarded the changes that we wanted. (Finkelstein 2001)

In 1972, Finkelstein helped found the Union of the Physically Impaired Against Segregation (UPIAS), one of a number of organisations dedicated both to the social model of disability and to the securing of these changes. For these organisations, the word 'impairment' referred to physical or mental condition, but the word 'disability' described society's (invariably neglectful) response to that impairment.

The third factor has been what Oliver and Barnes (2012: 144) charac-terise as the 'growing retreat from state-sponsored provision' for people with impairment, a retreat which they date from the late 1950s, but which quickened following successive economic crises in Western capitalism from the 1970s and the subsequent election of right wing governments such as the administrations of Margaret Thatcher in the UK between 1979 and 1990. During her first term as Prime Minister Thatcher com-missioned Roy Griffiths, deputy chairman of Sainsburys supermarket chain, to examine the management of the NHS; his report, delivered in 1983, recommended, perhaps predictably, that the NHS be run placed on more of a business footing, with the introduction of general man-agers. Griffiths, who received a knighthood in 1985, was subsequently invited to comment on community care and his second report, published in 1988, caught the political wind, which, following the public disquiet over findings such as Barbara Robb's, was blowing strongly against institutional care: his report therefore called for de-institutionalisation (Griffiths 1988). This meant that mentally and physically handicapped people should now be looked after 'in the community'. The phrase 'in the community' referred variously to the local authorities and, perhaps more than was acknowledged, to the families of the handicapped people con-cerned—in the latter instance, of course, it resonated strongly with Mrs Thatcher's widely quoted assertion the year before that 'there is no such thing as society: "There are individual men and women, and there are families"[2]. Crucially, Griffiths counselled that local social work departments should make maximum use of the voluntary and private sectors of welfare. Equally crucially, *Caring for People: Community Care in the next Decade and Beyond*, the government White Paper of the following year, identified the state as the 'enabler', rather than the provider, of services (see, for example, Langan 1990).

For the political left, which now included a large number of disabled activists, academics and campaigning groups, these changes were welcome since they represented some tacit acceptance of the social model and would take handicapped people out of the apparent ghetto of institutionalisation. For the political right, now increasingly reviving the small state, low tax ideologies of the nineteenth century liberals, public expenditure could be reduced along with state responsibility for welfare. In this regard the political memoirs of the Conservative politician Edwina Currie are instructive.

Currie became a junior minister in the Department of Health and Social Security (DHSS) in 1986. In September of the following year she confided to her diary that her view that John Moore, the cabinet minister for the DHSS, harboured party leadership ambitions and, to that end, had just made ' a strong speech on social security and getting away from the welfare state' (Currie 2002: 11). Currie then reflected on his prospects in this mission:

> There are only three groups/benefits Moore can tackle: pensioners (but they paid and there are ten million of them), child benefit (but the Tory ladies like it) and the disabled (he hasn't encountered them yet. Just watch!) Let it be recorded that the origin of his thinking is Charles Murray's *Losing Ground*, sent to him after the election [of 1983] by Keith Joseph [a government minister and advocate of free markets seen as a strong intellectual influence on the Thatcher government] [...] The book is fascinating and devastatingly accurate in its view of why more welfare makes things worse, particularly in its destruction of the status rewards of being respectable,

---

[2] Originally said in an interview with *Woman's Own* magazine, 31 October 1987; it was reproduced on Mrs Thatcher's death, by *The Huffington Post* website 4th August 2013: http://www.huffingtonpost.co.uk/2013/04/08/margaret-thatcher-dead-controversial_n_3037335.html Access 20th July 2015.

law-abiding, etc. Where it's hopeless is in suggesting what to do about it. (Currie 2002: 11)

Two important deductions can be made from Currie's comments. First, she clearly identifies welfare for the disabled as a target for a reduction in state responsibility, with pensions and child benefit being deemed too risky politically by comparison. Her 'Just watch!' is less easy to interpret. She may have meant that the disabled could 'just watch' and see what far-reaching changes to their welfare provision a radically right wing government was prepared to make. Similarly, she could have meant that political observers could 'just watch' and witness the angry resistance Moore (and his successors) would meet. In either case she would have been correct. Second, Charles Murray occupied a position on the far right of the Western political and philosophical spectrum and is sometimes described as a paleo-conservative—an atavistic creed based on limited government and free markets. His book *Losing Ground: American Social Policy, 1950–80* (Murray 1984) was written under the auspices of the Manhattan Institute, a right wing think tank based in New York and one of a clutch of far right bodies founded and funded by Charles and David Koch, who run one of the biggest industrial empires in the United States (see Mayer 2010; Schouten 2015). *Losing Ground* reiterated the long-held right-wing view that welfare spending created a culture of dependency and therefore did more harm than good; a review in the *New York Times* called it 'this year's budget-cutters' bible'[3]-words that would prove prophetic for disabled people in Britain. Moreover, if policy makers were being guided by literature this radical, then the cutting was unlikely to be piecemeal. Through the late 1970s and 1980s, meanwhile, the British popular press, as shown by the sociologists Peter Golding and Sue Middleton, had begun to revive the mediaeval notion of the undeserving poor—the 'sturdy beggars', now styled a 'scroungers', who were able to work, but chose not to, preferring to live off state benefits (Golding and Middleton 1982). It's fair to say that, in the 40 odd years since then, the British popular press have seldom relented in their pursuit of this broad narrative.

---

[3] 'Losing More Ground', unattributed, *New York Times* 3 February 1985. Available at: http://www.nytimes.com/1985/02/03/opinion/losing-more-ground.html (Access: 25 June 2015).

Although seldom the subject of mainstream political discussion, in the early 1990s, there was the occasional glimpse of what the state, as 'enabler', now had in mind for disabled people. At the Conservative Party conference of 1992, Peter Lilley, newly appointed Secretary of State for Social Security in the government of John Major, told delegates 'I'm closing down the "Something for Nothing" society', adding 'We are not in the business of subsidising scroungers'. Invoking an array of now-popular unentitled claimant myths (of 'so-called' new age travellers, 'bogus' asylum seekers, girls getting pregnant simply to obtain council housing) Lilley assured the audience that these people would no longer receive welfare. And this was just the beginning—there were 'scores of other frauds to tackle'[4]. If, however, Lilley was seeking to reduce welfare for the disabled—a group which could be assumed to command a high level of public sympathy—such a move would be difficult to sanction with this dismissive vocabulary alone.

In January 1994, Lilley presented the Social Security (Incapacity for Work) Bill to parliament; it became law the following year and entailed lower rates of benefit for certain groups and more stringent medical testing before qualifying for state support[5]. That same year (1995) the satirical magazine *Private Eye* carried an article by the investigative journalist Paul Foot which revealed that whereas disabled or sick people had hitherto been entitled to benefit if they could no longer do their job, from now on, they would be entitled to benefit only if they could do no work at all. It was forecast that this new dispensation would reduce expenditure by £410 million for the year 1995–96, by £1.2 billion for 1996–97 £1.2 billion, and for 1997–98 £1.7 billion. Ward Graffam, the chairman of American insurance company Unum, had described these measures as 'exciting developments' in the 'disability market', since many disabled people, who had previously relied on state benefits, would now be obliged to take out insurance. Dr John Le Cascio, second vice-president

---

[4] Lilley's speech can be seen at: https://www.youtube.com/watch?v=FOx8q3eGq3g (Access 27 June 2015).

[5] The bill can be read at: http://www.google.co.uk/url?sa=t&rct=j&q=&esrc=s&source=web&cd=1 &ved=0CCUQFjAA&url=http%3A%2F%2Fwww.parliament.uk%2Fbriefing-papers%2FRP94-13.pdf&ei=E8COVfLpO4Sd7gbHt6nYDA&usg=AFQjCNFvnT2L04hoBO2 VBVe-s70v3DIVIw&bvm=bv.96783405,d.ZGU (Access 27 June 2015).

of the Unum Corporation, had recently been seconded to the company's British operation and was in 1994 invited by the Department of Social Security (DSS) to help in the extensive training of doctors in the new techniques of testing claimant's eligibility for benefits/state support. Foot cannot have been the only person to view Dr Le Cascio's claim that this did not constitute a conflict of interest with scepticism (Foot 1995).[6] (At the time, the DSS made no public acknowledgement of their employment of Le Cascio; when asked in 2013, whether Unum had advised his department on testing disabled people for their benefits, Lilley replied enigmatically: 'When I was secretary of state? Maybe, I'm not denying it, but I certainly don't remember. If you tell me it happened, it happened, but I have absolutely no recollection' see Pring 2013).

Work Will Set You Free: The New Politics of Disablement

The Conservatives lost office in 1997 and made way for a Labour Party freshly re-branded as 'New Labour', a move widely seen as an acceptance of the market-oriented politics practised by the Thatcher and Major administrations. In an early appraisal of 'New' Labour's policies for the disabled, Robert F. Drake noted a greater emphasis on work and on employment-oriented programmes; however, the 'medical model', he argued, was still in play and there was little consideration of the many barriers that disabled people faced in seeking paid employment (Drake 2000).

But, far from accommodating the social model increasingly propagated by activists and researchers, 'New' Labour were soon suggesting that one of the main barriers in this regard was disabled people themselves. Significantly, in 2001 the work of the DSS had been absorbed into a new Department of Work and Pensions (DWP). That same year Malcolm Wicks, Parliamentary Under Secretary of State for Work, and Mansel Aylward, his Chief.

---

[6] Foot's article is reproduced at: https://beastrabban.wordpress.com/2013/08/12/paul-foot-on-the-insurance-company-unum-and-cuts-to-disability-benefit-in-private-eye-from-1995/ (Access 28 June 2015).

Medical Officer at the DWP, along with Dr Le Cascio of UnumProvident, attended a conference in Woodstock, Oxfordshire, on 'Malingering and Illness Deception', 'New' Labour having resolved to reduce the 2.6 million who were claiming Incapacity Benefit. The task of those present was, according to Jonathan Rutherford, to 'redefine the cultural meaning of illness'. Unum's contribution to this project was the facilitating of 'claims management'; to this end, in 2004, the company funded the UnumProvident Centre for Psychosocial and Disability Research at Cardiff University in the UK and appointed Aylward as Director following his recent retirement from the DWP (Rutherford 2007). The previous year, meanwhile, the DWP had instituted its *Pathways to Work* programme, the aim of which, according to a report of a survey conducted for the DWP, was 'to assist incapacity benefits claimants into, and towards, paid work'; the report refers to these claimants as 'customers' (Becker et al. 2010: 20, 14). In 2005, the Cardiff centre published its first monograph—'The Scientific and Conceptual Basis of Incapacity Benefits' by Aylward and Gordon Waddell (Waddell and Aylward 2005), an orthopaedic surgeon, which argued that recipients of incapacity benefits had trebled since 1979 'despite gradual improvements in objective measures of health'. Many of these recipients, they suggested, were capable at least of some work and wanted to do it. Waddell and Aylward, neither of whom declared their link to UnumProvident, therefore proposed a new model of disability, apparently the fruits of the Woodstock deliberations of four years earlier; this was the 'bio-psychosocial model' (BPS), which placed fresh emphasis on the subjective aspects of disability. Disability in this new formulation, which drew on earlier work on psycho-somatic illness and the American sociologist Talcott Parsons' concept of the 'sick role' (Parsons 1951), wherein individuals, in part, 'choose' to be ill and, similarly, authenticators, such as doctors, allow them to do so (Rutherford 2007), was increasingly seen as what the disabled person thought it was. Psychologist Peter Halligan, another proponent of this view, wrote: 'beliefs held by patients about their health and illness are central to the way they present, respond to treatment and evaluate their capacity for work' (Halligan 2007).

Needless to say, powerful critiques of this purportedly new model are now available (see, in particular, Rutherford 2007 and Thorburn 2012).

Thorburn has pointed out that (a) despite its multi-disciplinary title the biopsychosocial model is very largely 'psycho', with little attention paid to the 'bio' or the 'social' dimensions of disability, thus giving greater scope to the notion that the disability was all, or substantially, in the mind; (b) many proponents of the BPS model were professional psychologists and that some of them, Halligan in particular, stressed the human capacity for deception in their theorising: in 2003, Halligan helped to edit a book on 'malingering and illness deception' (Halligan et al. 2003); (c) BPS work tended to concentrate on certain disorders, such as musculoskeletal conditions, like back pain (although, in the tests instituted by the DWP, the arguments would have a much wider application); (d) BPS was, in effect, a closed discourse with its subscribers principally quoting each other in their work and together acting as the sole source of 'independent' research findings and academic advice to the DWP; and (e) illness was thus construed as a form of deviance, with many of those previously designated as disabled now liable to be seen either as malingering or falsely convinced that they were ill (Thorburn 2012). Crucially, work, for BPS writers, was seen as an abstractly 'good thing', with little sense that work itself could be the source of ill health (Rutherford 2007; Thorburn 2012). (It goes without saying that the social model has not gone away and it continues to inform much academic and activist writing and argument on the part, or on behalf of, the disabled. For a strong reaffirmation of this model, from a Marxian perspective, see Slorach 2015)

Adopting this carefully incubated new discourse, in 2006 Work and Pensions secretary John Hutton announced that the *Incapacity Benefit and Jobseekers Allowance* would be replaced by a single *Employment and Support Allowance* for those whose health affected their ability to work. Claimants, he said, had responsibilities, as well as rights. They would be encouraged to be 'actively engaged' in looking for a job, 'including attending regular interviews, completing action plans and work related activities': 'A new Personal Capabilities Assessment will be carried out for all claimants to judge their capacity to work, which Mr Hutton said would be "more than just a snapshot" and should look at fluctuating levels of disability' (BBC News 2006). The Welfare Reform Act, which was passed the following year, expressed this 'welfare to work agenda.' Mencap, the charity supporting those with learning disabilities, tried to explain this to those affected: the government, it said, 'aims to achieve an 80 % employment rate in the UK by moving one million people off incapacity benefits and into work in one decade from 2008'; the DWP's purpose, drawing on the title of the previous year's Green Paper, was said to be 'empowering people to work' (Mencap 2008). For a growing contingent of disabled activists, academics and politically sympathetic critics, this might be taking government policy dangerously at face value.

On 6 November 2007, six months after the passing of the Welfare Reform Act, BBC television news announced:

> A multinational insurance company accused of racketeering and cheating thousands of Americans out of welfare benefits, is giving advice to the British government on welfare reform. A BBC investigation has found that executives from Unum have held meetings with senior Whitehall officials to discuss changes to the benefit system

Reporter Mark Daly told viewers that US insurance commissioner John Garamendi had declared UnumProvident an outlaw company, which for years had operated 'in an illegal fashion.' In a settlement signed by all 50 US states, Unum had been fined $23million and ordered to reopen 300,000 denied claims totalling half a billion dollars. The transcript of this news item was removed from the BBC website but in 2011, it was re-posted on the website of the Black Triangle Campaign, a group formed to defend the rights of the disabled[7]. Black Triangle take their name from the badges, signifying 'arbeitsscheu' or 'work-shy', which disabled people were forced to wear in Nazi concentration camps.

The Labour government meanwhile contracted the private company Atos Healthcare to conduct its Work Capability Assessments (WCA). It may be surmised that the task was not given to the NHS (which would have been less costly) because of the prevalence of the neo-liberal 'private is better' ethos and/or because the real purpose of the WCA-ostensibly to filter claimants into the categories of 'work ready', 'having *some* barriers to work' and 'unlikely ever to be capable of working'—was to 'remove welfare from the sick' (Thorburn 2012).

Certainly, there was powerful evidence to support the latter argument. Atos were a multinational IT company, headquartered in France.

---

[7] A full transcript can be read at: http://blacktrianglecampaign.org/2011/08/20/bbc-10pm-news-6th-november-2007-transcript-news-article-unum/ (posted 20 August 2011; access 29 June 2015).

The system they employed to assess claimants was known as the Logical Integrated Medical Assessment (LIMA)—a computerised spreadsheet which generated questions on a screen along with boxes to tick according to the claimant's reply. LIMA had been devised by Unum (Rutherford 2007; Read 2011). Atos' training materials made clear to assessors 'that their role is not to act as the patient's advocate' (Thorburn 2012); similarly the DWP's 'Training & Development ESA [Employment and Support Allowance] Handbook' instructed assessors to play down the patient's disability: "Avoid making a statement such as "Can only walk 50 metres" as this may well be taken as fact by the Decision Maker or the Appeal Tribunal. Better would be 'Says he only walks 50 metres', then give an example of what the claimant actually does, as far as walking is concerned, on an average day: 'Walks to the shops and back (about 200 metres in all) but says he has to stop at least twice due to back pain'". (DWP 2008: 41, quoted in Thorburn 2012).

As Atos' assessment work progressed (for an extended critique, see Jolly 2011) a considerable controversy developed, in the press, on the internet, in parliament and on the streets. There is not sufficient space here to do justice to the apparent fear and anger provoked among the disabled by the new regime of testing. In 2010, a report commissioned by the government, by Prof. Malcolm Harrington said that the assessments were 'impersonal and mechanistic' and were failing those with mental illnesses and long-term disabilities (Harrington 2010; Ramesh 2010). The charge levelled by many critics was not only of cruelty, but also of bureaucratic confusion. In January 2011, a survey based on official DWP statistics by the left-wing campaign group Compass estimated that half a million people had been wrongly denied incapacity benefit: 300,000 had won it back on appeal and a further 200,000 had been refused it, only to have it restored (Stratton 2011). A few weeks later, for the same newspaper, Amelia Gentleman interviewed a family who had had this experience.

> The test did not identify the array of problems that make life complicated for Matthew Hutchings, 36, and awarded him just six points, not enough to qualify for the benefit. Smiling and shy, he sits quietly as his parents Ray, a retired accountant, and Diane, a teacher, explain that he has had lifelong speech and language problems, is probably autistic, and has always had difficulties with grasping simple tasks.

"People have problems understanding him, he has problems with his speech, with his understanding and with his confidence. But it is a hidden thing, until you know him well. He's very vulnerable," his mother says.

When he was called for his disability test, his father accompanied him. "The assessment was farcical for someone like Matthew—it was all: Can you sit down? Can you stand up? Can you pick something up? He can do all those things, but that's not what the problem is," Ray says. "They asked if he could cook. I explained he couldn't, but that he was able to heat something up in a microwave. In the report they said he could cook."

His father appealed immediately against the judgement and six months later the case went to appeal. The tribunal judge ruled that Matthew was not capable of work, and awarded him 15 points. (Gentleman 2011)

Disabled activist Lucy Glennon described the WCA as 'a genuine source of anguish', adding 'The computer software used by Atos is simplistic. A person may be able to hold a pen or pencil for a while, but may have trouble writing after a few minutes, yet such intricacies and complexities of a condition are not taken into account. The medical assessment makes little consideration of pain and the limitations it causes, and also the fluctuations in a condition that can vary wildly from day to day' (Glennon[8] 2011). The tight structure or 'tick-box tyranny' (Hunt 2011) of the test was such that many terminally ill claimants died shortly after being denied benefits: according to the DWP's own figures, 10,600 sick and disabled people died in 2011 within six weeks of their claim ending (Miller 2012). (The DWP later admitted that between December 2011 and February 2014, 2,380 people declared fit for work and disqualified from receiving the Employment and Support Allowance had died: Grice 2015[9]). A popular joke doing the rounds in 2012 was that the bones of the fifteenth-century English king Richard III, exhumed that year from beneath a council car park in the East Midlands city of Leicester, had been declared fit for work by Atos.

The government, which after the General Election of 2010 was a Conservative–Liberal Democrat coalition, continued to defend the WCA

---

[8] Lucy Glennon had epidermolysis bullosa, a condition that causes the skin to be very fragile. She died in 2015, aged 29.

[9] For further examples of the individual hardships visited on disabled people via these assessments, see Mendoza 2015: 61–8.

in the language of social inclusion. In 2011, Employment minister Chris Grayling said:

> It's all about saving lives not saving money. We have 2.1 million people on IB [Incapacity Benefit]: many of them have had little contact with the welfare state for a very long time. Nobody has ever talked to them, looked at their situation, understood whether they have the potential to do something else. There'll be some who'll be fit for work, there'll be others who can return to work with the right support, there'll be plenty who can't possibly work who will carry on receiving unconditional support. But unless you go through this exercise, you can never get to the point where you can say, right, how can we help these individuals. (Macrae 2011)

But the same could not be said of the right-wing press. Kayleigh Garthwaite noted the proliferation of stories of 'work-shy' disabled in the British *Daily Mail* and *Daily Express* in 2009 and 2010 (Garthwaite 2011). 'Three-quarters of those claiming to be too sick to work are fit and able to look for a job, figures reveal' said the *Daily Mail* in 2010, the phrase 'figures reveal' implying an official source (Chapman 2010). Five months later the same paper asserted: 'Nearly 1.5 million Britons are spending their fifth Christmas in a row on incapacity benefit. Figures released by the DWP expose the shocking degree to which a generation of Britons has abandoned work for a life "on the sick". The statistics show that almost £66 billion has been paid out in incapacity benefits alone over the past five years' (Shipman 2010). Such stories elicited a few stern comments of the 'Work or starve' variety on the web and some disability activists could see the attraction to readers of the anti-welfare rhetoric. 'Vicki', for example, responding to an article by Gill Thorburn (2012), wrote: 'In hard times people are perfectly willing to throw the weak against the wall, if it might make their own future a little less insecure, especially if someone will provide them with a glib phrase to cover their callousness. When they can be given the bonus of righteous indignation, then the package will sell itself'[10].

It was against this baleful political backdrop that the various parties to the controversy contemplated the London Paralympics of 2012.

---

[10] https://internationalgreensocialist.wordpress.com/illness-as-deviance-work-as-glittering-salvation-and-the-psyching-up-of-the-medical-model-strategies-for-getting-the-sick-back-to-work/ (Access 30 June 2015).

Atos, who had been Olympic and Paralympic 'IT Partners' since 2002, would be a lightning rod in this regard.

Impairment or Empowerment? Politics and Protest at the Paralympics of 2012

With the Paralympics of 2012 imminent, the event organisers and the liberal British media sought to affirm that these games now approached equal credibility with the established mega events of elite world sport. The event was thus explicitly rendered as offering social inclusion, by extension, to the country's disabled people. This was an important consideration, observers having for some time drawn attention to the comparatively low priority accorded to previous Paralympics: Goggin and Newell argued this for the Sydney Olympic coverage in Australia in 2000 (Goggin and Newell 2000), while an analysis of the 2002 Salt Lake City winter Paralympics 'discovered US journalists scrambling for the airport once the regular Games ended while many foreign journalists stayed on for the Paralympics' (Carroll 2012). Now the Paralympics would be broadcast in over 100 countries, 2.3 million tickets had been sold by late August and the British Channel Four TV channel would devote hundreds of hours to broadcasting events (Gibson 2012). Channel 4 presenter Georgie Bingham said in advance of the Paralympics 'Channel 4 has a target to make disability a bigger talking point. They hope that people with disabilities are not only inspired, but find a voice through our coverage. They want a fundamental change in attitude toward disabilities across the board' (Bingham 2012. In the same spirit, Ben Rushgrove, a British Paralympian sprinter with cerebral palsy, wrote:

> I'm hoping that the London Paralympics will deliver more than just sporting success. My dream is that these will be the breakthrough Games. When we can dispel so many myths and misapprehensions that people may have about disability. The number one myth I come up against all the time is that we can't do something. For me, there's no bigger incentive than that to prove people wrong. My philosophy is never to think what I can't do— only what I can. Disabled people, both athletes and non-athletes, are not so different from everyone else. (Rushgrove 2012)

Human interest stories might help in what Australian academic Katie Ellis has aptly called 'the media navigation of physical difference and social stigma' (Ellis 2008–2009) and these were certainly provided. The specific nature of this 'human interest', though, could not be guaranteed to show Paralympians as 'not so different from everyone else'. For example, the media spotlight settled on Martine Wright, who had lost both legs above the knee in the London bombings of 2005. She had since married, had a child and now played for the British sitting volleyball team. 'Sport healed me', Martine had told a reporter in 2010 (Hubbard 2010). Now she was framed as an innocent casualty of terrorism—**Terror victim finally puts 7/7 nightmare behind her in Paralympic volleyball debut... and is cheered on by the off-duty policewoman who saved her life** (*Daily Mail*, 31 August 2012)—who had heroically embraced its consequences: 'I don't know whether it's spiritual or it's fate, but I really truly believe that I was meant to do this journey' (Laville 2012). In similar vein, a short promotional film broadcast by Channel 4 invited viewers to *Meet the Superhumans*[11]. While laudable, this approximated to the 'Super Cripple' depiction warned of by Colin Barnes: emphasis upon the extraordinary achievements of some (wholly unrepresentative) disabled people lessens the relevance (and, perhaps, demeans) the experiences of the great majority of disabled people (Barnes 1992).

Disabled writers were well aware of this. Wheelchair user Paul Carter observed:

a real dichotomy in public and media attitudes towards disabled people and disability in general. We may be currently surfing a wave of Games-inspired media goodwill. For the last year or two, however, coverage of us disabled types has been increasingly hostile and negative; fuelled, or at least gently encouraged, by a government keen to push its thorny agenda of welfare reform. Reported hate crime against disabled people is increasing. [...]Don't get me wrong. I'm not bashing the Paralympics, quite the opposite. I love them and everything they stand for. I think we're on the cusp of achieving a real change in the British public's attitude towards disabled people for the better. However, I'm not superhuman. I'm normal. In the non-pejorative sense of the word. (Carter 2012)

---

[11] It can be seen at https://www.youtube.com/watch?v=tuAPPeRg3Nw (Access 2 July).

Writer-researcher Frances Ryan was less optimistic: '"Scrounger", "faker", or "genuine"—the government has responded to the economic downturn with lives reduced to soundbites, and taken much of the media with it. Perhaps it is refreshing, then, to witness the change brought about by the Paralympics, to see disabled people described, not as villainous scroungers but as heroic inspirations. It would be misguided, though, to think that this was progress. One set of comfy caricatures has just been replaced with another'. 'The Paralympics', she said, 'gets depicted less as a sporting event and more a feast of courage' (Ryan 2012).

It was, of course, the gulf in perceptions identified by Carter and Ryan that anti-Atos protesters were trying to close. Activism and political protest among the disabled and their supporters had been growing for over 20 years, initially to promote the 'social model' of disability, but latterly and largely in response to the changes in government policy. The Centre for Disability Studies at Leeds University in the UK had opened in 1990, the Disabled People's Direct Action Network (DAN) had evolved in the early 1990s out of the Campaign for Accessible Transport and Disabled People Against the Cuts (DPAC) had formed in the autumn of 2010, following a big demonstration in Birmingham against 'austerity' measures announced by the coalition government earlier that year (Williams-Findlay 2011).

Along with UK Uncut[12] DPAC had led opposition to reductions in public expenditure and together these two groups declared the week bridging the end of August and the beginning of September 2012, immediately prior to the Paralympics, a week of direct action and civil disobedience (Jolly 2012). *The Independent* reported: 'Protesters gathered at the French company's UK Head Quarters in central London this afternoon to commemorate the thousands of people who have died after being declared fit to work. DPAC activists also marched on Atos offices in Cardiff, Glasgow and Belfast as part of a week of direct action dubbed the Atos Games' (Lakhani and Taylor 2012). Demonstrator and lifelong

---

[12] UK Uncut formed in 2010 in response to 'austerity' measures announced by the newly formed coalition government in Britain. Their title implied resistance to cuts in public expenditure which the government styled as 'savings'. They specialised in direct action—usually the occupation of buildings—against prosperous organisations known to pay little tax. Companion organisations were formed in the USA and Portugal.

wheelchair user Micheline Mason expressed herself 'just so angry and so horrified at the demonising of disabled people. We're being used as an excuse for the government to take resources from the poorest, most vulnerable people' (Addley 2012a). Duncan Mackay reported: 'Among those involved in the protest were Tara Flood, a former swimmer who won seven Paralympic medals, including a gold at Barcelona in 1992, who has herself received a letter notifying her that she will be assessed about whether she is allowed to keep her disability living allowance. Last week, she took part in a spoof opening ceremony where she was stripped of her Paralympic medals after a mock Atos assessment' (Mackay 2012).

Remarkably, the *Daily Mail*, temporarily no longer the scourge of the work-shy, permitted columnist Sonia Poulton to praise the 'hardworking' protesters and to challenge the emergent 'superhuman' paradigm. While reminding readers that the WCA had been introduced by Labour, Poulton insisted: 'A Paralympian is no more like a regular disabled person than I am like [Jamaican sprinter] Usain Bolt or [British swimmer] Rebecca Adlington. Paralympians, just like Olympians, are unique in their field and should be regarded as such. So, to be clear, Paralympians are not a representation of the majority of disabled people. Even if [Prime Minister David] Cameron and Co. would like us to buy into that belief'. Nor did she absolve Atos:

> no-one has forced Atos to do it and rake in billions[13] while they are about it. Using the argument that they are only supplying a service and obeying orders is not dissimilar to the one trotted out by Nazi officers as they shepherded human beings into gas chambers. They were said to be merely conforming to requests, and they were, but it didn't clear them of their role in the barbaric treatment

Poulton's argument, best summarised in her headline, was that 'The Paralympics celebrate the strength of disabled people—as do all the protests that accompany them' (Poulton 2012).

However, other facets of these protests helped to open up the politics of the Olympic movement and, thereby, the rationale for Atos' involvement.

---

[13] At the time, Atos held £3 billion worth of contracts across ten British government departments (see Addley and Malik 2012).

During the opening ceremony of the Paralympics, observers noticed that the Atos branding on the passes worn by British athletes could not be seen and disability activists speculated, via Twitter, that this represented tacit support for the week of action.[14] This drew a swift and predictable denial from the British Paralympic Association (Addley and Malik 2012); whether the badges had been deliberately concealed or not, to admit that they had would have meant expulsion from the Games for the athletes concerned. Moreover, when Paralympic officials were pressed over the increasingly controversial partnership with Atos, they simply reiterated the indispensable contribution that the partner had made to the staging of the Games. Jackie Brock Doyle, Director of Communications and Public Affairs for the London 2012 Olympic and Paralympic Games, told the media: 'Without the sponsors there would be no Games' (see, for example, Lakhani and Taylor 2012). This is now a standard response. As I have argued elsewhere, there were contentious aspects to the business practices of most of the partners of the Olympics Games of 2012 and the inflexible restatement of the Without-Them-No Games argument, underscoring their role in facilitating a popular event, was part of what they were paying for. (See Wagg 2015: 119–127). Similarly, Lord Coe, chair of the London Organising Committee of the Olympic Games (LOCOG) readily dismissed the crowd's booing of various coalition government ministers who stepped forward to present medals at the Paralympics:

> There are 500 medal ceremonies, we require over 1,000 people, not just politicians, and from time to time, I know from my own personal experience, you do become the pantomime villain in politics. I don't think that we should read too much into that and I think it's really important that politicians have been seen supporting the two greatest sporting events in our lifetimes. Politicians are bold enough and brave enough to know that sometimes that is the landscape that they are in[15]

---

[14] See, for example, 'Did GB Paralympians choose to hide Atos branding?' on the Channel four website: http://www.channel4.com/news/did-gb-paralympians-choose-to-hide-atos-branding (Posted 30 August 2012; access 3 July 2015).

[15] See http://www.express.co.uk/news/uk/344172/Coe-plays-down-medal-ceremony-boos (Posted 6 September 2012; access 3 July 2015).

This Goes-With-the-Territory rendition emptied the controversy of its specifics, instead making it simply about 'politicians', characters in some ongoing pantomime.

At the beginning of the week, activist Debbie Jolly had suggested Atos' sponsorship of the Paralympics was 'beyond irony' (Jolly 2012). Public opinion may have been that there was at least a case to answer here, but, when challenged on this sponsorship, both Brock Doyle and Coe had, in effect, simply re-invoked the intellectually discredited, but still strategically advantageous, dictum that politics should not intrude upon sport.

Lifting the Clouds of Limitation? The Atos Games and Political Legacy

After the Games, in a widely reported eulogy, Lord Coe said: 'I really genuinely think we have had a seismic effect in shifting public attitudes. I don't think people will ever see sport the same way again, I don't think they will ever see disability in the same way again. One of the most powerful observations was made to me, by one of our volunteers, who talked about having lifted some of the clouds of limitation.'[16] Prime Minister David Cameron spoke in similar terms, revisiting the vocabulary of the 'super': 'It's been an absolute triumph. What it means for disabled people, the families of disabled children… it means a lot to an awful lot of people. You often think about what people can't do, but the Paras are about them being superhuman—about all the things they can do'[17]. And Richard Hawkes, Chief Executive of the disabled charity Scope, told the Conservative website *Politics Home*:

> We hoped the Games would have an impact beyond sporting success, and improve attitudes to disabled people. Some 84 % of disabled people told

---

[16] 'Lord Coe hails Paralympic's 'seismic effect in shifting public attitudes' to disability' *London Evening Standard* 9 September 2012 http://www.standard.co.uk/olympics/paralympics/lord-coe-hails-paralympics-seismic-effect-in-shifting-public-attitudes-to-disability-8120539.html (Access 4 July 2015).

[17] http://paralympics.channel4.com/competitions/london-2012/news-pictures/news/newsid=1237031/ (Access 4 July 2015).

us that greater public discussion of their lives would improve attitudes. For the last week-and-a-half disabled people have been everywhere. The focus is rightly on the sport, but disability has never been so consistently, openly and widely talked about. Attitudes don't change overnight but this could be the start. The legacy of the games should be a Britain where we focus on what disabled people can—rather than can't—do and where we have the support in place so that disabled people can achieve their aspirations— whether that's taking part in the 2016 Paralympics in Rio or simply being able to go to the pub with friends to watch the games[18]

This was, self-evidently, an empowering discourse, but free marketeers and disabled activists alike were quick to note its implications. In the *Daily Express*, columnist Leo McKinstry trod where even many other right-wing commentators had feared to tread:

Absurdly, Cameron and his Cabinet have been painted as reactionary demagogues.

Some have even drawn a parallel with the Third Reich. "Cameron and Iain Duncan Smith [Secretary of State for Work and Pensions] are little more than Hitler and Goebbels in suits, the chief architects of the Government- sponsored anti-disability pogrom," writes one protester. Such language is as offensive as it is misplaced. Despite all the emotional blackmail, it is ludicrous to suggest that the coalition is engaged in an ideological attack on the disabled. What ministers are trying to do is reform the welfare system so that it no longer rewards fraudsters and scroungers, but instead targets its funds at those most in deed. The real hypocrites at the Paralympics are not the Government but the professional protestors, whose mentality is the opposite of the spirit of the Games. The Paralympics is all about transcendence, where the athletes use their courage and resolution to conquer bodily constraints (McKinstry 2012)

For the protesters, 'professional' and otherwise, the honeyed rhetoric in which the Paralympians were now bathed obscured a continued retrenchment in the lives of most disabled people. Cameron's words, said the *Manchester Evening News*,

---

[18] https://www.politicshome.com/articles/opinion/scope/paralympics-showed-what-disabled-can-do (Posted 10 September 2012; access 4 July 2015).

came against a backdrop of uncertainty for many disabled people. In Greater Manchester, 200 people recently lost their jobs after the government decided to close down four Remploy factories, which employed those with disabilities. Thousands more face 'fit for work' tests which are aimed at cutting the numbers of those on benefits. Peter Williamson [an epileptic], who worked for Remploy for 26 years before it was shut, described Mr Cameron's comments as 'an insult'. He said: "It's a load of rubbish. He is just saying what he thinks people want to hear. It is heart-breaking to see him do this". (Keegan 2012)

At the same time, the Remploy factory in Springburn, Glasgow, which manufactured wheelchairs, was facing closure and its staff were on strike. Phil Brannan, the GMB union's shop steward for the factory, said: 'My members overcome their disabilities every week coming to work. They may not win gold medals at the end of it but they do create a world-class product.'[19]

Paralympians themselves noted the apparent disparity between rhetoric and political reality. A year on from the Paralympics, the *Daily Mirror* interviewed the widely venerated wheelchair athlete Tanni Grey-Thompson. 'I've lost track of the number of letters from disabled people who have been spat at in the street', said Thompson.

Letters from people being shouted at and abused. Instead of the deserving and undeserving poor, we have got deserving and undeserving disabled people.' As the first anniversary of the Games approaches and the word "legacy" is everywhere, it is a cruel irony that disabled people find themselves under sustained attack. Not just from the Government, but also from members of the public swept up in a war on welfare led from Westminster. 'One letter I received described how a disabled person was in a bus queue and someone came up and started asking them how many thousands in benefits they were costing,' Tanni says 'This was a working disabled person who takes nothing in benefits. There is suddenly a massive mismatch between how Paralympians and everyone else with disabilities are viewed. The irony is that of course there are Paralympians who are losing benefits under welfare reforms'. (Wynne Jones 2013)

---

[19] *Daily Record* 4 September 2012 http://www.dailyrecord.co.uk/news/scottish-news/two-faced-cameron-praises-disabled-athletes-1301957 (Access 4 July 2015).

Weeks earlier, it had been reported that the Prime Minister had refused to ratify Grey-Thompson's appointment as chair of the key national funding body *Sport England*. As Simon Hart, in the right-wing *Daily Telegraph*, put it:

> Grey-Thompson sits in the House of Lords as an independent cross-bencher, but she has been a vocal opponent of the Government's policies on disability benefits. It is claimed that a number of Tory peers sought private meetings with David Cameron to voice their concerns that she was too "political" to run the country's largest sports quango and that her appointment should be blocked. (Hart 2013)

In 2014, Atos extended its TOP (The Olympic Partner) partnership with the International Olympic Committee until 2020 and bought itself out of its DWP/WCA contract, citing death threats to its staff (Chorley 2014). As columnist Zoe Williams astutely observed, the company had acted as 'flak catcher' for government policy: 'These firms take the fall, and their reward is that they are lavishly paid to enforce other unpopular and/or ill-considered policies, for which they will again take the fall' (Williams 2012). The assessments, of course, continued, now administered by the American company Maximus (Gentleman 2015); the government, as Williams had observed, remained the organ grinder, having simply engaged a new monkey.

Nor was there any sign of a change in government policy. In April 2013, the under-occupancy charge-known almost universally as the 'bedroom tax'-was imposed, under which council or housing association tenants faced a 14 % reduction in their benefits if they were deemed to have a 'spare' bedroom, and of 25 % for two such rooms.[20] The following year, Mr Duncan Smith refuted claims that two-thirds of those affected were disabled, on the ground that their disablement was based on 'tenants' self-declarations' (Holehouse 2014). In 2015, the government claimed this charge had 'saved taxpayers around £1 billion' (Prince 2015). In June 2015, it was announced that the stars of the popular and long-running TV soap opera *Coronation Street* were among those

---

[20] http://england.shelter.org.uk/get_advice/housing_benefit_and_local_housing_allowance/changes_to_housing_benefit/bedroom_tax (Access 4 July 2015).

opposed to the closure of the Independent Living Fund, which helps the severely disabled to live on their own. Thousands of disabled people could be housebound or end up in care homes after another cutback this month. On June 30, the Tories will axe the Independent Living Fund which helps the severely disabled to live on their own. And campaigners fear it will cut off essential support for more than 17,500 disabled people. The Government plans to place the responsibility for the ILF on to local authorities. But councils are already struggling thanks to a 26 per cent cut in funding. And with £12bn in further welfare cuts to come, there is no spare cash to take on the extra expenditure (Mudie 2015)

Writer Rebecca Atkinson probably spoke for many when she said in the spring of 2015: 'Cameron said at the 2012 Paralympics, "When I used to push my son Ivan around in his wheelchair, I always thought that some people saw the wheelchair and not the boy." "He gets it," I thought. "He gets the way people with disability are looked through like a sheet of glass." Only he didn't. He doesn't. He can't see the humans for the deficit and debt. People with disabilities are just figures on the spreadsheet, books that would balance better if we all just went away' (Atkinson 2015). Moreover, as Dave King argued, the DWP tests were not actually about illness or impairment, but about the functionality of individual bodies: and in 'a world in which functionality rules, Atos as sponsors of the Paralympics makes perfect sense'. (King 2012)

Postscript

In mid-February of 2016, Justin Tomlinson, Minister for the Disabled in the Conservative government, announced a four-year stay of execution for the Independent Living Fund, implying to many disabled activists and their supporters that the protests described here had had some effect. See: http://www.disabledgo.com/blog/2016/02/government-agrees-four-more-years-of-ilf-transition-cash-for-councils/?utm_source=Disa bledGo+Blog+Update&utm_campaign=13aa93a2d6-RSS_EMAIL_CAMPAIGN&utm_medium=email&utm_term=0_444b4daa4c-13aa93a2d6-127135273

Acknowledgements I'd like to thank Viji Kuppan, Ashley Hardwell, Jon Dart, Gerard Goggin, Delva Campbell, Bob Williams-Findlay and Kayleigh Garthwaite for their help in the composing of this chapter.

References

Addley, E. (2012a). Paralympic sponsor Atos hit by protests. *The Guardian*, 31st August. Available at: http://www.theguardian.com/society/2012/aug/31/paralympic-sponsor-atos-hit-protests. Accessed 3 July 2015.

Addley, E., & Malik, S. (2012). Atos protesters clash with police in 'day of action' against Paralympics sponsor. *The Guardian*, 31st August. Available at: http://www.theguardian.com/society/2012/aug/31/atos-protest-paralympics-sponsor. Accessed 3 July 2015.

Atkinson, R. (2015). How David Cameron has betrayed people with disabilities. *The Guardian*, 15th April. Available at: http://www.theguardian.com/commentisfree/2015/apr/15/david-cameron-betrayed-disabled-coalition-retrogression. Accessed 4 July 2015.

Barnes, C. (1992). Disabling imagery and the media: An exploration of the principles for media representations of disabled people The British Council of Organisations of Disabled People/Ryburn Publishing. Available at: disability-studies.leeds.ac.uk/files/library/Barnes-disabling-imagery.pdf. Accessed 2 July 2015.

BBC News. (2006). "Radical" welfare reform detailed http://news.bbc.co.uk/1/hi/uk_politics/5211656.stm. Posted 26th July. Accessed 29 June 2015.

Becker, E., Hayllar, O., & Wood, M. (2010). Pathways to work: Programme engagement and work patterns. *DWP/Her Majesties Stationery Office*. Available at: https://www.gov.uk/government/uploads/system/uploads/.../rrep653.pdf. Accessed 29 June 2015.

Beveridge, Sir W. (1942). *Social insurance and allied service*. London: His Majesty'sStationery Office.

Bingham, G. (2012). I was scared about sounding patronising during the Paralympics, now I can't wait. *The Independent*, 24th August. Available at: http://blogs.independent.co.uk/2012/08/25/i-was-scared-about-sounding-patronising-during-the-paralympics-now-i-cant-wait/. Accessed 1 July 2015.

Brisenden, S. (1986). Independent living and the medical model of disability *Disability, Handicap and Society* (known as *Disability and Society* since 1993), *1*(2), 173–178. Available at: http://disability-studies.leeds.ac.uk/files/library/brisenden-brisenden.pdf. Accessed 24 June 2015.

Carroll, R. (2012). Paralympics: US finally engages with Games thanks to media interest. *The Guardian*, 27th August. Available at: http://www.theguardian. com/sport/2012/aug/27/paralympics-us-games-media-interest. Accessed 1 July 2015.

Carter, P. (2012). Will the Paralympics stop you staring at me? *The Guardian* 28th August. Available at: http://www.theguardian.com/society/2012/ aug/28/hope-paralympics-games-normalise-disability. Accessed 2 July 2015.

Chapman, J. (2010). 76 % of those who say they're sick 'can work': Tests weed out most seeking incapacity benefit. *Daily Mail*, 23rd July. Available at: http://www.dailymail.co.uk/news/article-1298192/76-say-theyre-sick-work. html. Accessed 30 June 2015.

Chorley, M. (2014). Sickness benefit tests firm pays the government to quit its £500million contract early after death threats to staff. *Daily Mail*, 27th March. Available at: http://www.dailymail.co.uk/news/article-2590590/ Sickness-benefit-tests-firm-pays-government-quit-500million-contract-death-threats-staff.html. Accessed 4 July 2015.

Currie, Edwina (2002) *Diaries 1987–1992* London: Little, Brown

Drake, R. F. (2000, November). 'Disabled People, New Labour, Benefits and Work'. *Critical Social Policy, 20*(4), 421–439.

DWP. (2008). *Training and development ESA handbook.* Office of the Chief Medical Adviser/Department for Work and Pensions.

Ellis, K. (2008–2009). Beyond the Aww factor: Human interest profiles of Paralympians and the media navigation of physical difference and social stigma. *Asia Pacific Media Educator, 19*, 23–35, June 2008/July 2009. Available at: core.ac.uk/download/pdf/11239826.pdf. Accessed 2 July 2015.

Finkelstein, V. (2001). *A personal journey in to disability politics.* Available at: http://www.independentliving.org/docs3/finkelstein01a.html. Accessed 25 June 2015.

Foot, P. (1995, June 16). Doctors on call. *Private Eye, 874*, 26.

Garthwaite, K. (2011). 'The language of shirkers and scroungers?' Talking about illness, disability and coalition welfare reform. *Disability and Society, 26*(3), 369–372.

Gentleman, A. (2011). The medical was an absolute joke. *The Guardian*, 23rd February. Available at: http://www.theguardian.com/politics/2011/feb/23/ government-reform-disability-benefits. Accessed 30 June 2015.

Gentleman, A. (2015). After hated Atos quits, will maximus make work assessments less arduous? *The Guardian*, 18th January. Available at: http://www. theguardian.com/society/2015/jan/18/after-hated-atos-quits-will-maximus-make-work-assessments-less-arduous. Accessed 4 July 2015.

Gibson, O. (2012). Paralympic TV deals break new ground as ticket rush contin- ues. *The Guardian,* 27th August. Available at: http://www.theguardian.com/ sport/2012/aug/27/paralympics-tv-deals-ticket-rush. Accessed 1 July 2015.

Glennon, L. (2011). The work capability assessment is a genuine source of anguish. *The Guardian,* 2nd March. Available at: http://www.theguardian. com/commentisfree/2011/mar/02/work-capability-assessment-anguish- disabled-people. Accessed 30 June 2015.

Goggin, G., & Newell, C. (2000). Crippling Paralympics? Media, disability and Olympism. *Media International Australia Incorporating Culture and Policy, 97,* 71–83.

Golding, P., & Middleton, S. (1982). *Images of welfare: Press and public attitudes to poverty.* Oxford: Basil Blackwell/Martin Robertson.

Grice, A. (2015). Thousands of disabled people declared "fit for work" died soon after, figures reveal. *The Independent,* 28th August, p. 4.

Griffiths, S. R. (1988). *Community care: Agenda for action. A report to the Secretary of State for Social Services.* London: HMSO.

Halligan, P. (2007, June). Belief and illness. *The Psychologist, 20,* 358–361. Available at: https://thepsychologist.bps.org.uk/volume-20/edition-6/belief- and-illness. Accessed 29 June 2015.

Halligan, P., Bass, C., & Oakley, D. (Eds.) (2003). *Malingering and illness decep- tion.* Oxford: Oxford University Press.

Harrington, M. (2010). *An independent review of the Work Capability Assessment.* London: The Stationery Office. Available at: https://www.gov.uk/govern- ment/uploads/.../wca-review-2010.pdf. Accessed 30 June 2015.

Harris, A., Cox, E., & Smith, C. R. W. (1971). *Handicapped and impaired in Great Britain.* London: Her Majesty's Stationery Office.

Hart, S. (2013). Tanni Grey-Thompson was offered Sport England role but appointment was vetoed by Number 10. http://www.telegraph.co.uk/ sport/10011572/Tanni-Grey-Thompson-was-offered-Sport-England-role- but-appointment-was-vetoed-by-Number-10.html. Posted 23rd April. Accessed 4 Feb 2016.

Holehouse, M. (2014). IDS doubts 'bedroom tax' disability figures. http:// www.telegraph.co.uk/news/politics/11032429/IDS-doubts-bedroom-tax- disability-figures.html. Posted 13th August. Accessed 4 July 2015.

Hubbard, A. (2010). Martine Wright: 'I lost both legs in the London bombings but sport healed me'. *The Independent,* 10th October. Available at: http:// www.independent.co.uk/news/people/profiles/martine-wright-i-lost-both- legs-in-the-london-bombings-but-sport-healed-me-2102529.html. Accessed 2 July 2015.

Hunt, T. (2011, February). Atos: Tick-box tyranny. *Red Pepper*. Available at: http://www.redpepper.org.uk/atos-tick-box-tyranny/. Accessed 30 June 2015

Jolly, D. (2011). The Billion Pound Welfare Reform Fraud: Fit for work? Available at: http://dpac.uk.net/2011/05/debbie-jolly-the-billion-pound-welfare-reform-fraud-fit-for-work/. Accessed 30 June 2015.

Jolly, D. (2012). The Atos Games will showcase disabled people's anger at Paralympic sponsors. *The Guardian*, 27th August. Available at: http://www.theguardian.com/commentisfree/2012/aug/27/atos-games-disabled-anger-paralympic-sponsors. Accessed 3 July 2015.

Keegan, M. (2012). 'Hypocrite' David Cameron is blasted by Manchester disabled campaigners after he praises Paralympians. *Manchester Evening News*, 11th September. Avaiable at: http://www.manchestereveningnews.co.uk/news/greater-manchester-news/hypocrite-david-cameron-is-blasted-by-manchester-694870. Accessed 4 July 2015.

King, D. (2012). Dave King: Down with the Paralympics! Down with Channel-4-liberalism! http://dpac.uk.net/2012/09/dave-king-down-with-the-paralympics-down-with-channel-4-liberalism/. Posted 8th September. Accessed 5 July 2015.

Lakhani, N., & Taylor, J. (2012, August 29). Hundreds protest against Paralympics sponsor Atos as anger about its role in slashing benefits bill intensifies. *The Independent*. Available at: http://www.independent.co.uk/news/uk/home-news/hundreds-protest-against-paralympics-sponsor-atos-as-anger-about-its-role-in-slashing-benefits-bill-intensifies-8092512.html. Accessed 3 July 2015.

Langan, M. (1990). Community care in the 1990s: The community care White Paper: 'Caring for People'. *Critical Social Policy, 10*(29), 58–70.

Laville, S. (2012, August 27). Martine Wright: From 7 July victim to Paralympic athlete. *The Guardian*. Available at: http://www.theguardian.com/sport/2012/aug/27/martine-wright-7-july-paralympics. Accessed 2 July 2015.

Mackay, D. (2012). Olympic sponsor targeted by angry disability groups over links to London 2012 Paralympics. http://www.insidethegames.biz/articles/18416/olympic-sponsor-targeted-by-angry-disability-groups-over-links-to-london-2012-paralympics. Posted 31st August. Accessed 3 July 2015.

Macrae, I. (2011). Grayling denies cuts and scrounger agenda. http://www.disabilitynow.org.uk/article/grayling-denies-cuts-and-scrounger-agenda. Accessed 30 June 2015.

Mayer, J. (2010, August 30). Covert operations: The billionaire brothers who are waging a war against Obama. *The New Yorker*. Available at: http://www.newyorker.com/magazine/2010/08/30/covert-operations. Accessed 25 June 2015.

McKinstry, L. (2012, September 6). The Paralympics show up a corrupt benefits system. *Daily Express.* Available at: http://www.express.co.uk/comment/columnists/leo-mckinstry/344254/The-Paralympics-show-up-a-corrupt-benefits-system. Accessed 4 July 2015

Mencap. (2008). Mencap briefing: An overview—The Welfare Reform Act 2007 and the new Employment and Support Allowance (ESA). Available at: https://www.mencap.org.uk/sites/default/files/documents/2008-08/Briefing_Overview_Welfare%20Reform%20Act%20and%20ESA_formatted.pdf. Accessed 29 June 2015.

Mendoza, K.-A. (2015). *Austerity: The demolition of the welfare state and the rise of the Zombie economy.* Oxford: New Internationalist Publications.

Miller, S. (2012). 10600 sick & disabled people died last year within six weeks of their claim ending. http://blacktrianglecampaign.org/2012/10/04/10600-sick-disabled-people-died-last-year-within-six-weeks-of-their-claim-ending/. Posted 4th October 2012. Accessed 30 June 2015.

Mudie, K. (2015). Independent living fund axe will see thousands of disabled people housebound or in care homes, campaigners fear. http://www.mirror.co.uk/news/uk-news/independent-living-fund-axe-see-5834070. Posted 6th June. Accessed 4 July 2015.

Murray, C. (1984). *Losing ground: American social policy, 1950–1980.* New York: Basic Books.

Oliver, M., & Barnes, C. (2012). *The new politics of disablement.* Basingstoke: Palgrave Macmillan.

Parsons, T. (1951). Illness and the role of the physician: A sociological perspective. *American Journal of Orthopsychiatry, 21*(3), 452–460.

Poulton, S. (2012, August 30). The Paralympics celebrate the strength of disabled people—As do all the protests that accompany them. *Daily Mail.* Available at: http://www.dailymail.co.uk/debate/article-2195885/The-Paralympics-celebrate-strength-disabled-people--protests-accompany-them.html. Accessed 3 July 2015.

Prince, R. (2015). Iain Duncan Smith: Government's controversial bedroom tax has saved taxpayers £1 billion. http://www.telegraph.co.uk/news/politics/conservative/11487387/Iain-Duncan-Smith-Governments-controversial-bedroom-tax-has-saved-taxpayers-1-billion.html. Posted 21st March. Accessed 4 July 2015.

Pring, J. (2013). Conservative conference: No regrets for Tory architect of 'fit for work' test. http://www.disabledgo.com/blog/2013/10/conservative-conference-no-regrets-for-tory-architect-of-fit-for-work-test/. Posted 16th October 2013. Accessed 29 June 2015.

Ramesh, R. (2010, November 23). Incapacity benefit faces overhaul after demining report. *The Guardian*. Available at: http://www.theguardian.com/politics/2010/nov/23/incapacity-benefit-test-review. Accessed 30 June 2015.

Read, C. (2011). ATOS: Notes on a Neoliberal Scandal. Available at: http://www.newleftproject.org/index.php/site/article_comments/atos_notes_on_a_neoliberal_scandal. Accessed 30 June 2015.

Robb, B. (1967). *Sans everything: A case to answer*. London: Thomas Nelson.

Rushgrove, B. (2012, August 29). Paralympics 2012: Why I hope this will be the breakthrough Games. *The Guardian*. Available at: http://www.theguardian.com/sport/2012/aug/29/paralympics-2012-hope-breakthrough-games. Accessed 2 July 2015.

Rutherford, J. (2007). New labour, the market state, and the end of welfare. *Soundings, 36*, 40–55. Available at: http://blacktrianglecampaign.org/2011/09/07/new-labour-the-market-state-and-the-end-of-welfare/. Posted 7th September 2011. Accessed 29 June 2015.

Ryan, F. (2012, August 28). Why the Paralympics won't challenge perceptions of disabled people. *The Guardian*. Available at: http://www.theguardian.com/society/2012/aug/28/why-paralympics-wont-challenge-perceptions-disabled. Accessed 2 July 2015.

Schouten, F. (2015, April 23). Who are the Koch brothers? *USA Today*. http://www.usatoday.com/story/news/politics/2015/04/23/koch-family-fred-frederick-charles-david-and-bill/26029549/. Accessed 25 June 2015.

Shipman, T. (2010). State-funded idleness: 1.5 m are spending fifth Christmas in a row on sick benefits. *Daily Mail*, 28th December. Available at: http://www.dailymail.co.uk/news/article-1342076/State-funded-idleness-1-5m-spending-fifth-Christmas-row-sick-benefits.html. Accessed 30 June 2015.

Slorach, R. (2015). *A very capitalist condition: A history and politics of disability*. London: Bookmarks.

Stratton, A. (2011). Up to 500,000 wrongly denied incapacity benefit, figures show. *The Guardian*, 3rd January. Available at: http://www.theguardian.com/politics/2011/jan/03/incapacity-benefit-compass-survey-dwp. Accessed 30 June 2015.

Sutherland, A. (2011). Vic Finkelstein: Academic and disability activist [Obituary]. *The Independent*, 16th December. Available at: http://www.independent.co.uk/news/obituaries/vic-finkelstein-academic-anddisability-activist-6277679.html. Accessed 25 June 2015.

Thorburn, G. (2012). Illness as 'Deviance', Work as glittering salvation and the 'Psyching-up' of the medical model: Strategies for Getting the Sick 'Back To Work'. Available at: http://blacktrianglecampaign.org/2012/07/25/illness-as-deviance-work-as-glittering-salvation-and-the-psyching-up-of-the-medical-model-strategies-for-getting-the-sick-back-to-work/. Accessed 30 June 2015 and https://internationalgreensocialist.wordpress.com/illness-as-deviance-work-as-glittering-salvation-and-the-psyching-up-of-the-medical-model-strategies-for-getting-the-sick-back-to-work/. Accessed 30 June 2015.

Waddell, G., & Aylward, M. (2005). *The scientific and conceptual basis of incapacity benefits*. London: The Stationery Office.

Wagg, S. (2015). *The London Olympics of 2012: Politics, promises and legacy*. London: Palgrave Macmillan.

Williams, Z. (2012). Atos is doing a good job—As the government's flakcatcher. *The Guardian*, 5th September. Available at: 'http://www.theguardian.com/commentisfree/2012/sep/05/atos-the-government-flakcatcher. Accessed 4 July 2015.

Williams-Findlay, R. (2011). Lifting the lid on disabled people against cuts. *Disability and Society, 26*(6), 773–778.

Wynne Jones, R. (2013). We've lost sight of the Paralympics legacy—Benefit reforms are victimising disabled. *Daily Mirror*, 23rd July. Available at: http://www.mirror.co.uk/news/uk-news/weve-lost-sight-paralympics-legacy-2082396. Accessed 4 July 2015.

'Messing About on the River.' Trenton Oldfield and the Possibilities of Sports Protest

Jon Dart

In April 2012 Trenton Oldfield, an Australian man in his mid-30s, disrupted the annual Boat Race between Cambridge and Oxford Universities by going for a swim in the River Thames. For some, Oldfield's timely swim in a public space was an imaginative and well-executed act of peaceful, civil disobedience which achieved maximum exposure and caused minimal damage. Live television coverage of the event and his use of social media allowed him to promote his manifesto 'Elitism leads to Tyranny' with Oldfield's actions an example of individual, autonomous political activity. This chapter considers the opportunities that a large sport event, here the Boat Race, offers to such individual autonomist protesters and how new forms of digital web-based media are changing the dynamic between sport, media and protest. The discussion focuses

J. Dart (✉)
Carnegie Faculty, Leeds Beckett University, Beckett Park Campus,
Leeds, LS6 3QU, UK

© The Editor(s) (if applicable) and The Author(s) 2016   **289**
J. Dart, S. Wagg (eds.), *Sport, Protest and Globalisation*,
DOI 10.1057/978-1-137-46492-7_13

on response to Oldfield's protest by sections of the English media and the UK government who, upset to see their sporting pleasures disrupted, sought to deport him from the UK.

In carrying out his direct action against what he saw as elitism and a widening division in British society, Oldfield targeted the highly symbolic event—commonly known as 'the Boat Race.' This event, held annually since 1856, is between the heavyweight eight-man crews from the elite universities of Cambridge and Oxford, is a non-ticket event which takes place between Putney Bridge and Chiswick Bridge (Mortlake) on the River Thames. Since its beginning, it has drawn huge crowds with the audience increasing when the BBC first broadcast the event in 1938. In 2012, Trenton Oldfield stopped the race for the first time in its 158-year history. He was immediately arrested, convicted of causing a public nuisance and sentenced to six months in prison. Following the direct intervention of a Home Office minister, he was then threatened with deportation (back) to Australia.

The discussion draws on a number of sources, primarily the British news media and other web-based material. It will show that there was an array of responses to Oldfield's action ranging from the centre-ground *Guardian* and *Independent* newspapers to the right-of-centre *Daily Mail* who gave extensive coverage to the event. It uses Trenton Oldfield's protest to examine the emerging predilection for autonomous individual protest. What was notable about Oldfield's protest is that it took place in a public space (i.e. the environs of the River Thames), rather than at a closed, ticketed event which often has its own increasing strict security measures in and surrounding the venue. By targeting this particular event, Oldfield increased the chances of gaining a significant TV audience and illustrates the changing relationship between sport, media and protest created by 'new' media tools and platforms. The chapter explores how concerns were raised that Oldfield's actions would inspire others to protest at the London 2012 Olympic and Paralympic Games. However, it did not, due in part to the legal and security measures that now surround most major sports events. The chapter concludes with consideration of the fragmented nature of the contemporary protest movements and the reluctance of the sporting audience to see their pleasures disrupted.

Sport as Platform for Political Protest

Protestors have long used sports events as a platform to protest. Activity can take place during, or surrounding, live events and so, securing sports events against unwanted intrusion has become increasingly important. The neophyte, lone protestor is potentially more able to avoid these security measures, particularly at non-ticket events, and use the element of surprise to their advantage. Arguably some of the most high-profile sporting protests have surrounded the Olympic Torch relay (Horne and Whannel 2010; Panagiotopoulou 2010; Tarantino and Carini 2010). One of the first post-World War II sports protests was in 1957 during the Wimbledon tournament when a woman called Helen Jarvis occupied the centre court during the men's double final waving a banner in her campaign for a new world banking system (Aldred 2012). In 1975, supporters of the 'Free George Davis' campaign destroyed the Headingley cricket pitch by digging holes and spreading engine oil with the final day of the final match abandoned, thus depriving England of the chance of winning 'the Ashes' (Wilkes 2011).[1] One of the most active disruptors of sports events in recent years has been Neil Horan, a laicised Irish Roman Catholic priest (BBC 2005; Mirror 2009; O'Connor 2010), although his actions have not been political protests, so much as a consequence of his poor mental health. In 2012, John Foley handcuffed himself to the goalposts during an English Premier League game at Everton's Goodison Park stadium in protest against the employment practices of Ryanair, whose Chief Executive was watching the match (Hough 2012). Described in the media as a 'known professional protester' Foley was arrested for 'pitch encroachment,' fined £665 and banned from attending Everton games for three years. He later received an anti-social behaviour order (ASBO)[2] banning him from attending sports grounds hosting an event to which the public had to pay to gain admission (Liverpool Echo 2012).

---

[1] The 'Free George Davis' campaign sought to increase public awareness of armed robber George Davis. Although his first conviction was overturned, he was later jailed for two other cases of armed robbery.

[2] An *ASBO* is an Anti-Social Behaviour Order issued by UK courts to people who behave in a persistently unacceptable manner. They were subsequently replaced by Criminal Behaviour Orders, and similar to ASBOs, aimed at stopping anti-social behaviour before it escalates.

Foley has protested at other sporting events including during the final day of England versus Australia cricket match (Brown 2009), and the Cheltenham horse racing festival where he ran onto the racecourse and waved a banner during the Ryanair Chase (Rossington 2011). However, none of these protesters attracted the level of media attention, or notoriety, achieved by Trenton Oldfield.

While traditional forms of resistance have not disappeared, as Harvey, Horne and Safai (2009: 399) suggest, 'new historical conditions are creating new forms of resistance.' Academic interest has been stimulated by the qualitative changes of protest which have been created by social media and their influence on the new social movements, such as that found around campaigns linked to sport and environmentalism (Harvey and Houle 1994; Stoddart and MacDonald 2011; Wheaton 2007, 2008). Informing these changes have been developments in digital web-based media which have altered the dynamic between sport, media and protest with the 'DIY' factor creating new opportunities to disseminate messages quickly, widely and cheaply (Earl and Kimport 2011; see also the chapter by Zervas in this book). The internet has created a medium for protestors to provide information directly to a wider public and thus reduce the traditional gate-keeping role performed by the mainstream mass media—something Oldfield used to his full advantage.

Trenton Oldfield and the 2012 Boat Race

At the time of his protest Trenton Oldfield was 35-year-old London resident. Educated at a private school in Australia he later transferred to a state school. Although Oldfield rowed whilst at the private school, he did not make mention of his status as a rower to frame his protest action. He moved to London where he studied for an MSc in Contemporary Urbanism at the London School of Economics and became a Fellow of the Royal Society of Arts. He then worked for the Thames Strategy group with responsibility to regenerate the river landscape from Kew to Chelsea (Pearlman et al. 2012).

The annual 'Boat Race' takes place on a Saturday afternoon around the end of March/beginning of April between the Boat Clubs from

Oxford and Cambridge University (specifically, the heavyweight men's eight). The first race took place in 1829 and was one of many sporting events organised between the universities. Neither the River Cam (in Cambridge) nor River Isis (in Oxford) were sufficiently wide or straight enough so the first events were held on the River Thames at Henley, beyond the tidal river. There are other boat races between the universities (i.e. women's, lightweight eights), but these take place during the Henley event (usually a week earlier). Since 1976 the race has been sponsored by a betting group, a gin manufacturer, an outsourcing company and a financial services business. Although the most recent sponsors did not disclose a figure for its five-year deal, Wigglesworth (1992) identified that during the 1990s, the two clubs received £70,000 each annually from their sponsors and the BBC.[3] Stereotypically, 'English' with the River Thames as the arena and the London skyline as the backdrop, the Boat Race is a well-established fixture in English sporting calendar along with Royal Ascot, the Ashes, the Grand National, Wimbledon and the FA Cup Final.[4] That it attracts a large spectator audience and significant television audience was one of the reasons why it was chosen by Trenton Oldfield.

According to press reports, Oldfield was first spotted approximately ten minutes into the 17–18-minute race by the assistant umpire, four-time Olympic Rowing medallist Sir Matthew Pinsent. The race was immediately stopped because Oldfield was swimming directly into the path of the boats and could have been severely injured if struck by an oar or part of the race boat. Oldfield submerged himself as the boats passed safely overhead. He was then pulled onto an official race boat and arrested by the river police. The event was restarted after a 25 minutes delay.

---

[3] Topolski (1990) states that in 1986 the commercial sponsor, Beefeater Gin, paid £330,000 to the two boat crews for three years.

[4] The Boat Race is not one of the so-called 'Crown Jewels' of listed sporting events which must be broadcast on free-to-air television. In 2004, organisers sold television rights to ITV after decades of the race being shown on the BBC, although BBC resumed coverage 2010 April. In 2015, BNY Mellon (men's) Boat Race and the Newton Women's Boat Race took place on the same afternoon over the same course.

The Manifesto: Elitism Leads to Tyranny

As suggested earlier, there has been a change in the dynamics of the relationship between sport, media and protest brought about by new forms of digital media with political activists successfully exploiting digital media. Given his protest was contentious, it was important that his Oldfield's statement was unmediated and directly and immediately available to the widest possible audience; he would therefore need to circumvent the traditional mainstream media to avoid his message being distorted or diluted. At the same time, it provided 'instant copy' for the mainstream media. Shortly before his carrying out his protest, Oldfield posted his manifesto online. His intention was to enable those using social media to find out about this protest by selecting key words that could be used by internet search engines. The full title of his manifesto *Elitism Leads to Tyranny* (*Boat Race Swimmers manifesto*) carried a subheading '*Performance upon Thames*' and was accessed via the Indymedia website (Oldfield n.d.). Once his manifesto was located, it was possible, via twitter, for his name to spread quickly and thus negate those who might seek to deny him what Margaret Thatcher in 1985 called 'the oxygen of publicity.'[5]

In his manifesto, Oldfield declares how *"This is a protest, an act of civil disobedience, a methodology of refusing and resistance"* before offering a brief history of capitalism, although that specific term is not used. He then identifies his use of guerrilla tactics: local knowledge, ambush, surprise, mobility, speed, detailed information and decisiveness. He explains the location of the protest as not only one of natural beauty, but also one replete with elitist establishments, including Fulham Palace (the residence of the Bishop of London), Chiswick House (historic seat of the Duke of Devonshire, now maintained by English Heritage), St Paul's School (and other prestigious fee paying schools), as well as being home to Nick Clegg (the Deputy Prime Minster). Oldfield explains this particular sporting event was chosen because it was a symbol of class elitism, partly informed by the event programme which lists the (private) schools

---

[5] In a speech to the American Bar Association, 15th July. See http://www.margaretthatcher.org/document/106096 (Access 14th July 2015).

the rowers attended before they 'went up' to Oxbridge. On his release from prison, Oldfield stated that he chose the Boat Race because it would have 'limited impact' on working people, affect a very small group of people, but have a profound and symbolic effect (Radio 5 Live 2012).[6]

In his manifesto, Oldfield states his desire to create a climate in which individuals are no longer 'victims' but set the agenda, placing the 'elite' on the back foot, increasing their costs, causing confusion, fermenting internal distrust and creating embarrassment and frustration. The current conditions in society were 'disorganised' and he offers a list of various suggestions on how protesters could exploit these conditions, including,

- Security guards setting off fire alarms in buildings where people work (at strategic times—for example, when there is a meeting in which cuts—for example, to jobs or public expenditure—are to be discussed).
- Work slowly, make mistakes, lose documents, send large emails to clog up accounts.
- Taxi drivers: take passengers on the slowest and most expensive route.
- Plumbers: create problems when called to office of a right-wing think tank.
- Tow truck drivers: park in front of Clegg/ [Prime Minister] Cameron's drive/tow their car away.
- Bike riders: chain your bike to the corporate bikes (in the absence of bike racks).
- Cleaners: don't put toilet paper in the toilet/bathrooms of the elite.
- Restaurant workers: serve the food cold—or the wrong food.

---

[6] This is not the place to consider whether rowing is elitist, but clearly, there is a widespread perception that it is, along with polo and golf. The public's perception of rowing was not helped when the 2012 Olympic rowing events were held at Eton Dorney, a private boat lake owned by one of Britain's most exclusive private schools. The composition of the Boat Race crews, between 1829 and 1992, contained 16 oarsmen (out about some 1600) from a non-private/independent school (Wigglesworth 1992). Similarly, as to whether Oxbridge is symbol of elitism divides opinion. Despite various initiatives, Oxbridge remains elitist in terms of its over-representation of students drawn from elite, private schools in comparison to state schools (The Sutton Trust). This over-representation is carried forward in terms of Oxbridge alumni entering senior levels in the civil service, finance, the judiciary and mainstream politics.

- Builders: repairing a house and bug it and share the footage/audio online.
- Call centre workers: find customers the best/cheapest deal possible.
- Let off stink bombs at events.
- Ask lots of questions to delay a decision/remove essential cables at conferences.

The list ends with a message to security guards: 'the elite depend on you.' Oldfield concludes his manifesto by inviting others to follow in his (and Emily Davison's[7]) footsteps by inviting them to disrupt and desta-bilise the forthcoming Olympics. The manifesto is linked to Oldfield's website 'This Is Not A Gateway,' a not-for-profit organisation calling for critical investigation into how modern cities are run.[8] Funded by Oldfield and his partner from private income, it has received financial support from the Arts Council, Tate Britain and the Institute for Contemporary Arts. Seeking to problematise agreed meanings and calling for critical pedagogy (Freire 1996, 2001), its steering group (comprising artists, architects, urban planners, curators, photographers and software devel-opers) organises salons and arts events, offering advice to charities and private businesses on how they may unwittingly cause, or perpetuate, inequalities.

Reponses to the Protest

According to media reports of his trial, Oldfield's protest caused great mental and physical stress to the rowers (Robinson 2012). Some of the rowers took to social media to vent their anger with the President of OUBC tweeting: '*my team went through seven months of hell. This was the culmination of our careers and you took it from us*' (Petro 2012). Oxford crew member William Zeng offered a philosophical series of tweets:

[7] Davison, a campaigner for women's suffrage, died in a collision with the King's horse during the Epsom Derby of 1913—see Carol Osborne's chapter in this book.

[8] The name of the organisation has been explained by Oldfield as a self-reflexive critique drawing on Magritte's painting 'This is not a pipe,' and Foucault's book of the same name. (See also Bauman's 2012 text 'This is not a diary').

When I missed your head with my blade knew only that you were a swimmer, and if you say you are a protestor then no matter what you say your cause maybe, your action speaks too loudly for me to hear you. I know exactly what you were protesting, you were protesting the right of 17 young men and one woman to compete fairly and honourably, to demonstrate their hard work and desire in a proud tradition. You were protesting their right to devote years of their lives, their friendship, and their souls to the fair pursuits of the joys and hardships of sport. You, who would make a mockery of their dedication and their courage, are a mockery of a man. (Petre 2012)

Matthew Pinsent (ex-Eton, Oxford and four times Olympian) who was acting as assistant umpire in the race offered a liberal response in stating how he wants "*to live in country where protest is possible. However unwelcome it was, I still value the freedom to do that*" (Pinsent 2012). This position uses the protest to exalt 'Western' freedoms, at the same time as these 'freedoms' are being severely eroded in order to 'protect' sports events (as will be discussed below).

The mainstream British press reported the disruption in a fairly predictable manner. *Daily Mail* newspaper columnist Melanie Phillips, well known for her right-wing, reactionary hyperbole, had evidently read his manifesto when she expressed concern that the Internet and social media had 'the potential to transform every crank, narcissist and bully into an instant celebrity.' Describing Oldfield as a 'grudge guerrilla' she chose to focus on his privileged upbringing rather than his 'unorthodox political views' (Phillips 2012). In the same newspaper, Jonathan Petre (2012) briefly mentioned the protest but concluded that 'it was not the only drama to affect the race.' Much of the article focused on a snapped oar, the collapse of a rower at the end of the race, the 'quick witted' reaction of Oxford's female cox Zoe de Toledo in alerting her crew to the presence of Oldfield, along with her education at fee-paying St Paul's Girls' School in London and her bit-part acting career.[9] By contrast, the centre-ground *Guardian* focused on the criminalisation of protest (Power 2012a) and on

---

[9] *Plus ca change…* After the Emily Davison incident, the King referred in his diary to 'a most disappointing day' whilst Queen Mary sent the jockey concerned a telegram wishing him well after his 'sad accident caused through the abominable conduct of a brutal lunatic woman.'

whether rowing was elitist (Cross 2012); those sympathetic to Trenton's protest had to go to social media and news sites such as Indymedia.[10]

Oldfield was initially detained under Section 5 of the Public Order Offence, namely, behaviour likely to cause harassment, alarm or distress. This was 'upgraded' from a minor public order offence (which had no custodial option and usually incurred a small fine) to one of 'public nuisance.' This 'upgrade' was instigated by Michael Ellis (Conservative MP and member of Home Office Select Committee) who wanted a more substantial punishment, given the forthcoming London Olympics and the perceived need to deter 'lone-wolf' protesters. This intervention by politicians in the legal system was seen by some as analogous to the 'Russian Pussy[11] riot' when the Russian government sought to influence the legal process (Criado-Perez 2012; Godwin 2012; see Lenskyj's chapter on the 2014 Sochi Winter Olympics in this book).

In October 2012, Oldfield faced a short court case. As outlined above, his protest was peaceful and non-violent and, after hearing the case, the jury asked for the judge to show leniency with the probation officer recommending a non-custodial sentence (Mair 2013). In spite of this, in what was clearly a 'deterrent sentence', the Judge stated that Oldfield's actions were dangerous and disproportionate and his *"decision to sabotage the race [was] based on the membership or perceived membership of its participants of a group to which you took exception."* She stated that every individual and group of society was entitled to respect and that in *"a liberal and tolerant society [...] no-one should be targeted because of a characteristic with which another takes issue."* This is not the place to question the Judge's clearly worded statement in which she claims there is no social class or group that does not have a label and therefore requires judicial protection from interference and possible offensive; however,

---

[10] One footnote to the typically reactionary coverage of the *Daily Mail* and the immediacy of social media was the online publication by the newspaper which linked Oldfield with one of the 'bête noires' of the paper, Abu Hamza. Maxfarquar (2012) identifies how the *Daily Mail* was the victim of a 'sting' when they published an online article that linked Trenton Oldfield to Abu Hamza. The origin of the information (i.e. their source) was the @TrentonOldfield twitter account, which turned out to be a spoof account. The original *Daily Mail Online* article was subsequently removed.

[11] 'Pussy Riot' is the name of a Russian feminist punk rock collective, who staged a protest against President Putin in a Moscow cathedral in February 2012. Several members were subsequently imprisoned, with Putin claiming they had insulted the church.

can Oldfield's actions properly be seen as a 'hate crime' against the elite? What the judgement does reveal is an increasing tendency to criminalise protest and undermine protest against legitimate causes, such as inequality. The Judge advised Oldfield that he should have limited his protest to the unfurling of a banner on the riverside and shown more clearly, what he was protesting against. Oldfield was convicted of causing a public nuisance and jailed for six months and fined £750. He was released with an electronic tag after serving seven weeks.

In June 2013, Oldfield was refused leave to remain in the UK with the Home Secretary, Theresa May, recommending his deportation to Australia.[12] Nine months earlier May had given a speech at her party conference in which she vowed to 'deport foreign criminals first, then hear their appeals' (Chorley et al. 2013). In December 2013, Oldfield went to the Immigration Appeals Tribunal to challenge against the Home Office's decision to deny him a spousal visa that would mean he could stay in the country.[13] During the court case, the Home Office argued that Oldfield had defied British law in what was a dangerous protest. Lawyers for the Home office stated that,

> The appellant, in an act of utter contempt for the law, abused his right to protest in a very public way, he endangered himself and other in the process. The whole nation saw this and there is a need to be firm against this type of behaviour. Those who come to the UK must abide by our laws. We refused this individual leave to remain because we do not believe his presence in this country is conducive to the public good. (Booth 2013)

Oldfield submitted a file containing over 100 letters highlighting the positive impact of his community work, including ones from an Oxford professor and the head of Central Saint Martins College of Arts and Design (Bland 2013). Oldfield argued that the visa refusal was disproportionate and that it breached article 8 of the European Convention on Human Rights, which guaranteed the right to a family life. The appeal judge agreed

---

[12] In 1788 the British Empire English state began to use Australia as a penal colony. Transportation of English prisoners officially ended in 1868 but this narrative continues to feature in Anglo-Australian relations.

[13] Oldfield had originally held a Tier One visa which is given to highly skilled migrants.

with this and stated there was 'a public interest in providing a platform for protest at both common law and the European Convention on Human Rights' (Mair 2013).

The threat of deportation experienced by Oldfield was one shared by thousands of migrants that year. However, most of those facing deportation did not have the cultural capital (white, male, educated, heterosexual, able-bodied) and support network which he was able to draw upon. He was punished because his was a high-profile case; but because it was high profile, he was able to garner wider support. The state's reaction to protest has, arguably, become increasingly repressive with a decision made to 'make an example' of Oldfield. This was seen by many, including Oldfield, as 'a vindictive decision, very political and very much an overreaction' (Mair 2013) and his successful challenge (to his deportation, if not his initial conviction) undermined the attempt to deter those, including migrants, who might consider engaging in political protest.

Individual Protest and New Social Movements

In this section, I suggest that Oldfield took inspiration for his protest from the Situationist movement of the 1960s and explore how his action chimes with the politics of contemporary protest. As a piece of theatrical political activity, Oldfield sought to achieve maximum exposure and cause the minimal amount of damage and, at the time of his protest, he was not linked to any wider political group. Although Oldfield did not explicitly claim inspiration from this movement, one can infer that he was aware of, and took inspiration from, the 1960s Situationist International and the US Yippies (Buiani and Genosko 2012). Indebted to Surrealism, Bakhtin, Lefebvre, the Dada-inspired Lettrist International (Gardiner 2000), and based around the cult/personality of Guy Debord (Barnard 2004), the Situationist International was an exclusive, theoretical, theatrical grouping that saw the contemporary 'spectacle' of global capitalism as a fake reality which masked the degradation of human life (Kaplan 2012; Rasmussen 2004). Based in Paris, their focus was art, architecture, urban planning, political theory and artistic radicalism (Debord 1995). All forms of social life, including sport, were viewed as becoming increasingly spectacularised, banal, passive and unfulfilling offering pseudo-fulfilment, something they sought to undermine by 'raising incendiary beacons heralding a greater game' (Barnard 2004: 115), a core tactic being acts of sabotage, pranks or publicity stunts designed to subvert/destabilise the capitalist system (St John 2008). However, as a project of purely propagandist politics it achieved limited success. Although claimed to be anti-hierarchical, it was based around the cult of Debord and, since it was unwilling to organise or build beyond their own intellectual elite group, it gained little widespread support and ultimately fragmented.

As noted earlier, recent protest movements have become adept at using social media to promote their cause, as demonstrated by Oldfield in promoting his manifesto. While there is criticism of the hyperbole surrounding how the internet is effecting social change (Morozov 2012, 2013), the *'clicktivism/slactivism'* nature of activity has informed changes in protest with an increase in internet-based activity, campaigns and petitions, exemplified by the 'Anonymous' collective of cyber activists and the Avaaz organisation of internet-based petitioners (Eordogh 2013; Gitlin 2013; Jones 2011; Kavada 2012; Sarigollu 2011). Extensive debate is taking place amongst those protesting against capitalism as to which tactics are appropriate, with much of the debate centred on the relationship between the individual and the organised political party. St John (2008) suggests that the 'protestival' has emerged as a creative response to traditional political rituals of the organised political left: of marching from A to B with permits and police escorts followed by a mass rally and speeches from the leaders.[14] More direct forms of action based on diversity, creativity, decentralisation and horizontal organisation have, depending on one's political position, sought to complement or replace the 'old-fashioned' style of protest with Hardt and Negri's (2000) text on autononism, anarchism and direct action an important reference point in the genealogy of protest against global capitalism, social and economic inequality (see also Johnston and Laxer 2003; Byrne 2012; Chomsky 2012; Mason 2012). Penny's (2010) individualistic, no hierarchy/no

---

[14] Although this form of protest is not entirely dead, as demonstrated by recent protest in southern Europe and North Africa the recent years.

leaders approach to political activity can be challenged by those who suggest that leaders will emerge (in the absence of a hierarchy), leaders who are unelected and potentially unaccountable, leading to a tyranny of the structureless.

Although not identifying himself with the Occupy movement, there are links between Oldfield, Occupy[15] and the 2012 Olympic Games. The London Organising Committee for the Olympic Games (LOCOG) were very concerned that protestors would 'spoil their party' with Lord Sebastian Coe noting how 'one man's protest can destroy someone else's dream' (Gibson and Walker 2012). LOCOG and the British state were primarily concerned with a 'terrorist attack' (deploying the Army and placing a missile system on residential tower blocks in East London); they were also conscious that the Games were taking place one year after extensive rioting in various London boroughs. In the intervening year, social and economic conditions deteriorated, with the authorities increasingly concerned that the Games might be subject to disruption. The custodial sentencing of 2011 rioters thus became opportunity to set a strong deterrent to potential Olympic protestors (Casciani 2011; Davey 2012). Illustrative of the UK state's response to any potential protest was the treatment of Occupy 'participant,' Simon More who received what was believed to be the first pre-emptive Olympic ASBO requested by the Metropolitan Police; the ASBO restricted him from going within 100 yards of Olympic venues, participants, officials, spectators and the torch relay route (Walker 2012).

Making a Splash?

Making an assessment of the effectiveness of Oldfield's protest depends on what criteria are used. In terms of gaining media attention and creating a profile, Oldfield was very successful and, echoing a Situationist tactic, carried out an imaginative, peaceful act of civil disobedience.

---

[15] A movement to occupy elite or privileged spaces, begun on Wall Street in New York in October 2011 and spreading to a number of other countries and cities, including London, by the following year.

His planning and execution were very effective and showed the potential media impact if the 'right' event is chosen. The use of social media to publicise his message to a wide audience was also effective. Oldfield called for protestors to 'do something similar to Emily Davison' during the London 2012 Olympics; however, his calls were not heeded. Oldfield differed radically from Davison in that she was part of a bigger movement, one that had a clear objective (i.e. votes for women); in contrast, Oldfield was not evidently part of anything bigger than himself and is more appropriately linked with the individualistic approach to contemporary political action.

All the ingredients were available for London 2012 to be a significant site of protest. The Games took place 12 months after serious civil unrest and a few months after a successful occupation in the financial district of central London, with an unpopular government continuing to implement economic policies that were hurting many people. However, when the Occupy movement dismantled their tents and physically disbanded, the lack of coordinated planning contributed to its failure to remerge during London 2012. Gitlin (2013) and Calhoun (2013) have argued that the Occupy movement produced clever theatrical events, but the absence of an extended strategy, experienced networks and a stabilising organisational structure, meant they could not parlay small victories into action with longer-term potential. They also suggest that its ability to operate horizontally was limited because too much frictional energy was spent in self-maintenance. Even so-called 'spontaneous organising' requires unglamorous groundwork, an ordering of priorities and tasks, planning and coordination if it is to be effective. 'Spontaneous' can, in practice, mean 'ephemeral,' with social media lending itself to a lack of accountability and commitment. Gitlin (2013) and Calhoun (2013) also note that direct action can drive a wedge between protestors and liberals sympathetic to many of the movements' messages, as can be seen in the field of sport. The 'new' individualistic (for some, unaccountable) approaches to protest can be criticised for being primarily about protesting ('against everything'), and lacking a coherent, forward-looking strategy. Those seeking a coordinated, centralist approach do not see this as elitist, but rather as facilitating accountability, responsibility and commitment.

Questions have been asked as to whether such theatrical events are the actions of a movement or a moment? (Gitlin 2013; Calhoun 2013), and in this vein, one can question whether the 'dramatic performance' such as that enacted by Oldfield was just that, and offered little in terms of a tangible legacy or forward momentum. As Calhoun (2013) and Gitlin (2013) have noted, Occupy may better be seen as a 'moment' rather than movement. For Calhoun (2013) the absence of centralised leadership made it hard to negotiate forward planning, with no agenda of tactics to take the place of the occupation. This left many seeing Occupy as a gesture, the idealist moment, the performance rather than permanence, less an organisational effort (i.e. a movement), more a dramatic performance. However, Oldfield, Occupy, and those who did protest in during 2012 have contributed to the debate on capitalism. Consciousness-raising is an important element in any political action; one can point to how the Greenham Camp peace camp against US cruise missiles in the 1980s drew constant attention to the missile presence, even if the camp was dismantled long before the missiles were removed. Similar consciousness-raising was an important element within the British suffragists' campaign, the US civil rights movement and the South African anti-apartheid struggle.

Conclusion

What this chapter has shown is that although daring and resourceful gestures like Oldfield's are still possible, their likelihood is increasingly being restricted by the legal and security measures that surround major sports events, coupled to the fragmented nature of contemporary protest. Demonstrating at sports events is unlikely to engender sympathy from the wider public with the mainstream media reporting such actions (and, by default, their motivation) as little more than 'silly stunts'. Such actions are effective publicity stunts, but can remain just that if they are not connected to a wider political movement. Although his protest was against elitism, it was perhaps too random for many people to identify with. Therefore, care needs to be exercised to avoid enacting the politics of the spectacle, which does little to encourage others to get involved and become part of a wider, more inclusive movement.

Atomising protest and leaving individuals to decide what stunts/action they should take, risks undermining the tactic of coordinated unity, strength and accountability. At the heart of the matter is one's political philosophy and how this informs protesting against an organised, coordinate 'enemy'—in short one's views on the role of the individual and a centralised organisation.

Prior to London 2012, the organisers were very concerned the events would be an attractive site for protest, be this large or small. In addition to the visible deterrent, the British state sent a clear message to potential protesters in the sentences given to those associated with the 2011 summer disturbance and the pre-emptive use of ASBOs. What is evident is the increasing criminalisation of protest in the UK (Gilmore 2012; Power 2012b), and this is not restricted to major sports events (Graham 2012). In considering sporting events, what is manifested is how comparatively rarely they have been the site of, and for, protest. Although Trenton's protest was 'successful', the response of the mainstream media and UK government, coupled to the lack of activity around London 2012, identifies a reluctance amongst the sport-media audience to see their pleasures disrupted. However, the low levels of protest at sporting events cannot be due to the security and state jail sentences alone. Little protest takes place at sports events because it is more likely to alienate people and forfeit public sympathy. What is evident is the power of the corporate world and media to promote and frame sports events, and undermine those with a different agenda. Disrupting a sports event can also have a direct, negative impact upon individual athletes rather than the organisers, sponsors or the wider system. Coordinated action between the media, sponsors, organisers and athletes creates a powerful sporting ideology in sport is seen as a non-political arena and an activity that is 'off-limits' for protest. In this chapter, I have suggested that even though sports events are potentially significant sites for protestors to gain media attention, there is very little public support for those using sports events as a platform for protest. This is especially the case when the protest is conducted by an individual unaffiliated to a wider political organisation and lacking a clear agenda; disrupting sports events with a seemingly random, ad hoc interruption is likely to undermine the cause and leaving one looking, literally, like a spoilsport.

References

Aldred, T. (2012). Laura Robson's rainbow hair-band follows a long tradition of political statements in the sporting arena. *The Telegraph*. http://www.telegraph.co.uk/sport/tennis/australianopen/9021286/Laura-Robsons-rainbow-hair-band-follows-a-long-tradition-of-political-statements-in-the-sporting-arena.html

Barnard, A. (2004). The legacy of the situationist international: The production of situations of creative resistance. *Capital & Class, 28*(84), 103–124.

BBC News. (2005). Sport disrupting priest defrocked. *BBC News*. http://news.bbc.co.uk/1/hi/england/london/4193285.stm

Bland, A. (2013). Trenton Oldfield deportation: Boat Race protester allowed to stay in UK. *The Independent*. http://www.independent.co.uk/news/uk/politics/boat-race-protester-trenton-oldfield-wins-unprecedented-case-against-deportation-as-judge-overturns-theresa-mays-decision-8993322.html

Booth, R. (2013). Boat Race protester Trenton Oldfield wins appeal against deportation. *The Guardian*. www.theguardian.com/world/2013/dec/09/boat-race-protester-trenton-oldfield-wins-appeal-deportation

Brown, A. (2009). Airline protesters interrupt test. *ESPN Cricket Info*. http://www.espncricinfo.com/engvaus2009/content/story/413910.html

Buiani, R., & Genosko, G. (2012). Little brothers break the fifth commandment. *Cultural Studies, 26*(6), 934–955.

Byrne, J. (2012). *The occupy handbook*. New York: Little, Brown and Company.

Calhoun, C. (2013). Occupy wall street in perspective. *The British Journal of Sociology., 64*(1), 26–38.

Casciani, D. (2011). Riot convictions: How tough are they? Deterrent in the sentencing of last summer rioters. *BBC News*. http://www.bbc.co.uk/news/uk-14595102

Chomsky, N. (2012). *Occupy*. London: Penguin.

Chorley, M., Slack, J., & Chapman, J. (2013). 'Immigration system is like a never-ending game of snakes and ladders': Theresa May vows to kick out illegal migrants BEFORE they get chance to appeal. *Mail Online*. www.dailymail.co.uk/news/article-2438130/Theresa-May-Ill-kick-illegal-migrants-BEFORE-chance-appeal.html#ixzz3Dfj9FkQc

Criado-Perez, C. (2012, October 21). Is Trenton Oldfield our Pussy Riot? *New Statesman*. http://www.newstatesman.com/politics/2012/10/trenton-oldfield-our-pussy-riot

Cross, M. (2012, April 9). Rowing is elitist, but not in the way Trenton Oldfield thinks. *The Guardian*. http://www.guardian.co.uk/commentisfree/2012/apr/09/rowing-trenton-oldfield-boat-race

Davey, E. (2012). London England riots one year on: Culprits jailed for 1800 years. *BBC News.* http://www.bbc.co.uk/news/uk-england-london-19111720

Debord, G. (1995). *The society of the spectacle* (trans: Nicholson-Smith, D.). New York: Zone Books [Original edition: Debord, G. (1967). *La Societe du Spectacle* (Editions Buchet-Chastel) Paris].

Earl, J., & Kimport, K. (2011). *Digitally enabled social change: Activism in the internet age.* Cambridge, MA: MIT Press.

Eordogh, F. (2013). How anonymous have become digital culture's protest heroes. *The Guardian.* http://www.guardian.co.uk/commentisfree/2013/apr/15/anonymous-digital-culture-protest

Freire, P. (1996). *Pedagogy of the oppressed* (trans: Ramos, M. B., 2nd ed., Rev. ed.). London: Penguin.

Freire, P. (2001). *Pedagogy of freedom: Ethics, democracy and civic courage.* Lanham: Rowman & Littlefield.

Gardiner, M. (2000). *Critiques of Everyday Life.* London: Routledge.

Gilmore, J. (2012). Criminalizing dissent in the 'war on terror': The British State's reaction to the Gaza war protests of 2008–2009. In S. Poynting & G. Morgan (Eds.), *Islamophobia: Muslims and moral panic in the West.* London: Ashgate.

Gitlin, T. (2013). Occupy's predicament: The moment and the prospects for the movement Occupy's predicament: The moment and the prospects for the movement. *British Journal of Sociology, 64*(1), 3–25.

Godwin, R. (2012). Parallels between Trenton Oldfield trial and Pussy Riot... wife of Boat Race protestor speaks out. *Evening Standard.* http://www.standard.co.uk/lifestyle/london-life/parallels-between-trenton-oldfield-trial-and-pussy-riot-wife-of-boat-race-protester-speaks-out-8254705.html

Graham, S. (2012). Olympics 2012 security: Welcome to lockdown London. *The Guardian.* http://www.guardian.co.uk/sport/2012/mar/12/london-olympics-security-lockdown-london

Hardt, M., & Negri, A. (2000). *Empire.* London: Harvard University Press.

Harvey, J., & Houle, F. (1994). Sport, world economy, global culture, and new social movements. *Sociology of Sport Journal, 11*(4), 337–355.

Harvey, J., Horne, J., & Safai, P. (2009). Alterglobalization, global social movements, and the possibility of political transformation through sport. *Sociology of Sport Journal, 26,* 383–403.

Horne, J., & Whannel, G. (2010). The 'caged torch procession': Celebrities, protesters and the 2008 Olympic torch relay in London, Paris and San Francisco. *Sport in Society, 13*(5), 760–770.

Hough, A. (2012). Everton fan handcuffs himself to goalpost during 1-0 victory over Manchester City. *The Telegraph*. http://www.telegraph.co.uk/sport/football/teams/everton/9053340/Everton-fan-handcuffs-himself-to-goalpost-during-1-0-victory-over-Manchester-City.html

Johnston, J., & Laxer, G. (2003). Solidarity in the age of globalization: Lessons from the anti-MAI and Zapatista struggles. *Theory and Society, 32*(1), 39–91.

Jones, J. (2011). Social media and social movements. *International Socialism, 130*. http://www.isj.org.uk/?id=722

Kaplan, R. (2012). Between mass society and revolutionary praxis: The contradictions of Guy Debord's society of the spectacle. *European Journal of Cultural Studies, 15*(4), 457–478.

Kavada, A. (2012). Engagement, bonding, and identity across multiple platforms: Avaaz on Facebook, YouTube, and MySpace. *MedieKultur. Journal of media and communication research, 28*(52), 28–48.

Liverpool Echo. (2012). Ryan air protestor John Foley found guilty at Liverpool community justice centre of invading pitch at Everton—Man City game. http://www.liverpoolecho.co.uk/liverpool-news/local-news/2012/04/24/ryanair-protester-john-foley-found-guilty-at-liverpool-community-justice-centre-of-invading-pitch-at-everton-man-city-game-100252-30823617/

Mair, H. (2013). Judge rules Home Office's bid to deport Boat Race protester 'an overreaction.' *The Guardian*. http://www.theguardian.com/uk-news/2013/dec/23/judge-rules-home-office-deportation-boat-race-trenton-oldfield

Mason, P. (2012). *Why it's kicking off everywhere: The new global revolutions*. London: Verso.

Mirror. (2009). Britain's got talent exclusive: Irish dancer Neil Horan is defrocked priest jailed for Grand Prix stunt. *Daily Mirror*. http://www.mirror.co.uk/3am/celebrity-news/britains-got-talent-exclusive-irish-dancer-394598#ixzz2Q3txhZTP

Morozov, E. (2012). *The net delusion: How not to liberate the world*. London: Penguin.

Morozov, E. (2013). *To save everything, click here: Technology, solutionism, and the urge to fix problems that don't exist*. London: Allen Lane.

O'Connor, A. (2010). *The dancing priest: The story of Fr Neil Horan*. Dublin: Londubh Books.

Oldfield, T. (n.d.). Elitism leads to tyranny (Boat race swimmers manifesto). *Indymedia*. http://www.indymedia.org.uk/en/2012/04/494640.html?c=on

Panagiotopoulou, R. (2010). Greece: The Olympic Torch relay in ancient Olympia—An ideal showcase for international political protest. *International Journal of the History of Sport, 27*(9/10), 1433–1452.

Pearlman, J., Evans, M., & Hough, A. (2012). Trenton Oldfield: Boat Race protester's privileged Australian education. *Daily Telegraph.* http://www.telegraph.co.uk/news/worldnews/australiaandthepacific/australia/9193388/Trenton-Oldfield-Boat-Race-protesters-privileged-Australian-education.html

Penny, L. (2010). Out with the old politics. *The Guardian.* http://www.guardian.co.uk/commentisfree/2010/dec/24/student-protests-young-politics-voices

Petre, J. (2012, April 8). 'We went through seven months of hell and you wrecked it': Fury at protestor who tried to derail Oxford and Cambridge Boat Race. *Daily Mail.* http://www.dailymail.co.uk/news/article-2126695/Boat-Race-protest-2012-Swimmer-Trenton-Oldfield-tried-derail-Oxford-Cambridge-event.html

Phillips, M. (2012). A Boat Race rebel without a clue and the politics of narcissism. *Daily Mail.* http://www.dailymail.co.uk/debate/article-2127037/Trenton-Oldfield-A-Boat-Race-rebel-clue-politics-narcissism.html

Pinsent, M. (2012). Olympic rower Matthew Pinsent on Trenton Oldfield's Boat Race protest. *The Guardian.* http://www.guardian.co.uk/sport/2012/dec/23/protestor-trenton-oldfield-boat-race

Power, N. (2012a). The criminalisation of protest is part of the elite's class war. *The Guardian.* http://www.guardian.co.uk/commentisfree/2012/oct/19/boat-race-protest-class-war. Friday 19.

Power, N. (2012b). Dangerous subjects: UK students and the criminalization of protest. *South Atlantic Quarterly, 11*(2), 412–420.

Radio 5 Live. (2012). BBC Radio 5 Live's Nicky Campbell interviewing Boat Race Protester Trenton Oldfield. http://www.youtube.com/watch?v=qNzN8Gw1QwQ

Rasmussen, M. (2004). The situationist international, surrealism, and the difficult fusion of art and politics. *Oxford Art Journal, 27*(3), 367–387.

Robinson, M. (2012, October 19). Protestor who stopped Oxford-Cambridge race by swimming into paths of boats is jailed for six months. *The Daily Mail.* http://www.dailymail.co.uk/news/article-2220098/Trenton-Oldfield-Protester-stopped-Boat-Race-swimming-River-Thames-jailed-6-months.html

Rossington, B. (2011). Liverpool Ryanair protestor storms track at Cheltenham horse racing festival. *Liverpool Echo*. http://www.liverpoolecho.co.uk/liverpool-news/local-news/2011/03/18/liverpool-ryanair-protestor-storms-track-at-cheltenham-horse-racing-festival-100252-28358579/

Sarigollu, B. (2011). Avaaz—The possibility of a transnational public sphere & new cosmopolitanism within the networked times: Understanding a digital global utopia: "Avaaz.org" and a global media event: "Freedom Flotilla". *Online Journal of Communication & Media Technologies, 1*(4), 150–170.

St John, G. (2008). Protestival: Global days of action and carnivalized politics in the present. *Social Movement Studies, 7*(2), 167–190.

Stoddart, M. C. J., & MacDonald, L. (2011). 'Keep it wild, keep it local': Comparing news media and the internet as sites for environmental movement activism for jumbo pass, British Columbia. *Canadian Journal of Sociology, 36*(4), 313–335.

Tarantino, M., & Carini, S. (2010). Authoritarianism, opacity and proxies: The 2008 Olympic Torch relay in the Italian media. *International Journal of the History of Sport, 27*(9/10), 1473–1489.

The Guardian. (2012). Topless feminists protest against International Olympic Committee in London. http://www.guardian.co.uk/sport/video/2012/aug/02/topless-feminists-protest-olympic-london-video

Topolski, D. (1990). *True blue, the Oxford Boat Race mutiny*. London: Bantam Books.

Walker, P. (2012). Protester receives Olympics asbo. *The Guardian*. http://www.guardian.co.uk/society/2012/apr/17/protester-receives-olympic-asbo

Wheaton, B. (2007). Identity, politics, and the beach: Environmental activism in surfers against sewage. *Leisure Studies, 26*(3), 279–302.

Wheaton, B. (2008). From the pavement to the beach: Politics and identity in 'surfers against sewage'. In M. Atkinson & K. Young (Eds.), *Tribal play: Subcultural journeys through sport* (pp. 113–134). Bingley: JAI Press.

Wigglesworth, N. (1992). *A social history of English rowing*. London: Frank Cass.

Wilkes, D. (2011). George Davis IS innocent: He punches the air with joy after winning appeal against his conviction for 1974 robbery. *The Daily Mail*. http://www.dailymail.co.uk/news/article-1390400/George-Davis-punches-air-joy-winning-appeal-conviction-1974-robbery.html

Sochi 2014 Olympics: Accommodation and Resistance

Helen Jefferson Lenskyj

In 2007, when the International Olympic Committee (IOC) announced that it had awarded the 2014 Winter Olympics to Sochi, critics denounced the organization that defines itself as the 'moral authority for world sport' for rewarding a country guilty of widespread human rights violations. Developments in the next six years, culminating in Russia's 2013 anti-gay laws—broadly worded legislation banning the distribution of 'gay propaganda' to minors—demonstrated that critics' concerns were well justified, as the persecution of gay activists, environmentalists and other dissidents escalated. President Vladimir Putin's military interventions in Crimea and Ukraine, commencing a few weeks after Sochi's closing ceremony, shifted world attention from Russia's domestic human rights abuses to the broader Eastern European context. In March 2015, the assassination of opposition politician and Sochi whistle-blower Boris

H.J. Lenskyj (✉)
University of Toronto, Toronto, Canada

© The Editor(s) (if applicable) and The Author(s) 2016   **311**
J. Dart, S. Wagg (eds.), *Sport, Protest and Globalisation*,
DOI 10.1057/978-1-137-46492-7_14

Nemtsov was emblematic of Russia's status as a failed democracy, or, as some have argued, a successful totalitarian state.

IOC Innocence: The Olympics Aren't Political

International outrage at Russia's anti-gay laws prompted some Olympic officials to express mild concern, acting as if it came to them as a complete surprise to witness the resurgence of traditional conservatism and homophobia in Russia. In reality, the suppression of dissident voices, attacks on civil society, the influence of the Russian Orthodox Church and widespread political corruption had characterized Putin's leadership since 1999. These trends were well documented by international human rights groups such as Amnesty International, Human Rights Watch (HRW) and the International Lesbian and Gay Association—Europe.[1] For years, non-governmental organizations (NGOs), particularly lesbian/gay/bisexual/transgendered (LGBT) and environmental activist groups, had been subjected to surveillance and harassment.[2] Laws on NGOs became even tougher in 2012, requiring that organizations involved in 'political activities' must register as 'foreign agents' and submit to annual audits, with severe penalties for non-compliance (Lanskoy and Suthers 2013).

In light of developments during Putin's presidency, I argue that IOC should have known that human rights were at risk in Russia by at the latest 2007, the year when it named Sochi as host of the 2014 games. However, the Olympic industry relies on the 'magic' of the sporting spectacle to divert public attention from its less attractive underbelly, in this case, IOC's long-standing practice of ignoring human rights abuses in host countries, most notably Berlin, Germany, in 1936 and Beijing, China, in 2008.

---

[1] Human Rights Watch, among other groups, had monitored these developments since 2000. For a review, see Human Rights Watch (24 April 2013b) *Laws of Attrition: Crackdown on Russia's Civil Society after Putin's Return to the Presidency;* Russia: Human rights activists: Voices from the ground (3 October 2013) *Amnesty International.*

[2] Lomagin, N. (2012) Interest groups in Russian foreign policy: The invisible hand of the Russian Orthodox Church, *International Politics* 49:4, 498–516.

As the Olympic year of 2014 progressed, Putin's annexation of Crimea, the invasion of Ukraine, and Russia's presumed culpability for the Malaysian Airline crash over Ukraine led some commentators to predict resumption of the Cold War, and, with these events dominating the international media, the plight of sexual minorities in Russia was virtually ignored. In fact, by November 2014, even the United Nations seemed to have forgotten the human rights violations associated with the Sochi games. According to Russian Olympic Committee president Alexander Zhukov, as quoted in Russian media sources *RIA Novosti* and *Sputnik News*, the Sochi Olympics 'were assessed highly by the UN General Assembly'.[3]

Moreover, the same UN General Assembly passed a resolution calling for 'the independence and autonomy of sport'. Like IOC president Thomas Bach, Zhukov interpreted this as a call to keep politics out of sport, a popular position among Olympic industry officials.[4] Clearly, the Olympics are by definition political: they involve politicians, multinational corporations, taxpayers and public money. Those who object to the so-called politicizing of the Olympics are often the individuals and organizations that have the most to lose from protests and boycotts. Interestingly, when politicians and Olympic boosters try to sell an Olympic bid to taxpayers, or when they co-opt Indigenous children to sing and dance for IOC members, these initiatives are not labelled 'political'. Nor is it called political when organizing committees lobby politicians to pour unlimited tax dollars into Olympic infrastructure budgets, or, in the case of the Sydney 2000 games, when the head of the organizing committee happens to hold the portfolio of Olympic minister in the state parliament. Yet when anti-Olympic activists protest their government's misplaced spending priorities and use the games to focus world attention on local and global injustices, they are accused of politicizing

---

[3] It is significant that no reference to the UN's positive assessment of Sochi could be found through an online search of non-Russian media. UN praise of Sochi Games testifies to success: Russian Olympic Committee (November 10, 2014) www.sputniknews.com/sport/20141107/1014365848.html

[4] UN praise of Sochi Games testifies to success: Russian Olympic Committee (10 November 2014) www.sputniknews.com/sport/20141107/1014365848.html. Wharton, D. (3 November 2014) Olympic leaders welcome United Nations resolution, *Los Angeles Times.* www.latimes.com/sports/sportsnow/la-sp-sn-olympic-un-resolution-20141103-story.html

and contaminating something pure and holy, as if the Olympics were a religion, or a social movement, or an extended family—all favourite metaphors in Olympic industry circles (Lenskyj 2000).

In its inadequate response to yet another human rights crisis in Olympic host cities and countries, the IOC relies on the myth that sport is apolitical. Olympic officials, like Putin, exhorted the world to keep politics out of sport, a self-serving position that exploited the appeal of international sporting competition and its numbing effects on political consciousness. Once again, this strategy worked. In fact, newly elected president Bach falsely claimed that IOC had no power over governments (BBC Sports 2013), ignoring his organization's clearly demonstrated capacity to confer legitimacy on eligible member-countries, including, historically, those with contested political status such as Taiwan and Palestine.

Gender and Sex in the USSR and Russia

Long-standing controversies over issues of gender and sexualities coalesced in Sochi in 2013, when anti-gay legislation in Putin's Russia not only threatened the safety and security of LGBT athletes and visitors during the Olympics, but, more significantly, had long-term implications for all sexual minorities and their allies living in Russia. The country's long history of sexual repression continued to shape public attitudes and government policies in the post-Soviet era, with the majority of Russians agreeing with Putin and Russian Orthodox Church leaders in their condemnation of homosexuality as a contaminating western import, a practice that was alien to the values of 'normal' Russians, as the captain of the Russian women's track and field team famously asserted in August 2013 (Buckley 2013; Majendie 2013).

Writing in 1998, Russian social scientist and gay activist Igor Kon provided a chronology that demonstrated shifts in government policy and practice following the 1917 Revolution:

1. 1917–1933: decriminalization of homosexuality, relative tolerance and homosexuality officially labelled a disease.
2. 1934–1986: homosexuality recriminalized and severely dealt with by prosecution and discrimination and silence.

3. 1987–1990: beginning of open public discussions of the status of homosexuality from a scientific and humanitarian point of view by professionals and journalists.
4. 1990 –May 1993: gay men and lesbians themselves take up the cause, putting human rights in the forefront, resulting exacerbation of conflict and sharp politicization of the issue.
5. June 1993: decriminalization of homosexuality; the homosexual underground begins to develop into a gay and lesbian subculture, with its own organizations, publications, and centres; continued social discrimination and defamation of same-sex love and relationships. (Kon 1998; Halper 2013)

With the end of the Soviet era came fresh hope of the acceptance of sexual minorities. However, by 2002, there had been unsuccessful attempts by Russian politicians to recriminalize sodomy and criminalize lesbianism. In 2006, the region of Ryazan passed laws banning the promotion of homosexuality to minors, and, in 2010, the Constitutional Court of the Russian Federation backed the Ryazan law. The same year, the European Court of Human Rights ruled that Russia had violated the Convention for the Protection of Human Rights and Fundamental Freedoms when lawmakers prevented gay pride parades in Moscow in 2007, 2008 and 2009. In 2012, laws banning 'homosexual propaganda to minors' were enacted in St Petersburg and six other regions, and finally, in June 2013, a new law to that effect encompassed the entire country (Halper 2013).

Official rationales for the anti-gay laws, as well as reactions inside and outside Russia, reflected real and symbolic exclusionary practices, most notably the pathologizing of homosexuality as 'western decadence' and antagonism towards Russia's growing Europeanization movement. Like Putin, elected officials and church leaders equated homosexuality with paedophilia, and claimed to be protecting Russia's children from 'contamination' when they condoned the public humiliation and punishment of homosexuals, even suggesting public whipping.[5] Lest these attitudes be interpreted as evidence of Russia's unique conservatism, it should be

---

[5] Roberts, S. (20 June 2013) Russia MP calls for law allowing gays to be whipped in public squares, *Pink News* www.pinknews.co.uk/2013/06/20/russia-mp-calls-for-law-allowing-gays-to-be-whipped-in-public-squares/. *Foreign Correspondent*, (October 29, 2013) Interview with St Petersburg mayor, Russian Orthodox deacon Vitaly Milonov ABC (Australia).

noted that in 2014, Dr Ben Carson, an American seeking nomination as the Republican presidential candidate for the 2016 election, justified his opposition to gay marriage on the grounds that it led to bestiality and paedophilia, a position shared by many western religious fundamentalists (Brinker 2014).

In the months after the anti-law was passed, there was a well-documented increase in homophobic violence carried out by vigilante groups, as well as growing numbers of LGBT refugees who have sought asylum in other countries, with the USA reporting a 15 per cent increase in Russian applicants in 2013 and 2014.[6] For his part, Putin invoked his 'demographic crisis' rationale when defending the anti-gay law to western audiences. In view of the country's low birth rates and high mortality rates, he claimed that exposure to the mere idea of homosexuality would lead youth to become gay or lesbian, with resulting threats to the next generation. Speaking to domestic audiences, he called on Russian women to reproduce for the sake of the country's future, pointing to nations that protected sexual minorities as examples of western decadence and declining populations (Shlapentokh 2005; Rivkin-Fish 2010).

Dissidents, Activists and Potemkin Protest Zones

In the years following the dissolution of the USSR in 1991, LGBT activists had struggled to develop a movement that was relevant to the Russian context and not necessarily following western models or evaluating progress by western markers. Along with other Russian dissidents, they faced a conservative majority suspicious of all things western and antagonistic to Europeanizing trends in their country. Following the example of protests mounted at earlier Olympics, activists sought to use the event

---

[6] *Foreign Correspondent*, (October 29, 2013). Interview with St Petersburg mayor, Russian Orthodox deacon Vitaly Milonov. ABC (Australia). Shreck, C. (15 October, 2014) Number Of Russian asylum seekers to US spikes in wake of 'antigay' law, *Radio Free Europe Radio Liberty* www.rferl.org/content/russia-asylum-us-gay-law/26639402.html. Gessen, M. (27 April 2014) Salon of the Exiled, *New York Times*www.nytimes.com/2014/04/28/opinion/gessen-salon-of-the-exiled.html?smid=tw-share

to draw international attention to human rights abuses, the best-known historical examples being the student protests (see Celeste González de Bustamante's chapter in this book) and the American Black Power demonstrations at the 1968 Mexico City Games. Decades earlier, American civil liberties and religious groups, among others, had unsuccessfully lobbied the American Athletic Union to boycott the 1936 Berlin Olympics because of the growing Nazi persecution of Jews (for an analysis of the latter, see the chapter by Large and Large in this book). More recently, following the Soviet invasion of Afghanistan in 1979, only 65 out of 145 countries boycotted the 1980 Moscow Olympics, and these figures are often invoked to support the argument that boycotts are ineffective.

Russia's anti-gay law was passed with no opposing votes in June 2013, just eight months before the Sochi Olympics began. The timing was not coincidental; in fact, it could be seen as a deliberate provocation on Putin's part vis-à-vis the west, carried out with the sure knowledge that the IOC would never strip Sochi of the games at that late stage in the preparations. Several sources reported that Putin and other Russian officials believed there was 'an organized campaign to discredit Russia over Sochi', and that this subsequently prompted Putin's uncompromising position vis-à-vis Crimea and Ukraine.[7] Indeed, his Crimean intervention began just weeks after the Sochi Olympics had ended.

In August 2013, Putin signed off a law banning any and all protest in Sochi between January 7 and 21 March 2014, a move followed up by Russia's deputy prime minister who sent a letter to the IOC assuring them that Russia would comply with the Charter's anti-discrimination provisions. Bach announced that he was 'satisfied' with Putin's assurances regarding LGBT rights and spoke approvingly of Russia's decision to create so-called protest zones (even if they were 15 km away from Sochi's major sport venues). These zones were no doubt a concession that the spineless IOC president—a personal friend of Putin's—felt obliged to applaud. *Wall Street Journal* reporter Jane Buchanan felt no such compulsion, pointing out that IOC should be speaking out about

---

[7] Walker, S. (17 December, 2014) The Sochi Olympic legacy, *The Guardian* www.theguardian.com/sport/2014/dec/17/sochi-olympics-legacy-city-feels-like-a-ghost-town. *Russia Times* (February 15, 2014) Sochi 2014: Invisible security is the best security, says Putin's chief of staff, http://rt.com/news/sochi-legacy-sergey-ivanov-220/

Russia's 'crushing free-speech restrictions', the crackdown on LGBT activists, environmentalists and journalists, and the planned surveillance of all foreign visitors' phone and internet use. Instead, as she pointed out, IOC merely used the occasion to remind athletes of the Olympic Charter's rule banning demonstrations or political propaganda (Buchanan 2013).

Russia: Open and Democratic?

Olympic boosters routinely rely on 'miracle of sport' rhetoric, including its alleged powers to stop wars, prevent teen pregnancy and unite the world in peace and harmony, as the old Coca-Cola jingle so aptly put it.[8] Vague references are made to 'behind the scenes' negotiations and 'soft diplomacy' aimed at persuading leaders to lift their game, particularly in the area of human rights, when the eyes of the world are focused on their countries. In the case of Beijing 2008, the suppression of Tibetan protesters, among numerous other examples, amply demonstrated that western 'persuasion' was largely ineffective. Even in a democracy like Australia, concerted efforts to keep up the appearance of Aboriginal inclusion in the cultural events and on the playing field belied the fact that hosting the 2000 Summer Olympics had no impact whatsoever on the numbers of Aboriginal incarcerations and deaths in police custody, and the ongoing and shameful aspects of that country's race relations (Lenskyj 2002).

In 2007 when Sochi was awarded the 2014 games, President Putin and Dmitry Chernychenko, head of Sochi's bid committee, proclaimed that all was well on Russia's human rights front. According to Putin, Russia's governance was based on 'free elections [and] free expression',[9] while

---

[8] The alleged connection to pregnancy prevention was made in some American sources in the 1990s and popularized in Nike advertisements, with little evidence that there was a causal relationship. The concept of the Olympic truce, commonly misinterpreted as the equivalent of a cease fire in modern warfare, originated at the time of the ancient Olympics, when athletes travelling to Greece were given safe passage through battle lines. Sport as a force for character-building and peace is a tenet of the UN Inter-Agency Task Force on Development and Peace, Right to Play, and numerous other NGOs. See United Nations Inter-Agency Task Force on Sport for Development and Peace (2003), *Sport for Development and Peace: Towards Achieving the Millennium Development Goals* (New York: United Nations).

[9] Statement in Sochi bid documents cited in Nemtsov and Martynyuk (2013) Winter Olympics in the subtropics: Corruption and abuse in Sochi, *The Interpreter* (trans. C. Fitzpatrick).

Chernychenko predicted that 'Russia will become even more open, more democratic' as a result of hosting the 2014 games (Toronto Star 2007). But, as opposition politicians Boris Nemtsov and Leonid Martynyuk documented in their 2013 report on corruption and abuse related to Olympic construction, 'these statements were lies then, and all the more so now' (Nemtsov and Martynyuk 2013). A key allegation concerned oligarchs' attempts to profit from government-backed loans by inflating construction costs and demanding lower interest rates. In fact, several oligarchs were quite forthright in sharing their concerns with western media sources, for example, in an Associated Press special report published in February 2013 (Grove 2013). Putin had 'invited' these oligarchs to fund Sochi's major construction projects, which some viewed as a 'social project', with the mutual understanding that lucrative state contracts would come their way in the future. However, they claimed that so much extra work was required that they were demanding compensation through interest rate subsidies with the government.

Critics pointed to the cost of constructing the 48 km Krasnaya Polyana combined motorway and railway ($US8.7 billion), an amount that, as a *Business Insider* article noted, exceeded the total budget of the Vancouver 2010 Winter Olympics by $1 billion (Manfred 2014). Even government-affiliated sources reported that at least $506 million of the $50 billion spent on Olympic construction had been skimmed off through graft and inflated overhead costs, one consequence being the firing of the Russian Olympic committee's vice president because of delays and cost overruns in building the ski jump (Lanskoy and Suthers 2013). Although these 'political capitalists' were not guaranteed any profits from Olympic-related contracts, they recognized the importance of complying with Putin's requests as a 'kind of tax' for doing business in Russia (Fox News 2013). After all, they were already quite rich; on Putin's watch, the number of Russian billionaires grew from 8 in 2000 to 110 in 2013 (Russia Times 2013).

By early 2015, as western sanctions, plummeting oil prices and the looming recession took their toll, investors sought to recoup some

---

www.interpretermag.com/winter-olympics-in-the-sub-tropics-corruption-and-abuse-in-sochi/.
See also Milov, V. and Nemtsov, B. (2010) Putin: What 10 years of Putin have brought. www.putin-itogi.ru/putin-what-10-years-of-putin-have-brought/#13

of their losses, and Associated Press journalist Nataliya Vasilyeva (Vasilyeva 2015a, b) uncovered several examples of behind-the-scenes deals which were now struck. In February, Sochi investor Viktor Vekselberg handed over a hotel to Putin, who agreed to compensate him for construction costs. In doing so, Vekselberg's company 'quietly' dumped its 'toxic assets': a $450 million loan. Using a similar strategy, Sberbank, which had taken out a $1.7 billion loan to fund the ski jump and ski resort, handed over these properties to the regional government in exchange for Sochi's media centre ... again, dumping uncollectable loans on the state.

In her 2014 book exposing the power of Russia's oligarchs, *Putin's Kleptocracy*, historian Karen Dawisha (2014) argued that Putin's Russia should not be viewed as a failed democracy but as a successful totalitarian regime, a not so hidden agenda embraced by Putin and his inner circle from the outset.

Boycott or No Boycott?

In response to the 2013 anti-gay law, there were calls for an Olympic boycott from progressive organizations around the globe. However, well-rehearsed anti-boycott arguments on the part of Olympic industry officials, sponsors, athletes, and even LGBT organizations prevailed, and the games were carried out as planned (Aguiar 2013; Gillespie 2013; Plank 2013; *Russian LGBT Network* 2013).

In 2012, members of the feminist, pro-gay punk group Pussy Riot had been sentenced to two-year prison terms for the crime of 'hooliganism' following their performance of an anti-Putin protest song in Moscow's Russian Orthodox cathedral. The January 2014 publication of dissident journalist Masha Gessen's *Words Can Break Cement*, an account of their horrific experiences in labour camps, should alone have been sufficient grounds for a boycott. In fact, conditions in Russian penal colonies resembled those experienced by Australian convict women living in factory prisons in the nineteenth century, as well as those in Soviet gulags. As Pussy Riot member Nadia Tolokonnikova's diary recorded, she survived 16-hour work days on a diet of 'stale bread, generously watered down milk, exceptionally rancid millet and ... rotten potatoes.'[10] Female prisoners were subjected to regular gynaecological examinations, permitted only one bath per week and forced to work in sub-zero temperatures. Magical thinking that sport is an apolitical enterprise that unites the world led many to reject proposals for an Olympic boycott. Those opposed, including most world leaders, claimed that boycotts were ineffective and only served to penalize athletes—portrayed as pure and innocent victims of political meddling—who had been training and making sacrifices for years in order to represent their countries. Such statements implied that the rights of elite sportsmen and women were more important than the rights of millions of lesbians and gays. It was disturbing to see a world leader such as President Obama appearing to justify his anti-boycott position on these grounds, thereby implying that his country's foreign policy was influenced by a mere 230 American athletes (Johnson 2013). There was some support for a boycott of Olympic sponsors and for other low-key forms of protest, such as wearing a rainbow pin to signify solidarity with sexual minorities. More radical critics compared Putin's Russia to Nazi Germany in its treatment of homosexuals, and pointed to the missed opportunity to boycott the 1936 Berlin Olympics, a position that was widely dismissed as overreacting to the situation in Russia.

An analysis of rationales used to support the anti-boycott position demonstrates the Olympic industry's continued success in perpetuating the 'sport-as-special' myth: magical thinking that sport transcends politics and, as I've termed it, 'the myth of the pure Olympic athlete and pure Olympic sport' (Lenskyj 2000: 102–5). One popular position held that it was inappropriate for LGBT activists to 'politicize' sport, that it was the responsibility of sport leaders and politicians to resolve any issues of discrimination and that it was impolite for visitors to challenge another country's laws.

Views expressed by significant numbers of western LGBT organizations revealed similar blind spots in relation to the brutality of Putin's regime.

---

[10] Excerpt from Nadya Tolokonnikova's journal, written in a Mordovian penal colony, in M. Gessen (2014) *Words Will Break Cement: The Passion of Pussy Riot* (New York: Riverhead) 295, 301. For experiences of Australian women convicts, see, for example, A. Summers (1975) *Damn Whores and God's Police* (Ringwood, VIC: Penguin).

For such groups to agree with Russian politicians' calls to 'keep politics out of sport' demonstrates the wide reach of 'sport-as-special' propaganda as well as the limitations of liberalism. Many LGBT groups also invoked the 'sport as LGBT proving ground' argument, claiming that winning performances by LGBT athletes would speak for themselves, and echoing Gay Games founder Tom Waddell's 'heteronormalizing' rationale for promoting high performance sport for gays and lesbians (Waddell and Schaap 1996: 147). Obama took a similar stance, predicting that LGBT athletes would bring home medals and show the world their level of commitment (Daily Beast 2013).

In a persuasive counter-argument, drawing parallels with the 1936 'Nazi Olympics', political commentator Adam Goldenberg (2013) stated that 'showing up at Putin's games would allow the Olympics to overshadow his government's human rights abuses. Not showing up would ensure that the opposite happens.' His predictions were borne out in the subsequent months as western attention focused on the sport mega-event (and Sochi's stray dogs, which a private company was hired to kill), to the virtual exclusion of Putin's crackdown on any form of political protest.

The Games Begin… and the Media 'Play the Game'

In December 2013, Putin made the token gesture of freeing some political prisoners, including Greenpeace protesters and Pussy Riot members, in order to mollify western opinion. It appeared, however, that he had difficulty maintaining any illusion of tolerance. In a television interview a few weeks before the games began, after assuring gay and lesbian visitors that they would be welcome, he repeated the popular conservative/Orthodox view that equated male homosexuality with paedophilia (Berry 2014). Similarly, police and security forces seemed to forget the omnipresent international television cameras, and the viewing world witnessed Cossack security personnel attacking Pussy Riot members with whips, batons and pepper spray during an anti-Putin protest in Sochi.

In the lead-up to the opening ceremony, western activists continued their futile attempts to shame IOC into action on Russia's anti-gay laws by invoking the so-called 'Olympic values' and erroneously claiming that the Olympic Charter guaranteed freedom from discrimination. In fact, a closer reading shows that the anti-discrimination clause of the Charter applies specifically to the *practice of sport*, and is not violated as long as *athletes* are not disqualified because of 'race, religion, politics, gender or otherwise' If a country guilty of systemic racism, sexism, homophobia, or other forms of discrimination were disqualified, the Olympics would not have survived into the twenty-first century.

Nine months after the Sochi Olympics, the IOC added a new anti-discrimination clause, specifically including sexual orientation, to the Charter and the host city agreement, a move that may produce some positive outcomes in the future, but only if adherence to the terms of this agreement is strictly monitored—and in light of IOC's past failures to do so on other fronts, this seems unlikely. For example, in February 2013 a HRW report on migrant workers in Sochi noted that, although IOC had made a commitment to address human rights abuses in 2009, it had failed to address HRW's concerns. IOC had merely requested the Sochi organizing committee for more information about the abuse cases that HRW had already thoroughly documented. Moreover, IOC failed to use its coordination committee visits to Sochi to monitor and enforce human rights protections (Human Rights Watch 2013a). The urgent need to do so should have been obvious to IOC: between 2010 and 2014, almost 18,000 Uzbek migrant workers died on Olympic construction sites in Russia (Shaw et al. 2014).

Throughout the Sochi Olympics, there was an unmistakable reduction in western media coverage of anti-Olympic and anti-Putin protest, as well as LGBT issues, arrests of protesters and police brutality. The public whipping of Pussy Riot members briefly gained western media attention, but did not generate the degree of international outrage that might have been expected if the Olympic sporting competition had not eclipsed all such protests. In a clear example of selective reporting throughout 2014, more western media attention focused on the rescue of Sochi's stray dogs than on the arrests of protesters and their likely prison sentences. For example, a photo of a dog rescue appeared on the same page as a (rare) critical opinion piece titled 'Russia's anti-gay stance has spoiled the Sochi Olympics' in a Toronto newspaper during the games (Mullins 2014).

A Google search for *Sochi dogs* on 13 January 2015 produced 597,000 hits, compared to 439,000 hits for *Sochi protesters*.

Not coincidentally, shortly after the games were over, Russia escalated its military action in Crimea and Ukraine. These events effectively diverted world attention from LGBT issues, which, by the end of 2014, appeared to be largely forgotten in mainstream media. Even the supposedly progressive UK newspaper, *The Guardian*, failed to comment on any political aspect of the Sochi games in its December 2014 'Sport Year in Review' article, focusing instead on Team GB's medal count and a 'heartbreaking fall' by a Russian skater. In contrast, a full paragraph on FIFA in the same article dealt with that organization's corruption problems, so the failure to comment on Russia's anti-gay laws could not be attributed to a 'keeping politics out of sport' position (Ronay 2014).

Future Olympics: Lessons from the Past

In 2001, when the IOC selected Beijing as host of the 2008 Summer Olympics, several critics expressed cynicism. The IOC's decision, one *National Review* commentator observed, 'furthers the perception that the People's Republic of China – with its sprawling gulag system and often murderous repression – is a normal country, if with some peculiar characteristics.'[11] A *Washington Post* columnist concluded, 'If Beijing is marred by corruption, burning with commercial fever and eager to have a chance at self-promotion, maybe that makes it the perfect place for what the Olympics have become' (Mufson 2001).

The accuracy of these views was confirmed by events of 2014, when only two candidate cities—Beijing and Almaty (Kazakhstan)—were left in the running for the 2022 Winter Olympics, thereby guaranteeing that these games would be hosted by a non-democratic country. Earlier European contenders—Krakow, Lviv, Stockholm, St Moritz, Munich and Oslo—had all dropped out of the race by October 2014, with

---

[11] *National Review Editor (24 July, 2001) China: A dangerous decision*, National Review. http://en.minghui.org/html/articles/2001/7/24/12511.html

negative referendum results and concerns about mounting costs largely responsible for Moritz's, Munich's and Oslo's withdrawals. In fact, with the escalating security presence and the suppression of peaceful protest in every recent host city, consistent with or exceeding Olympic industry requirements, it seems that dictatorships can more readily meet these needs than democracies.

As *Business Insider* (Manfred 2014) reported at the time of Oslo's withdrawal, a major Norwegian newspaper's exposé of IOC's 'arrogance', as reflected in its long list of demands of the 2022 host city, made the organization look 'particularly aristocratic' at a point when it was struggling to attract bids from democratic countries (or indeed any countries, in the case of Winter Olympics)

Throughout most of 2014, the IOC had been working on Agenda 2020, a 'roadmap' comprising 40 proposals for the future of the Olympics, with stakeholders, 'external experts and the public' contributing to the process (Macur 2014). Not coincidentally, one of the recommendations approved at IOC's December meeting involved changes to the bid process to make it (somewhat) less expensive: the IOC would cover costs of evaluation committee members' visits as well as bid delegates' travel and accommodation at various IOC meetings, and bid documents were to be prepared in electronic format only. Another change allowed some events to be held outside the host country, if warranted by geography and sustainability—not an unprecedented arrangement since equestrian events had been held in Stockholm during Melbourne's 1956 Olympics because of Australia's quarantine rules.

A more substantive change involved the inclusion of sexual orientation in the 6th 'Fundamental Principle of Olympism' as well as in the host city contract. There was, however, no dearth of proposals catering to IOC self-interest, including extending members' mandatory retirement age to 74, and launching an Olympic television channel. Despite an emphasis on environmental sustainability, there were two references advocating the use of temporary or demountable facilities 'where no long-term legacy need exists or can be justified'—an inconsistent provision in light of bid boosters' long-standing reliance on promises of permanent facilities as a 'legacy'. Further, if a bid city lacks the mandatory 'state-of-the-art' facilities (as required in Recommendation 2:3 of the document), it is unlikely to succeed.[12] Of course, such contradictions are not new: one of the long-standing criteria used to evaluate bids refers to the number of pre-existing facilities, which by definition cannot be 'state-of-the-art' if they exist seven or more years before the event. IOC's internal politics, as well as wider geopolitical considerations, play a bigger part in host city selection than a careful technical analysis of bids (Lenskyj 2010).

Events surrounding the Sochi Olympics undoubtedly contributed to the relative lack of interest in bidding for the 2022 winter games, particularly sponsors' concerns about boycotts and the Sochi's unprecedented costs, at a time when the global economy was faltering. Empty rhetoric about Olympic legacies had little credibility when mainstream media were documenting countless examples of crumbling, unwanted 'white elephant' venues and underused facilities that constituted an ongoing drain on public money (Lobo 2014; Mackin 2013).

Throughout 2014, even conservative business sources were pointing to the economic risks of hosting the games, with rising security costs and the uncertain currency of the Olympic image and brand, and these kinds of critiques flourished in February 2015 on the anniversary of the Sochi Olympics (Gibson 2014; Clarey 2014; Panja 2014; Manfred 2014). Yet, as early as September 2000, while Sydney was hosting the Olympics, an article in *Report on Business Magazine* had identified the main source of misleading calculations in Olympic budgets: whether the accounting methods included or excluded infrastructure expenses, otherwise known as the *Olympic legacy* (Grange 2000). Such details are routinely obscured or rationalized in Olympic bids, while the power of the brand and the 'legacy' hype continue to sway public opinion. However, when it serves Olympic industry interests, boosters and politicians remind taxpayers that infrastructure costs are excluded. Putin and Russian officials repeatedly emphasized this fact when challenged about the $51 billion price tag, even claiming that the games made a $US261 million profit (Moscow Times 2014).

Russia's hosting of international events continued to attract media attention throughout 2014 and into 2015. The first scheduled event was

---

[12]IOC (December 2014) Olympic Agenda 2020: 20+20 Recommendations. www.olympic.org/ Documents/Olympic_Agenda_2020/Olympic_Agenda_2020-20-20_Recommendations-ENG.pdf

to be the G8 Summit in June 2014, but Russia was ejected from that meeting after its annexation of Crimea, no doubt an embarrassing start to Putin's plan to put Sochi on the world political map. In October, after years of negotiations, Russia signed a contract with Formula One and Sochi hosted the F1 Grand Prix. F1 chief Bernie Ecclestone, unsurprisingly putting business above all else, praised Putin for his ability to get things done and for his stance on 'gay propaganda' (Whatt 2014a, b). In a *San Francisco Chronicle* opinion piece, a Russian consular official made the exaggerated claim that the Formula One track was 'an Olympic Games legacy project'—thus implying that auto racing was a winter Olympic sport (Avdoshin 2014). And, in what looked like yet another effort to rehabilitate Sochi, Putin agreed to consider the option of creating a gambling zone in the region, despite his earlier opposition on the grounds that Sochi should continue to appeal to family holiday-makers. It is interesting to note that he had not hesitated to propose Crimea as a gambling zone in April 2014, mere weeks after the Russian annexation (GGRAsia 2014).

In January 2015, the *Moscow Times* reported that the regional government would pay $46m to have the stadium roof removed in preparation for Sochi's hosting of the 2018 FIFA World Cup (since FIFA required a roofless stadium for soccer tournaments), thereby demonstrating a typical limitation of legacy promises: bid committees' failure to disclose costs of maintaining and/or 'repurposing' expensive, unwanted facilities (Moscow Times 2015). Despite IOC's stated commitment to sustainability, Sochi organizers' 'repurposing' agenda included converting the speed skating arena into an exhibition centre and tennis academy, dismantling the temporary freestyle skiing venue and using the Ice Cube (curling) as a concert venue.

What About the Dissidents?

Dissidents took significant risks in the period leading up to the Sochi Olympics. For years, environmental activists had been targeted and subjected to ongoing harassment by security services, and several were arrested in the months before the games began. In February 2014,

Evgeni Vitishko, a prominent member of Environmental Watch on North Caucasus (EWNC), was sentenced to three years in a penal colony on charges of defacing a construction fence that encroached on protected forest and Black Sea beach access in the Sochi region (Digges and Levchenko 2014). Fellow activist David Khakim received 30 hours of 'correctional labour' for holding a picket in support of Vitishko without a permit, (The Ecologist 2014) thereby exemplifying Russian officials' pattern of overzealous policing and excessive penalties for Olympic-related infractions.

Although IOC subsequently requested the Russian authorities to 'clarify' Vitishko's situation, an IOC spokesman claimed that his sentence involved 'a matter that was not in relation to the Olympics' (Associated Press 2014a)—a patently false assertion in view of the fact that activists had set up EWNC specifically to monitor environmental impacts on the Sochi region and official corruption related to the Olympic project. Furthermore, Russian authorities clearly believed that the group's activities were 'in relation to the Olympics', as evident in the fact that they had accused its members of seeking to 'trash' the games.[13] In an equally transparent attempt to protect the Olympic brand from the contaminating effects of Russian dissidents, IOC spokesman Mark Adams claimed that police detention of Pussy Riot members, who vocally supported a boycott and had been protesting in front of an Olympic banner in Sochi, was not linked to the Olympics (CBC News 2014).

In St Petersburg and Moscow on 7 February, the opening day of the Olympics, police arrested 14 pro-gay protesters. In a typical example of media bias, these events were delegated to page 22 of the broadsheet, *The Toronto Star*, occupying a mere 7 column inches, in contrast to the adjacent longer piece and two photographs of the funeral of Hollywood actor Philip Seymour Hoffman, which took up almost half the page (Associated Press 2014b). And, on the day after Sochi's closing ceremonies, Pussy Riot members Nadezhda Tolokonnikova and Maria Alyokhina were among the 234 peaceful protesters arrested outside a Moscow court where the landmark Bolotnaya Square protest trial resulted in seven of

---

[13] *RIA Novosti* (12 March 2010) Russian ecologists 'trashing' 2014 Sochi Olympics—deputy PM. http://en.rian.ru/sports/20100312/158174256.html. Walker, ibid.

the eight Bolotnaya defendants being sentenced to prison terms of 2.5–4 years.[14] A few days earlier, in Sochi, Khakim and another environmental activist, Olga Noskovets, had been arrested in order to prevent them from protesting during the closing ceremonies.[15] These patterns of harassment, arrests, release and repeat arrests of environmental activists and political dissidents became routine in the lead-up to the Sochi games. Except for human rights groups, few western sources reported on these trends, most of which were forgotten when the games left town.

Future Challenges

In the conclusion to my first Olympic book, *Inside the Olympic Industry* (2000), I called on those who enjoy sport and value democracy to direct their energies towards alternative international sporting competitions that were conducted in more ethical, less exploitative ways (Lenskyj 2000: 195). However, having carefully monitored the bid process and preparations for Toronto's 2015 PanAm Games, I now question whether *any* sport mega-events, particularly those like the PanAms with direct links to the Olympic industry, can meet these criteria (Lenskyj 2014; Dutkiewicz and Trenin 2011; Healey 2001).

Well-organized watchdog groups in some recent host cities have managed to minimize or prevent some of the most severe damaging impacts on vulnerable populations and environments. Successful resistance strategies include peaceful protest, non-violent direct action, support

---

[14] On the eve of Putin's 2012 inauguration, thousands of protesters had assembled in Bolotnaya Square in a rally that had been sanctioned by the police. However, police closed one exit, resulting in violent confrontations between trapped protesters and police. Arrests and violence at overflowing rally in Moscow (6 May, 2012), *New York Times* www.nytimes.com/2012/05/07/world/europe/at-moscow-rally-arrests-and-violence.html. Putin's repression resumes, now that the Olympics have ended (25 February, 2014), *Washington Post.* www.washingtonpost.com/opinions/putins-repression-resumes-now-that-the-olympics-have-ended/2014/02/25/071f010a-9e5a-11e3-b8d8-94577ff66b28_story.html

[15] Amnesty International (24 February, 2014) Russia: Hundreds more protesters arbitrarily arrested outside Bolotnaya trial sentencing, *Amnesty International Press Release.* www.amnesty.org/en/for-media/press-releases/russia-hundreds-more-protesters-arbitrarily-arrested-outside-bolotnaya-tria. Sochi court fines Olga Noskovets by 1000 roubles (25 February, 2014) *Caucasian Knot.* http://eng.kavkaz-uzel.ru/articles/27393/

of whistle-blowers, boycotts of sponsors, and coalitions that unite progressive voices across gender, race/ethnicity, sexuality and social class. Social media and citizen journalism represent effective tools in organizing resistance campaigns and countering Olympic industry's control of mainstream media. Admittedly, these are David-and-Goliath encounters, but events surrounding the 2008 Beijing Olympics and the 2014 Sochi Olympics amply demonstrate that to do nothing is to condone the abuse of power.

References

Aguiar, R. (2013, August 28). Coca-Cola defends sponsorship of 2014 Sochi Winter Games, Towleroad. www.towleroad.com/2013/08/coke-responds-to-sochi-olympics-controversy-and-protests.html

Associated Press. (2014a, February 13). IOC presses Sochi officials over jailed activist. www.wintergames.ap.org.wfaa/article/ioc-presses-sochi-officials-over-jailed-activist

Associated Press. (2014b, February 8). Russians arrest 14 gay rights activists. *Toronto Star*, A22.

Avdoshin, E. (2014, December 15). What the Games meant to Sochi. *San Francisco Chronicle*. www.sfchronicle.com/opinion/openforum/article/What-the-Olympic-Games-meant-to-Sochi-5958842.php?t=f13b17cd86&cmpid=twitter-premium

BBC Sports. (2013, September 11). New IOC President Thomas Bach wants 'evolution not revolution'. www.bbc.com/sport/0/olympics/24044915

Berry, L. (2014, January 19). Russian President Putin links gays to pedophiles. *Huffington Post*. http://www.huffingtonpost.com/2014/01/19/putin-gays-pedophiles_n_4627438.html

Brinker, L. (2014, December 22). Ben Carson stands by claim that gay marriage leads to pedophilia and bestiality, Salon. www.salon.com/2014/12/22/ben_carson_stands_by_claim_that_gay_marriage_leads_to_pedophilia_and_bestiality/

Buchanan, J. (2013, December 20). Sochi's Potemkin protest zones. *Wall Street Journal*. www.wsj.com/articles/SB10001424052702303773704579265800387636682

Buckley, N. (2013, September 19). Putin urges Russians to return to values of religion. *Financial Times*.

CBC News. (2014, February 19). Pussy Riot attacked with whips by Cossack militia at Sochi. www.cbc.ca/news/world/pussy-riot-attacked-with-whips-by-cossack-militia-at-sochi-1.2542843

Clarey, C. (2014, October 2). A winter games few care to host. *New York Times.* www.nytimes.com/2014/10/03/sports/olympics/-a-winter-games-few-care-to-host-.html?_r=0

Daily Beast. (2013, August 15). Comments. www.thedailybeast.com/articles/2013/08/15/boycott-putin-not-the-sochi-olympics.html

Dawisha, K. (2014). *Putin's Kleptocracy: Who owns Russia?* (p. 5). New York: Simon and Schuster.

Digges, C., & Levchenko, V. (2014, December 18). Environmentalists locate illegal fence than landed Sochi critic Vitishko in prison, Bellona News.

Dutkiewicz, P., & Trenin, D. (Eds.) (2011). *Russia: The challenges of transformation.* New York: New York University Press.

Fox News. (2013, May 20). Russian oligarchs foot much of the bill. www.foxnews.com/sports/2013/05/20/russian-oligarchs-foot-much-bill-for-2014-olympics-in-sochi-as-price-doing/

Gessen, M. (2014). *Words will break cement: The passion of Pussy Riot.* New York: Riverhead.

GGRAsia. (2014, June 13). Putin now in favour of gambling zone in Sochi. www.ggrasia.com/ putin-now-in-favour-of-gambling-zone-in-sochi/

Gibson, O. (2014, October 14). Sochi continues the rise of non-democratic host nations. *The Guardian.* www.theguardian.com/sport/blog/2014/oct/14/sochi-non-democratic-host-nations-russia-f1-winter-olympics-world-cup

Gillespie, K. (2013, August 16). Sochi: Canadian Olympians weigh in on boycott plan. *Toronto Star*, S2.

Goldenberg, A. (2013, July 26). Why Canada's Jews should stand up for Russia's gays. *Macleans.* www2.macleans.ca/2013/07/26/why-canadas-jews-should-stand-up-for-russias-gays/

Grange, M. (2000, September). Summer Olympic red alert, Report on Business Magazine, pp. 25–32.

Grove, T. (2013, February 21). Russia's $50 billion Olympic gamble, Reuters Special Report. www.reuters.com/article/2013/02/21/us-russia-sochi-idUSBRE91K04M20130221

Halper, K. (2013, August). Putin's war on gays: A timeline of homophobia. Policy Mic policymic.com/articles/58593/putin-s-war-on-gays-a-timeline-of-homophobia.html

Healey, D. (2001). *Homosexual desire in revolutionary Russia.* Chicago: University of Chicago Press. http://bellona.org/news/russian-human-rights-issues/

2014-12-environmentalists-locate-illegal-fence-landed-sochi-critic-vitishko-prison

Human Rights Watch. (2013a). Race to the bottom exploitation of migrant workers ahead of Russia's 2014 Winter Olympic Games in Sochi. www.hrw. org/sites/default/files/reports/russia0213_ForUpload.pdf

Human Rights Watch. (2013b, April 24). Laws of attrition: Crackdown on Russia's civil society after Putin's return to the Presidency. www.hrw.org/reports/2013/04/24/laws-attrition-0

IOC. (2014, December). Olympic Agenda 2020: 20+20 Recommendations. www.olympic.org/Documents/Olympic_Agenda_2020/Olympic_Agenda_2020-20-20_Recommendations-ENG.pdf

Johnson, L. (2013, September 9). Obama opposes Olympic boycott, criticizes Russian anti-gay law, Huffington Post. www.huffingtonpost.com/2013/08/09/obama-olympic-boycott_n_3733275.html

Kon, I. (1998). Moonlight Love, Gay Russia. http://english.gay.ru/life/history/moonlightlove/SovietHomophobia.

Lanskoy, M., & Suthers, E. (2013, July). Outlawing the opposition. *Journal of Democracy, 24*(3), 75–87.

Lenskyj, H. (2000). *Inside the Olympic industry: Power, politics and activism.* Albany: SUNY Press, Chapter 3.

Lenskyj, H. (2002). *Best Olympics ever? Social impacts of Sydney 2000.* Albany: SUNY Press.

Lenskyj, H. (2010). Olympic power, Olympic politics: Behind the scenes. In A. Bairner & G. Molnar (Eds.), *The politics of the Olympics* (pp. 15–26). London: Routledge.

Lenskyj, H. (2014, December). Sport mega-events and leisure studies. *Leisure Studies, 33*, 1–7.

Lobo, R. (2014, August 13). From London to Sochi, whose Olympic games paid off? Business Destinations. www.businessdestinations.com/move/travel-management/from-london-to-sochi-whose-olympic-games-paid-off/

Lomagin, N. (2012). Interest groups in Russian foreign policy: The invisible hand of the Russian Orthodox Church. *International Politics, 49*(4), 498–516.

Mackin, B. (2013, July 2). Ten legacies of the Vancouver Olympics. *The Tyee.* http://thetyee.ca/News/2013/07/02/Vancouver-Olympics-Legacies/

Macur, J. (2014, November 19). At glacial pace, Olympics weigh change. *New York Times.* www.nytimes.com/2014/11/20/sports/lack-of-suitors-for-winter-olympics-prompts-ioc-to-change-bidding-process.html?smid=tw-share&_r=0

Majendie, M. (2013, August 15). We are normal Russians. *The Independent.* www.independent.co.uk/news/world/europe/we-are-normal-russians-pole-vaulter-yelena-isinbayeva-defends-antigay-laws-8764393.html

Manfred, T. (2014, October 24). The bidding for the 2022 Olympics is a disaster. *Business Insider*. www.businessinsider.com/sochi-olympics-cost-reform-2014-10

Moscow Times. (2014, June 19). Sochi Olympic organizers report $261 million profit. www.themoscowtimes.com/business/article/sochi-olympic-organizers-report-261-million-profit/502223.html

Moscow Times. (2015, January 20). Russia to spend $50 million taking roof off Sochi Olympic stadium. www.themoscowtimes.com/business/article/russia-to-spend-50-million-taking-roof-off-sochi-olympic-stadium/514657.htm

Mufson, S. (2001, July 2). Politics of the Games finds China deserving. *Chronicle*. www.chron.com/opinion/editorials/article/Politics-of-the-Games-finds-China-deserving-2045243.php

Mullins, A. (2014, February 13). Russia's anti-gay stance has spoiled the Sochi Olympics. *Metro News*, p. 9.

Nemtsov, B. and Martynyuk, L. (2013, June 7). Winter Olympics in the subtropics: Corruption and abuse in Sochi, The Interpreter (trans: Fitzpatrick, C.). www.interpretermag.com/winter-olympics-in-the-sub-tropics-corruption-and-abuse-in-sochi/

Panja, T. (2014, October 30). Sochi Olympics $51 billion price tag deters host cities. *Bloomberg News*. www.bloomberg.com/news/articles/2014-10-30/sochi-olympics-51-billion-price-tag-deters-host-cities

Plank, E. (2013, August 8). Want to stand with gay athletes? Then don't boycott the Sochi Olympics, Policy Mic. policymic.com/articles/58651/want-to-stand-with-gay-athletes-then-don-t-boycott-the-sochi-olympics

Rivkin-Fish, M. (2010). Pronatalism, gender politics, and the renewal of family support in Russia. *Slavic Review, 69*(3), 701–724.

Ronay, B. (2014, December 19). World Cup overshadowed the worst efforts of Fifa's elite. *Guardian Weekly*, p. 63.

Russia Times. (2013, October 11). Gap between rich & poor in Russia among world's biggest. http://rt.com/business/russia-world-household-wealth-944/

Russian LGBT Network. (2013, July 30). Russian LGBT Network on the Olympics in Sochi. http://lgbtnet.ru/en/content/winter-olympics-we-should-speak-not-walk-out

Shaw, C. Anin, R. and Vdovii, L. (2014, December 10). Ghosts of Sochi. *The Black Sea*. http://theblacksea.eu/index.php?idT=88&idC=88&idRec=1184&recType=story

Shlapentokh, V. (2005). Russia's demographic decline and public reaction. *Europe-Asia Studies, 57*(7), 951–968.

The Ecologist. 2014, February 17). Russian eco-activist gets 3 years in penal colony. http://www.theecologist.org/News/news_round_up/2285706/russian_ecoactivist_gets_3_years_in_penal_colony.html

Toronto Star. (2007, July 5). Winter Olympics are headed to Russia, with love, in 2014, S2.

United Nations Inter-Agency Task Force on Sport for Development and Peace. (2003). *Sport for development and peace: Towards achieving the millennium development goals.* New York: United Nations.

Vasilyeva, N. (2015a, February 5). The most expensive Olympics ever is now a major headache for Putin. *Business Insider.* www.businessinsider.com/the-most-expensive-olympics-ever-is-becoming-a-major-headache-for-putin-2015-2#ixzz3RrSOBCDw

Vasilyeva, N. (2015b, February 9). Putin offers to take vast Sochi hotel from oligarch, Salon. www.salon.com/2015/02/09/putin_offers_to_take_vast_sochi_hotel_from_oligarch/

Waddell, T., & Schaap, D. (1996). *Gay Olympics: The life and death of Dr. Tom Waddell* (p. 147). New York: Knopf.

Whatt, B. (2014a, February 21). Bernie Ecclestone backs Vladimir Putin on gay rights issue. *CNN News.* http://edition.cnn.com/2014/02/20/sport/motorsport/ecclestone-putin-sochi-formula-one/index.html?hpt=isp_t3

Whatt, B. (2014b, February 22). Sochi's formula for the future: 'Fantasy' becomes reality. *CNN News.* www.edition.cnn.com/2014/02/21/sport/sochi-formula-one-winter-olympics-ecclestone-tilke-f1/index.html

An Anatomy of Resistance: The Popular Committees of the FIFA World Cup in Brazil

Christopher Gaffney

Introduction

The 2013 FIFA Confederations Cup was characterized by massive public protests in hundreds of Brazilian cities. Such were the scale and intensity of the social unrest that the successful realization of the tournament was threatened. Violent police reactions to demonstrations exposed their lack of preparation and training, as well as the pervasive modes of conflict resolution in Brazil. While the protests themselves called into question the channelling of public resources into state of the art sports facilities, tax exemptions for FIFA and their corporate partners, and the lack of decent public services in Brazilian cities, protest-space was also contested territory. Black blocs, anarchists, right-wing anti-government organizations, and undercover police mixed with lower- and middle-class protesters, student groups, syndicalists, political parties, and newly formed civil

C. Gaffney (✉)
Department of Geography, University of Zurich, Zürich, Switzerland

© The Editor(s) (if applicable) and The Author(s) 2016   **335**
J. Dart, S. Wagg (eds.), *Sport, Protest and Globalisation*,
DOI 10.1057/978-1-137-46492-7_15

society organizations. All of these groups were often present at the same protests, creating a melange of messages, tactics, and interpretations of dissent.

The protests that erupted in June of 2013 continued and shifted, gaining renewed impetus with the approach of the 2014 FIFA World Cup before petering out during the tournament itself. As public resources and global attention returned their focus to Rio de Janeiro with the city's preparations for the 2016 Olympic Games, the social movements leading the fight against these massive public outlays again took centre stage. The recent decline in Brazil's economy and a series of corruption scandals created a crisis of legitimacy for the Rouseff government that may bring new social actors into the global spotlight of the Olympics. In anticipation of the impending media frenzy, it will be important to understand the actors involved, their mechanisms of formation and action, and to contextualize their "performance" within the field of urban social movements in Brazil and elsewhere (Atkinson and Dougherty 2006).

In the years leading up to the protests of 2013, groups known as Popular Committees of the World Cup (Comitês Populares da Copa) formed in each of the 12 host cities of the 2014 World Cup. These committees were among the primary articulators of the counter-discourses that have framed social resistance to sports mega-events in Brazil. This chapter will discuss the formation, political structure, social composition, and strategic actions of the first such group in Brazil—the Comitê Popular da Copa e das Olimpíadas do Rio de Janeiro (Rio de Janeiro Popular Committee of the World Cup and Olympics—CPRJ). Based on the author's personal involvement with the CPRJ, interviews with activists and an examination of the historical trajectory of urban social movements within the Brazilian socio-political conjuncture, the chapter will contribute to an understanding of the increasing role of protest in the planning and realization of sports mega-events (Boykoff 2014; Lenskyj 2000).

This research and its findings emerged from the author's activity as a member of the Comitê Popular in Rio de Janeiro. From 2009 to 2014, I attended weekly meetings of the CPRJ, helped in the conceptualization and organization of protests and contributed to the production of dossiers, press releases, and other content. While I did not engage in these activities as a participant observer, I was aware that I would eventually compile my reflections about the experience and practice of activism and resistance to sports mega-events. I used my multiple roles as a foreign observer in the global media spotlight, a blogger, a member of the foreign press corps, and a Visiting Professor at the Universidade Federal Fluminense to gain access to information about the hosting of mega-events in Rio de Janeiro and Brazil in order to forward the political goals of the CPRJ. I broadly place these activities under the rubric of "activist scholarship" (Hale 2008), wherein I use the study of sports mega-events as a site of political engagement.

In what follows, I give a brief history of the formation of the Comitês Populares (CPs) in the context of sports mega-events in Brazil. I then describe the ways in which the CPRJ functioned in the daily context, explaining some of the mechanisms of meetings, communications, and social action. I describe how the CPRJ targeted specific urban areas for intervention and the means through which they made their demands as an emergent counter-hegemonic organization (de Omena 2015, 15–18). I then place the CPRJ within the national context of mega-event resistance, delving into some of the key moments that were reflective of a particular political philosophy. In the conclusion, I reflect upon the role of the CPRJ in fomenting debate and articulating resistance to sports mega-events while assessing its role within Rio's Olympic conjuncture.

Brief History of the Comitês Populares

For the Brazilian population, observers and mainstream media, the size and strength of the 2013 protests came as a surprise, but their sudden emergence was partly influenced by years of work on the part of activist groups that had targeted the hosting of sports mega-events. These groups, while small, had consistently contributed to a discursive stream that challenged directly the hegemonic media, economic, and political coalitions that generate and manufacture a consensus view of the events.

The first group to resist sports mega-events in Brazil was called the Comitê Social do Pan (CSP). The CSP coalesced "from below" in

2005 in response to a series of urban and social interventions being pursued by the Rio de Janeiro city and state governments in preparation for the 2007 Pan American Games (PAN). Foremost among the concerns of the activists, syndicalists, social workers, academics, middle-class professionals, and low-income residents that comprised the CSP's membership was the threat of forced removals to make way for PAN-related infrastructure. The varying degrees of success experienced by the CSP were largely determined by the level of organization that the group encountered in the communities slated for removal (Freire 2013). For instance, a large mobilization to prevent the forced removal of the Canal do Anil favela was ultimately unsuccessful because many residents accepted the government's offer of indemnity, which weakened community bonds. In the case of the Vila Autódromo, a community proximate to some of the PAN venues, collective actions were more successful in preventing forced removal due to community solidarity, effective organization, and stronger claims to legal land tenure. In addition to the resistance to forced removals, a middle-class organization helped to prevent substantive reforms to and privatization of the Marina da Gloria in the centrally located Flamengo Park. Organizations in more peripheral neighbourhoods such as Engenho do Dentro and Deodoro were less successful in preventing major urban interventions such as stadium building projects. The wide-ranging interventions of the PAN and the CSP's ability to react to them were indicative of an ability to articulate social capital through networks that extend beyond the immediate community.

In the two years leading up to the PAN, the CSP engaged in small-scale protests, media campaigns, and social organization to draw public attention to the projects and policies being instituted through the realization of the event. This "multiform social movement" (Boykoff 2014, 26) was working through an unfamiliar process of framing and counter-framing of political discourse to resist a political coalition that had formed around the staging of the PAN itself. While the CSP was able to draw upon many of the tactics used in the long struggle for "The Right to the City" in Brazil (McAdam et al. 1996; Mayer 2009), the social and political contexts of sports mega-events were new, and the relatively predictable interactions between urban social movements, mass media, and public authorities in Brazil were unable to influence significantly public opinion (read: middle-class opinion) regarding the PAN or to achieve many of the stated aims of the CSP.

Following the hosting of the PAN in 2007, the CSP lost impetus. Even though there was no on-going agenda, the strong ties that had formed through direct political action were significant enough to allow the association to continue in a very loose form even though the PAN itself was no longer a political object. When I arrived in Rio in mid-2009, I began to frequent the CSP meetings, which were held in the offices of the Regional Economic Council of Rio de Janeiro (CORECON-RJ). Even though there had been an accumulation of experience in resisting the PAN, the membership of the CSP declined precipitously as its former members re-joined their previous social movements, or were discouraged from their activism because of the time lag until the 2014 World Cup. For the 2016 Summer Olympics announcement party staged on Copacabana Beach in October of 2009, the CSP brought together 15 people to protest in front of Copacabana Palace. An estimated 300,000 people celebrated Rio's "victory."

Even though FIFA had selected Brazil as the host of the 2014 World Cup in October of 2007, the CSP had not reformulated its agenda to meet the impending urban, security, political, and economic interventions that the event would eventually bring. One reason for this delay was that the Local Organizing Committee of Brazil 2014 did not finalize its host city choices until May of 2010 and no particular suite of projects had been announced for Rio de Janeiro. In the midst of this uncertainty and with public opinion largely favouring the hosting of the World Cup and Olympics, the CSP tried various mechanisms to rearticulate its core messages. In early 2010, it rebranded itself as the Forum Social de Megaeventos Esportivos (FSME—Social Forum of Sports Mega-events) and then, building on the momentum gained through the realization of the Urban Social Forum, at a conference that ran parallel to the United Nation's World Urban Forum in Rio de Janeiro in March of 2010, it reformulated as the Rede de Megaeventos Esportivos (REME—Sports Mega-events Network).

The clear focus on the processes and impacts of the World Cup and Olympics would gain national traction in late 2010 once the 12 World Cup host cities announced their Responsibility Matrices, documents that defined the projects that each city would pursue in World Cup preparations. Sensing that there was a way of connecting social actors across Brazil to draw attention to human rights violations and shifting paradigms of urban governance, REME was reconstituted in early 2011, this time as the CPRJ.

For a number of legal (liability) and practical (cost) reasons the CPRJ never sought formal status with the Brazilian government. Previous experiences in sports-related social mobilization had shown the difficulties and problems inherent in assuming an institutional form under Brazilian law (Gaffney 2013). Similar to the way that the CSP organized around core people and institutions, the CPRJ counted on the financial support of NGOs such as Global Justice, Witness, Amnesty International, Jubileu Sul, StreetNet, Political Action for the Southern Cone (PACS), Henrich Boll Stiftung, and the networks of academic institutions, progressive political parties, and pre-existing social movements. These same national scale actors were key to the formation of popular committees in all of the World Cup host cities. In particular, the strong pre-existing relationships between urban planning and geography departments at Brazilian Federal Universities helped to bring academics and their local political networks into the CPs.

Because Rio de Janeiro had previous experience with organizing to confront sports mega-events and because it would host the final of the World Cup as well as the Olympics, the CPRJ was the first and most active Comitê Popular in Brazil. Incorporating the football-based assemblages of organizations such as the Associação Nacional dos Torecedores (ANT) and the Frente Nacional dos Torcedores (FNT), the CPRJ model served as an impetus for the formation of similar groups in most of the World Cup host cities.[1] Once established, local CPs formed an umbrella organization called the Articulação Nacional dos Comitês Populares da Copa (ANCoP—National Articulation of Popular Committees of the World Cup).

Through ANCoP financing, each CP was able to send representatives to bi-annual national meetings and ANCoP took on the responsibility

---

[1] The exceptions to this were Cuiabá, Manaus, and Brasília, which never developed strong networks of resistance.

for directing national policy, communication through the ANCoP website (first post on 10 December 2011, last post on 19 November 2014), maintenance of a Facebook page (first post on 1 March 2012, last post on 26 February 2015: 5476 likes), Twitter account (first post on 25 April 2012, last post on 26 February 2015 and linked to Facebook: 356 followers), the collection of national-scale data for publications, documentation, and strategies of resistance for all of the member committees. It was then left up to the local committees as to whether or not they would take local actions based on the national-scale decisions, though there was an expectation that ANCoP's directives would be followed.

The challenge of organizing on a national scale in Brazil reflected the different compositions, strengths, experiences, and interests of the regional CPs. The strength of representation from the large urban centres of São Paulo, Rio de Janeiro, Fortaleza, Salvador, and Porto Alegre guaranteed that their CPs would bring their particular agendas (sometimes tied to individual political goals) to bear on the national debate. Many of the members that chose to partake in the national-scale political articulation had previous experience with organizing social movements in Brazil and they inevitably brought their regional and ideological interests to the table. As the meetings of ANCoP rotated throughout Brazil, a familiarity and dialogue developed between the ANCoP activists and the regional CPs, which created a sense of shared national struggle.

One of the initial projects undertaken by ANCoP was the #copapraquem? [World Cup for Whom?] campaign.[2] This nationwide media initiative was intended to bring as many social movements into the ANCoP framework as possible. The objective of the campaign was to, "[e]xpose the human rights violations suffered as a consequence of the mega-event and to question the real legacy that will remain for the country after the games." The campaign had eight primary agenda items that were arrived at through the deliberation of local CPs and other organizations at a national meeting in 2012:

(1) The end of forced removals and housing demolitions, (2) Guarantee the homeless population access to food, shelter, and personal hygiene, as

---

[2] http://www.portalpopulardacopa.org.br/index.php?option=com_k2&view=item&id=536:#copa paraquem?

well as work and social assistance, (3) Immediate revocation of the exclusive areas set aside for FIFA in the General Law of the World Cup, (4) Creation of campaigns to combat sexual abuse and human trafficking, (5) Overturn the installation of extraordinary judicial tribunals in and around the football stadiums, (6) Revoke the law that grants FIFA and its commercial partners tax-free status and the implementation of a "popular audit," (7) The immediate end of legal processes that define the crime of terrorism, (8) Demilitarization of the police and the end of the repression of social movements. (Author translation, original text available at www.portalpopulardacopa.org.br).

Despite this initial consensus regarding a national platform, the internal politics of ANCoP were fraught with tensions as particular CPs tried to assert control over the group. For instance, in 2013, the adoption of the slogan "Não vai ter Copa" (There will be no World Cup) for all CPs was rejected by the CPRJ because its members felt it to be an absolutist and defeatist political position. This rupture between the CPRJ and ANCoP was problematic for both organizations. However, as the political philosophy of ANCoP, as well as the CPRJ, was based on consensus, and there was no way of arriving at a consensual position with such a polarizing campaign, the only solution was for the CPRJ to exit the national organization. The wholesale rejection of the World Cup was seen by the CPRJ as a negation of popular culture, which would impinge upon the re-appropriation of fandom as a political act in its own right. As the CPRJ left the national-scale debate, the politics of the CPSP (São Paulo) and CPBA (Salvador) began to dominate ANCoP.

Composition, Organization, Strategies, and Actions of the CPRJ

The CPRJ defined itself as an *espaço de articulação* (space of articulation) that allowed for multiple stakeholders to converge around two principal themes: (1) the opening of a public debate to prevent the privatization of the Maracanã sporting complex and its associated sporting and social infrastructures and (2) the immediate cessation of forced removals of *favelas/comunidades*, particularly those threatened with removal for

World Cup and Olympic-associated infrastructure projects. As World Cup and Olympic projects began to take physical form in the city, the CPRJ mobilized to call public attention to the wave of forced removals, privatizations, and radical re-makings of urban space that were underway. The processes through which the CPRJ worked to identify and meet its goals shifted over time in response to current events and the composition of the CPRJ itself. In what follows, I offer a brief explanation of the meetings and their structure before turning to the targeted interventions and the specific actions taken.

Typically, meetings were announced via the CPRJ list served for Tuesday evenings at 1830 hours at an available space in the city centre. The locations varied between institutions such as the Teachers' Union Headquarters, NGO offices (Global Justice), the Brazilian Press Association (ABI), the Manuel Congo Occupation (a nine-storey squat next to City Council Chambers), or other institutional spaces. The lack of consistent meeting space was a perpetual concern and the downtown location of the meetings was a frequent topic of conversation as it tended to privilege access for those who lived in the wealthier South Zone. This however, was where the majority of CPRJ members lived, even though the forced removals that the CPRJ was targeting were taking place in the periphery.

At the beginning of each meeting, everyone in the room was required to identify himself or herself and to declare their institutional affiliations or reason for their presence. As the work of the CPRJ gained more traction in the media and the protests began to draw thousands of people to the streets, this interpersonal identification at the beginning of meetings became increasingly important. At the larger plenary sessions designed to attract new members and to explain the goals of the CPRJ to a wider audience, there were instances in which undercover police were suspected to have infiltrated the meeting. Some members of the CPRJ had had their email and twitter accounts hacked and their phones tapped. One prominent CPRJ member received a phone call from a Brazilian Army General as he was marching in a protest during the UN's Rio+20 Environmental Conference in 2012 with the ominous message: 'We are watching you, right now.' There were also frequent foreign visitors, typically students or journalists, who were trying to get 'inside' information on the CPRJ.

I myself brought several journalists to meetings so that they could make connections with key actors. The increasingly delicate nature of the discussions surrounding protest actions in the lead up to the Confederations Cup in 2013 began to shift what had once been an almost completely open dialogue into something much more guarded and suspicious of outsiders. Once the political stakes had become higher and there was not enough time for introductions, the meetings would begin by asking if everyone in the room knew everyone else by name. If that were not the case, a full round of introductions would occur. In this way, only known individuals could participate in the meetings. It was expressly forbidden to record audio or video of any meeting and still photography was also prohibited.

Following the introductions, one of the members would begin the meeting by listing the items for discussion (*a pauta*). While there was no official leader of the CPRJ, a core group of individuals took it upon themselves to schedule the meetings, organizing the discussion points, calling for order when needed, timing individual speeches, and trying to push through the agenda. At the beginning of the meeting, time was always allowed for individual reflections on recent CPRJ events, news items, and updates from the field sites. During debates, only one person would or could speak at a time and the opportunity to speak was mediated by attracting the attention of the meeting coordinator, who would write the person's name on a list. When individuals tried to argue a point out of turn they were quickly reminded that they had to put their name on the list and wait for their chance to take the floor. When individuals were not known by their first names, they were typically referred to as *companheiro/companheira* until greater familiarity developed.

It was extremely rare for a vote to be taken either at the plenary sessions or in the working groups. If consensus amongst all of the members could not be reached, an issue was simply dropped or put to one of the working groups for further analysis. Arriving at a consensus among so many different stakeholders was frequently frustrating and always time-consuming, but this process formed the backbone of the CPRJ's political philosophy. Everyone, no matter his or her standing within the committee's informal hierarchy, was able to voice their opinion. When objections were raised to a proposed motion that had already been in discussion, the debate would continue until consensus was reached. If consensus could not be reached, no action (*encaminhamento*) would be taken and the motion was shelved. These debates would frequently be taken up after the meeting at a bar in Rio's Centro. It was during these informal discussions that many of the CPRJ's internal political tensions and external actions were talked through and resolved. Indeed, these informal meetings at the bar were as important in forming friendships and strong bonds as the formal meetings and direct political action.

The longer that I participated in the CPRJ, the more I began to understand that the sourcing and management of finances were never going to be made explicit to the membership as a whole. There was not necessarily secrecy regarding these operations, but neither was there transparency. The same small core group of people ran meetings, controlled the media communications, and organized the finances. This group articulated with NGOs, universities, and international development agencies to accumulate money for the production of banners, rental of sound cars, and the production of published material in addition to financing travel, staging events, and so on. The core functions of the organization were primarily shared between individuals whose "day jobs" allowed them to use institutional space for CPRJ projects. Members of political offices, for example, could use the printers and copiers of their offices to produce pamphlets for distribution at rallies and events whereas members of NGOs used their work computers and time for CPRJ communication and labour. Graduate students could leverage their activities with the CPRJ into their coursework. From the perspective of an academic within the CPRJ, the explicitly recognized and overlapping nature of the political project of the CPRJ with many of the affiliated NGOs and academics helped to concretize the CPRJ as a space of articulation between various stakeholders.

The following is a partial list of organizations that were represented in meetings between February 2011 and January 2013. This list was generated from collecting the sign-in sheets that are distributed in every meeting. Some members did not declare an institutional affiliation and some of these were foreign "visitors" who were participating in the plenary sessions (*plenárias*) as part of a research project or as members of the media. The data are not robust enough to indicate the length of time or frequency with which individuals or institutions were present at meetings or the degree to which they participated. During this time the list of members that had attended at least one meeting in the previous two months grew from 41 in July of 2011 to 98 in January 2013.

Universities and Departments

Universidade Federal Fluminense (Economics; Architecture and Urbanism; LeMetro)
Universidade Federal do Rio de Janeiro (Urban Planning and Regional Development; Psychology; Communications; Electrical Engineering; Production Engineering)
Pontifica Universidade Catolica Rio de Janeiro (History)
UniGranRio
Fundaçaõ Getulio Vargas
University College London (Geography)
Yale University (USA)
University of Osnabruck (Germany)
Roskild University (Denmark)
Rutgers University (USA)
University of Lyon (France)

NGOs

Amnesty International
Witness
Political Action for the Southern Cone (PACS)
ActionAid
FASE
StreetNet
Associação Brasileiro de Imprensa
Laboratório DHs Manginhuos
INKA
SEPE-RJ
NEA/BR
Viva Hoje

SOS Estado do Remo
ECO Museu Neg Vilma
Rio Antigo
Meu Rio
Moviemento Nacional para a Luta de Moradia
Mais Democracia
Mup. Conselho Popular
Juventudes CDP
Amigos da Praça
MUCO
CTOII
Catalytic Communities/Rio on Watch
Projeto Legal
PMS
ANEL

Social Movements

Comitê of Moviementos Populares (RJ)
AVAT-RJ
Centro Unido dos Camelôs
Favela não se cala

Neighborhoods/Communities/Associations

Aldeia Maracanã
Vila Autódromo
Comunidade Arroio Pavuna
Comunidade Metrô-Mangueira
Comunidade Pico Santa Marta
Comunidade Vila União de Curicica
Comunidade Asa Branca
Centro Indígena
Escola Friendenrich
Associação de Atletas Celio de Barros

Political Offices

Office of Marcelo Freixo (PSOL)
Office of Renato Cinco (PSOL)
Office of Eliomar Coelho (PSOL)
Office of Paulo Pinheiro (PSOL)

This list is perhaps less extensive than it appears. While there were in fact many organizations that sent representatives to meetings, the degree to which any one of these institutions/people functionally participated in the CPRJ is difficult to gauge. From my observations, approximately 10–15 people carried out the fundamental, daily work of the organization. Some individuals would come in for a time and then leave for personal reasons, or like myself, would be called away for several weeks of travel for academic conferences. There was always significant flux in the numbers of individuals that attended meetings, but the typical size was between 20 and 50. In general terms, a small, core group was the first to arrive at meetings, the last to leave, and always assumed organizational responsibilities. They were also responsible for much of the behind-the-scenes work of organizing for protests, making banners, communicating with police, editing publication material, and the innumerable small tasks of communication and organization that make effective social mobility possible. The CPRJ became a social world as much as a political project and because the core group of people spent at least five hours meeting on Tuesdays and communicated extensively through email over many years, the "deep ties" of the group created a sense of shared urgency as well as momentum that kept the CPRJ active, even after the failure to mobilize during the 2014 World Cup.

The incompleteness of the above data is indicative of the ad-hoc nature of the CPRJ and many of the organizations that were formed to meet the challenges presented by sports mega-events in other places (Boykoff 2014; Lenskyj 2000; Shaw 2008). Within the Comitês and ANCoP, there was a constant struggle to reach a level of coordinated, secure communication that would allow for more structured discussion and action. As the CPs became more visible, the communication policies became more complex.

In early 2013, the CPRJ divided communications between a "general list" for passing along information about meetings and public actions and an "operational list" for those who consistently took part in meetings and organization. In order to qualify for the Operational List, one had to be present for several months of meetings and take up a position of responsibility within the organization. The General List became a means of keeping individuals with an interest in what was happening with the CPRJ up to date about current actions and deliberations.

In addition to maintaining tighter control of internal communication, there was also a continual search for the professionalization of press releases and printed material that would give both ANCoP and the local Comitês legitimacy with national and international media, which in turn (it was hoped) would help to swing global public opinion against the mega-events. Contrary to the examples of Vancouver, London, Chicago, and Toronto, Brazil's anti-Games movements were not able to count on a critical mass of highly motivated and engaged upper-middle-class professionals to contribute their time, skills, and resources. While the CPRJ and ANCoP did eventually produce high-quality publications, on-line content, and effectively organized resistance actions, these were not easily realized.

Actions and Strategies

Public Debates

The CPRJ organized a series of public debates that treated various aspects of the physical, financial, political, and social transformations occurring in the city. These debates were typically expositions representing one point of view though on occasion a member of one of the mega-event organizing committees or the government could be induced to attend. Universities, public schools, residents' associations, and civil society organizations all hosted these events. The CPRJ received a number of invitations to participate in these events and also did a significant amount of outreach. CPRJ members who were available would volunteer to participate as debaters. Importantly, no individual was permitted to speak *for* the CPRJ, but could only express his or her opinion *as a member of* the CPRJ. This distinction was rigidly enforced and those who were quoted in the press or in public as speaking *for* the CPRJ were always reminded of the necessity of not attributing individual speeches to a collective that based its decision-making processes on consensus. CPRJ members were trusted to represent faithfully the political position of the CPRJ, but as word choice and syntax were always subject to interpretation, the individual as *member* and not as *representative* was an important distinction to maintain.

Manifestos and Documents

In addition to the imperatives of putting bodies in the physical space of the city as a means of signalling strength in numbers and organization, the struggle for the "right to the city" also plays out in the field of ideas and public consciousness. While public manifestations are typically ephemeral, the circulation of accumulated knowledge and, importantly, the deconstruction of hegemonic discourses, was a major objective of the CPRJ and ANCoP. To this end, the CPRJ's annual dossiers of human rights violations occurring in the context of the World Cup and Olympics in Rio de Janeiro are important markers in the history of resistance to global mega-events. The first dossier was published in 2012 and quickly became a reference document for researchers and journalists.[3] These dossiers largely depended on the research conducted by academics and graduate students into the impacts of the World Cup and Olympics in Brazilian cities. The inclusion of hard data as well as the translation of the language of social science into a professionally produced, glossy document was (and continues to be) one of the major achievements of the CPRJ. These dossiers were launched in large institutional settings and drew significant national and international media attention.

---

[3] http://www.slideshare.net/RosaneGafa/dossi-megaeventos-violaes-dos-direitos-humanos-no-brasil

http://www.cartacapital.com.br/sociedade/o-rio-que-viola-os-direitos-humanos

http://global.org.br/programas/edicao-2014-do-dossie-megaeventos-e-violacoes-de-direitos-humanos-e-lancada-no-rio-de-janeiro/

http://br.boell.org/pt-br/2014/06/13/edicao-2014-dossie-megaeventos-e-violacoes-dos-direitos-humanos-no-rio-de-janeiro

Actions in Public Space

As we shall see in the context of the CPRJ's working groups, despite the rise in electronic communication and high levels of social media adherence, the physical presence of human bodies in urban space is still the surest way to demonstrate discontent. Much as corporations can determine the success of their product through the amount of income received, the true currency of social movements in urban contexts are the numbers of people that come to the street. In this context, the #vemprarua hashtag (come to the street) that emerged during the June 2013 protests can be understood as a way of bridging the gap between social media and political action. Public actions were key to the strategic framework of the CPRJ and were generally scheduled for weekends, when more people would be able to participate, or for large media events associated with FIFA or the International Olympic Committee (IOC). Sometimes, ad-hoc protests with smaller numbers were sufficient in order to register a voice of discontent with the media or with public authorities. In other instances, larger numbers (and more planning and investment) were necessary to make as visible and audible an impact on public consciousness as possible. Regardless of the action, the CPRJ was continually searching for strategic political targets and significant openings in the media landscape in which it could bring its coalition to the streets.

Internal Strategies and Division of Labour

In 2011–2012, as the fights over the privatization of the Maracanã and the elimination of favelas in key tourist and competition areas became more central to public debates over mega-events, the CPRJ opted to have fewer general meetings and to divide tasks between *Grupos de Trabalho* (GTs, or Working Groups). From June 2011, these were divided between the Maracanã GT, Removals GT, and Communications GT. The latter GT was absorbed into the other two once a website had been established and there were some more clear directives for email communications. The GTs met after abbreviated plenary sessions and then on alternate weeks. They were tasked with devising action strategies and for bringing proposals to the plenary sessions for approval. The GTs also developed internal communications apart from the general email list that the CPRJ maintained through Google groups.

The meetings and deliberations of the GTs were key to developing the direct actions that would allow the political project of the CPRJ to move into the spaces of the city. As part of my personal interest and previous activity in Brazilian social movements, I participated in the Maracanã GT, which took as its subject the Maracanã stadium and its immediate environs. I also participated in some of the Removals GT sessions. Interestingly, the division of the CPRJ into these groups was largely along gendered lines: the majority of the Maracanã GT participants were male and the majority of the Removals GT members were female. This reflects some of the persistent gendered divisions in Brazilian society between spheres of social action.

Strategies and Actions of the Maracanã GT

As part of the State of Rio de Janeiro's proposals for the World Cup, the Maracanã, which had undergone a R$330 million refit for the 2007 Pan American Games, would receive an additional R$900 million of public funding to make the stadium "FIFA-standard." Additionally, the stadium would be taken out of the public sphere, being transferred to a private consortium as a 30-year concession. As part of this concession, the Maracanã stadium, the Maracanãzinho gymnasium, the Célio de Barros athletics stadium, the Julio Delamare aquatics centre, and the historic Museu do Índio would be handed over to the concessionaire to do with as they pleased. As part of this process, which had occurred without due process or engaging relevant stakeholders, the Rio de Janeiro State government announced plans to demolish all of the facilities except the Maracanãzinho. The demolitions extended to one of Rio's best performing public schools (Escola Friedenrich). In response to these plans the CPRJ mobilized civil society actors and users of the facilities in an attempt to impede the realization of the reform and privatization of the Maracanã complex.

Each GT had different strategies for community outreach and publicity, yet were remarkably similar in their general patterns. In the case of the Maracanã GT, the initial step was to search out and bring together as many stakeholders as possible. This included the Indigenous community encamped at the Museu do Índio, the Olympic athletes and coaches that trained at the Celio de Barros and Julio Delamare installations, the parents and teachers at the Escola Municipal Friendenrich, and football fans. While each of these groups had different goals, the Maracanã GT was able to help articulate a message against the destruction of public facilities that each of the stakeholders could identify with and contribute to. The outreach work by a small number of extremely motivated individuals was able to create conversations between these groups, resulting in the formation of a group external to the CPRJ called *O Maraca é Nosso* (The Maracanã is Ours). However, there were no attempts to engage in dialogue with SUDERJ, the Rio de Janeiro state superintendence for sport, principally because this institution was closed to conversation and was the primary state agency engaged in the privatization and destruction of the Maracanã complex.

As Brazilian society has become increasingly articulated through the use of electronic communications, so too have Brazilian social movements have become dependent upon the use of social media to communicate their messages. In this regard, the Maraca é Nosso movement was exemplary in that the group quickly developed a blog, Facebook page, and a twitter feed.[4] As the campaign to preserve the Maracanã was accompanied by the on-going destruction of the stadium, the CPRJ and the Maraca é Nosso movement opened the kind of public consultation process that the state itself did not, or would not provide. The "Popular Consultation for the Maracanã" provided a series of diagnostic problems that could be addressed through public input into a website (http://consulta.omaracaenosso.org.br/). This process of formulating questions and mediating responses was one of the most difficult and delicate proposals of the Maraca é Nosso movement for three principal reasons. One, it was difficult to imagine what a democratically managed stadium could look like and how it would function, as there were no Brazilian and very few foreign models to draw from. Two, there could not be an expressed desire to maintain the status quo ante, which was quite bad, but there

---

[4] http://maracanosso.blogspot.com.br/ and https://www.facebook.com/OMaracaENosso

had to be a counterproposal. Three, the Maracanã sporting and cultural complex does not function in an isolated context but has to be integrated into other urban services, transportation in particular. Because of this the issues associated with the stadium were more elaborate than those that could be easily accessed through an on-line consultation platform. The lack of experience in thinking through the cooperative management of large public spaces such as parks and stadiums is an issue that continues to plague Rio de Janeiro as it deals with the aftermath of hosting mega-events.

Even though the Brazilian public at large is highly proficient in social media, one of the most important elements of public resistance continues to be the occupation of public space and marking a presence at live events. In this regard, the CPRJ acted in concert with the Maraca é Nosso movement to engage in a series of marches that began in a public square near the Maracanã (Saens Pena) and ended with symbolic gatherings in front of the Maracanã's eastern entrance, also known as the Bellini Gate (because there is a statue of Brazil's 1958 and 1962 World Cup captain, Bellini). These choreographed events varied in size from 250 to many thousands of people and in retrospect were important moments in preparing for the larger public protests that would happen in 2013. Unfortunately, some of the smaller protests were also used as training exercises for Rio's notoriously violent Military Police, which took the opportunity to practice moving in formation, launching tear gas canisters, and "clearing" space around the stadium.

Because Rio de Janeiro had become the effective centre of mega-event production for the Brazilian government and FIFA (and later, the IOC), it was also an increasingly large platform upon which acts of resistance could be performed. The CPRJ and other groups were very agile in their movements to use the global media spotlight of the 30 July 2011 FIFA World Cup Draw held in Rio de Janeiro to call for the ousting of then-CBF (Brazilian Football Confederation) president and FIFA vice-president Ricardo Teixeira (since implicated in innumerable corruption scandals). A "Fora Teixeira!" (Teixeira Out!) march was organized to draw the attention of the national and international media that were covering the FIFA event and exposed some of the internal ruptures of social mobilization as well as the possibilities for effective resistance. In the former case, the FNT (National Fans Front) were expelled from participation with the CPRJ after they declared to the media that they alone were responsible for organizing the event. A similar rupture would happen the following year with the NGO Meu Rio after they had declared unilateral responsibility for the preservation of the Escola Friendenrich. The success of the Fora Teixeira event in drawing more than 25,000 people to a public protest drew attention to the corrupt practices of the CBF and FIFA and was one of the first instances in the history of the World Cup event cycle that global media attention was drawn towards public criticism of football governance and the realization of the World Cup in Brazil.

In addition to public protests, enduring campaigns in stadiums, and advocacy on behalf of the users of the Maracanã complex, the members of the CPRJ and the Maraca é Nosso movement engaged critically in the few public hearings about the fate of the stadium. The most notable of these occurred in November of 2012, when the government resolved to present the public with a pre-approved privatization scheme for the Maracanã. CPRJ and Maraca é Nosso activists interrupted the farcical charade from the beginning; some of them threw rotten fruit and faeces at the officials on the stage. Determined to bring the proceedings to a conclusion so that the contracting process could go forward, security guards stood over the government officials with raised umbrellas while they read through their prepared speeches and reluctantly opened the floor to the public. At the end of the meeting, the Governor's chief of staff, Regis Fichtner declared that, "everything would continue as it had been planned before the meeting" (Carpes 2012).

While directly pressing public authorities, members of the Maracanã GT (as with other GTs) participated in formal political debates in various institutional and informal settings. Personally, I gave lectures in private and public universities, in public high schools, and in favelas under threat with removal from mega-event-related projects. Along with many other members of the Maracanã GT, I also participated in radio talk shows, engaged in local, national, and international media outreach, and created a mutual feedback loop between sympathetic politicians and their constituents. While ultimately unsuccessful in preventing the destruction and privatization of the Maracanã stadium, the Maracanã GT and the Maraca é Nosso movement were successful in creating wider public consciousness about the absurdities of destroying some of the few Olympic training facilities in a city that was planning on hosting the 2016 Olympics. Importantly, the preservation of the Escola Friendenrich and the physical (but not social) integrity of the Museu do Índio owe much to the organization and actions of these groups.

Strategies and Actions of the Removals GT

The processes for working with favela communities threatened with forced removal were somewhat different to the Maracanã GT, though they followed similar patterns. Identifying and bringing on board stakeholders was slightly more difficult as the communities were socially and geographically diverse and working to bring these stakeholders to meetings was not always successful. Negotiating the internal politics of communities was always highly sensitive and the political project of guaranteeing the right to housing had to move beyond the mere permanence of the favelas in question. While present in the Maracanã GT discussions, in the Removals GT, broader issues of human rights were a central concern, especially in the face of the expansion of the UPP program (an "pacification" of select favelas by Rio State Military Police), gentrification dynamics, and the removal pressures coming from the city government.

In the early years of the CPRJ, five favelas in particular were central to the CPRJ's attentions: Vila Autódromo, Santa Marta, Metrô-Mangueira, Arroio Pavuna, and Indiana-Tijuca. Since 2014, the Vila União de Curicica, slated for removal because of the Transolimpica BRT line (associated with the 2016 Olympic project), has also been a site of intervention. Some favelas such as Recreio II and Camorim came to the attention of the CPRJ too late to intervene meaningfully in the removal processes that were justified by the city government as essential to the installation of the BRT Transoeste line. Once the forced removals had occurred, the CPRJ organized "missions" to the former sites to catalogue what had been done with the land as well as to strengthen individual relationships with the CPRJ. The CPRJ also invited members of the displaced communities to tell their stories in public meetings. In reality, key members of these communities were always present at the CPRJ general meetings as well as in the Removals GT, so the invitations were a matter of availability and mobility.

The arrival of sports mega-events accelerated pressures that were already impinging upon lower-income residents in wealthier areas of the city. While gentrification was also occurring in middle-class neighbourhoods, these dynamics were being felt more acutely in the wealthy Zona Sul favelas than in other areas of the city. It was impossible for the CPRJ to reach out to all of the favelas targeted for removal by the city government and real-estate speculators, so it focussed on those that were under direct and immediate threat (initially, Metrô-Mangueira and Vila Autódromo and later Santa Marta, União de Curicica, Indiana) as well as in those in which previous actions had had some resonance.

Similar to the Maracanã GT, the removals GT articulated with academic institutions and civil society organizations to create policy documents that would point the way forward for more inclusive, more democratic forms of governance. Following the announcement that the British–Australian firm AECOM had won the design competition for the 2016 Olympic Park and that within that design the Vila Autódromo had been included in a long-term urbanization project, a movement started to create an alternative, community-centred urbanization project. Even though the spatial parameters of the 2016 Olympic Park competition did not include the Vila Autódromo, the AECOM design gave impetus for the formation of a team of urbanists, architects, activists, and residents to start developing what would become an award winning *Plano Popular de Urbanização* (Popular Urbanisation Plan). The multi-disciplinary group worked with residents over a year to map and catalogue community needs within the environmental parameters established by the city government and the Olympic Park plans developed by AECOM. This plan, given a $80,000 USD award by the London School of Economics and Deutche Bank,[5] was presented to the city government, but was summarily ignored. The Plano Popular was an innovative piece of proactive, community-based consultation that if taken on board by the city government, would have been less costly than the eventual forced removals and resettlement that have occurred.[6]

There were moments when the two GTs were able to work together on specific events. As the Confederations' Cup kicked off in June of 2013,

---

[5] http://www.alfred-herrhausen-gesellschaft.de/en/3748.html

[6] http://www.portalpopulardacopa.org.br/vivaavila/index.php/joomla-forums

the CPRJ hosted the *Copa Popular* (Popular Cup) in Rio's Gamboa neighbourhood. Imagined as a football tournament that would bring together teams from communities under threat of forced removal, the tournament was most successful in attracting international media attention to the wave of human rights violations occurring in Brazilian cities as a result of mega-event hosting. The Copa Popular was a marginal success in terms of building relationships and cross-city solidarities as many of the places under threat were many dozens of kilometres apart and creating new formal networks of resistance in the context of a competitive event proved difficult, without specific interventions that would politicize the act of playing football. In the months leading up to the 2014 World Cup, a series of small tournaments were programmed in favelas around the city, but they ultimately failed to stimulate further resistance or to develop cross-favela solidarities, or contribute to the CPRJ as a city-wide social movement (Table 1).

The massive public protests of 2013 that defined the Confederations' Cup in Brazil may have forever changed the way that the world perceives the impact of these events on urban and social environments. It is unquestionable that the CPRJ and ANCoP had important roles in constructing and maintaining a counter-discourse in public consciousness in the months and years leading up to the "Vinegar Revolt"—a nationwide protest against Brazilian government priorities, during which demonstrators soaked their clothes in vinegar to protect themselves from police tear gas (Gaffney and Michener 2013). However, it is not just a matter of maintaining this counter-discourse, it is the attempt to construct an alternative politics that both gives legitimacy to the words and also creates the

Table 1 Actions of the CPRJ, 2011–2013. Used with permission from Omena de Melo (2015, 30)

Activities	2011	2012	2013
Street protests	6	7	14 (6 before June)
Organization of public debates	1	7	3 (2 before June)
Publishing of documents and manifestos	3	18	7 (6 before June)
Meeting with government representatives and other institutions	2	7	3 (2 before June)
Educative actions in poor communities	3	2	4 (3 before June)

spaces for this politics to be realized. The above table describes the consistent actions that the CPRJ undertook in the years and months leading up to the 2013 protests. To be sure, there were many other social actors involved and the CPRJ did not have the organizational capacity or social network to draw hundreds of thousands of disaffected Brazilians into the streets. Yet, many of the themes regarding corruption, transparency in governance, and the end of police violence that had defined many of the CPRJ's campaigns were also present on the streets in 2013. And the CPRJ banners were there as well, backed by growing numbers of people associated with the organization.

The CPRJ and ANCoP updated their dossiers for the 2014 World Cup and worked diligently to get them into the hands of journalists (Articulação Nacional dos Comitês Populares da Copa 2012). The CPRJ dossier engaged all of the themes of the events and was based on research conducted by professors and graduate students at the major public universities in Rio de Janeiro who were members of the CPRJ. The principal sections of these documents were housing, labour, public space, mobility, sport, security, environment, budgeting and finances, access to information, and participation in decision making. The comprehensive overview used specific case studies to identify where and how human rights violations were occurring, identified larger trends in urban governance, and explicitly sought to de-legitimize the narratives of the events put forward by the organizing committees and government.

Significant to the discussion regarding social activism in Brazil is the rapid expansion of access to cell phone technology, the Internet and social media networks. When the CSP formed, Internet usage across Brazil was around 21 %, with a strong concentration in large urban centres. By 2014, this figure had grown to 53 % and Brazil was the country with the second largest number of Facebook, Twitter, and WhatsApp accounts (Holmes 2013). An estimated 79 % of those with access to the Internet in Brazil use some form of social media to communicate and this diffuse usage pattern has changed the scope and impact of social mobilization. The quality and quantity of the media coverage of dissent is critical to its success (Boykoff 2014, 30) but so is its internal communication structure.

Conclusion

Following the 2014 World Cup, many of the local comitês dissolved and ANCoP ceased its activities. While there were undoubtedly new social formations that emerged from these temporary organizations, a much more thorough investigation into the historical trajectories of individuals, their relationships to social movements, and the permanence or effectiveness of those movements will have to be undertaken in each city to gain a deeper understanding of their significance and impact. Because Rio de Janeiro will host the 2016 Olympics, the CPRJ is still active, though the two-year time-lag between the mega-events has somewhat reduced the urgency of social action and increased the difficulty in retaining the focus of the general public. This condition, coupled with a continuing crisis of legitimacy of the governing Worker's Party because of revelations about systemic corruption and poor economic management, has deepened an already profound crisis on the Brazilian left. The difficulty of attracting new members to the CPRJ and to articulate messages that have to do with social justice and progressive politics in a city that is notoriously unjust and conservative has been compounded by the corrupt politics of a nominally left government. A new wave of protesters hit the streets in early 2015, though this time they were demanding impeachment of the president and some even called for the return of the military dictatorship. It is in this difficult political climate the CPRJ continues to function as a space of articulation. The question as to whether or not the CPRJ as a social movement has been a failure or a success is to perhaps miss the importance of its role in raising global consciousness about the trauma that the business model of sports mega-events brings to cities and populations. Though the movement is somewhat reduced in size as the 2016 Olympic Games approach, the major accomplishments of the CPRJ were (and remain) the following:

- The CPRJ gave the national and international media a place to go to for information and counter-discourse.
- The CPRJ created and kept alive space for the kinds of public action that erupted in 2013, although they themselves were surprised by the strength and breadth of the protests.

- The expert information of the CPRJ was legitimized through their presence on the ground and their ability to deconstruct dominant narratives.
- The CPRJ brought together a constellation of social actors that created a hybrid political platform that responded effectively to the shifting and contradictory politics of city, state, and federal governments. The ability of the CPRJ to articulate differently with different social actors, in widely varying urban geographies suggests that the forms of flexible accumulation that Rio's mega-event cycle has unleashed, may be met with flexible resistance. Whether this resistance is ultimately successful or not may be beside the point as at the very core of the CPRJ's political philosophy is the creation of horizontal political solidarities.

Relative to the generalized failures of the CPRJ to stop or to interfere significantly in the announced projects of the government and mega-event organizers, we may assert the following:

- The failure of the CPRJ to generate effective actions *during* the World Cup could have been a condition of their success *before* the World Cup. Once the stunning and successful delivery of the message had occurred in 2013, the core individuals of the CPRJ lost focus and energy over the subsequent nine months, and with the increased media noise around the World Cup as a football tournament, it was increasingly difficult to find new ways to deliver the same message. The political battles inside of ANCoP, particularly in relation to the "Não vai ter Copa" message and the activities of Black Bloc and Anarchist elements that favoured vandalism in protests, diluted the consistency and effectiveness of a national movement.
- The CPRJ eventually developed alternative proposals for the management of urban space and for the opening of dialogue with the general public regarding mega-event related projects, but could not construct channels for dialogue with the mega-event coalition (mayor, governor, FIFA, LOC, etc.). The CPRJ was not seeking a place at the table, but was committed to establishing counter-discourses, educating and informing the public, and pushing the state to change its course of action. Success can be measured in part by the effective delivery of a

message of discontent with mega-events to national and global audiences. The global terrain of resistance to mega-events was given new contours in Brazil and the *comités populares* played a decisive role in this.

The CPRJ continues to meet and organize around local issues such as the repression of informal street vendors and the increasingly violent and repressive security apparatus being installed for the Olympics. Other localized resistances, such as the Ocupa Golf (protesting the development of an Olympic Golf course in an environmental protection zone), SOS Estado do Remo (fighting for preservation of the rowing facility), and SOS Parque do Flamengo (trying to preserve the Flamengo Park from illegal development projects), as well as issues associated with environmental degradation are also on the agenda. The nascent local alternative media and the global media, when they return to Brazil in 2016 to cover the Olympics will be able to use an updated human rights violations dossier to question the narrative of the Olympic spin doctors and help to expose the costs of hosting these events in a radically unequal and increasingly violent society. The actions of the CPRJ will continue to enhance "the possibility of high-profile dissident citizenship on the global media terrain" (Boykoff 2014, 40).

References

Articulação Nacional dos Comités Populares da Copa. (2012). Dossiê Megaeventos E Violações de Direitos Humanos. *Rio de Janeiro*. http://www.portalpopulardacopa.org.br/index.php?option=com_k2&view=item&id=198:dossi%C3%AA-nacional-de-viola%C3%A7%C3%B5es-de-direitos-humanos

Atkinson, J., & Dougherty, D. S. (2006). Alternative media and social justice movements: The development of a resistance performance paradigm of audience analysis. *Western Journal of Communication, 70*(1), 64–88. doi:10.1080/10570310500506953.

Boykoff, J. (2014). *Activism and the Olympics: Dissent at the games in Vancouver and London, Critical issues in sport and society*. New Brunswick: Rutgers University Press.

Carpes, Guilander. (2012, August 11). Audiência Do Maracanã É Alvo de Boicote, Ovos E Até Fezes. Terra. http://esportes.terra.com.br/futebol/copa-2014/audiencia-do-maracana-e-alvo-de-boicote-ovos-e-ate-fezes,191816b19 c6fa310VgnCLD200000bbcceb0aRCRD.html

Freire, Leticia de Luna. (2013). Mobilazações Coletivas Em Contexto de Megaeventos Esportivos No Rio de Janeiro. *O Social Em Questão, XVI*(29), 101–128.

Gaffney, C. (2013). Virando O Jogo: The challenges and possibilities for social mobilization in Brazilian football. *Journal of Sport & Social Issues, 37*(4), 1–20. doi:10.1177/0193723513515887.

Gaffney, C., & Michener, G. (2013, June 27). Explaining Brazil's Vinegar Revolt. *Al Jazeera English*. http://www.aljazeera.com/indepth/opinion/2013/06/201362411928841390.html

Hale, C. R. (Ed.) (2008). *Engaging contradictions: Theory, politics, and methods of activist scholarship, Global, area, and international archive.* Berkeley: University of California Press.

Holmes, R. (2013, December 9). The future of social media? Forget about The U.S., look to Brazil. *Forbes.* http://www.forbes.com/sites/ciocentral/2013/09/12/the-future-of-social-media-forget-about-the-u-s-look-to-brazil/

Lenskyj, H. J. (2000). *Inside the Olympic industry: Power, politics, and activism.* Albany: SUNY Press.

Mayer, M. (2009). The 'right to the city' in the context of shifting mottos of urban social movements. *City, 13*(2–3), 362–374. doi:10.1080/13604810902982755.

McAdam, D., McCarthy, J. D., & Zald, M. (Eds.) (1996). *Comparative perspectives on social movements: Political opportunities, mobilizing structures, and cultural framings, Cambridge studies in comparative politics.* Cambridge/New York: Cambridge University Press.

Omena de Melo, E. (2015). *Only because of 20 cents?: the 'demonstrations cup' and the processes of urban transformation in Brazil.* Working paper.

Shaw, C. A. (2008). *Five ring circus : Myths and realities of the Olympic Games.* Gabriola Island: New Society Publishers.

Index

Note: Locators followed by 'n' refers to notes

© The Editor(s) (if applicable) and The Author(s) 2016
J. Dart, S. Wagg (eds.), *Sport, Protest and Globalisation*,
DOI 10.1057/978-1-137-46492-7

Printed in the United States
By Bookmasters